READING
STATISTICS
AND
RESEARCH

READING
STATISTICS
AND
RESEARCH

SCHUYLER W. HUCK
University of Tennessee

WILLIAM H. CORMIER
West Virginia University

WILLIAM G. BOUNDS, JR.
University of Tennessee

HarperCollins*Publishers*

Sponsoring Editor: Michael E. Brown
Project Editor: Cynthia Hausdorff
Designer: Michel Craig
Production Supervisor: Will Jomarron

Library of Congress Cataloging in Publication Data
Huck, Schuyler.
    Reading statistics and research.
    1. Statistics. 2. Research. 3. Experimental design. I. Cormier, William, joint author.
II. Bounds, William G., joint author. III. Title.
HA29.H799   001.4'22'014   74–1417
ISBN 0–06–042976–3

CONTENTS

PREFACE

With few exceptions, both undergraduate and graduate students approach courses in statistics and research with great anxiety and trepidation. If they cannot find some way to circumvent the requirement for these courses, most students put them off for as long as possible. When they finally get around to enrolling in the mandatory "stat" or research course, their fears are justified—the course turns out to be a painful, distasteful experience. It is no secret that statistics and research are usually the least popular courses and are generally considered a boring, unnecessary, and irrelevant portion of the academic program.

There is no a priori reason why students should view statistics and research courses as being valueless compared to the other courses they take. The problem lies in the lack of relevancy. If designed to meet the students' needs, these courses would be considered interesting, helpful, and valuable to the students' professional careers. This book has been written in an attempt to improve the relevancy and, therefore, the reputation of these highly unpopular courses.

Some students who sign up for courses in statistics and research fully intend to become involved in their own research activity. Many will limit their involvement to a master's thesis or a doctoral dissertation. A few will remain actively involved in research long after their academic programs have been completed. These students need courses in statistics and research that lead to proficiency in how to conduct research, how to set up experimental designs, and how to use various statistical techniques. For the future researcher, applied courses in statistics and research are highly appropriate.

However, not all students plan to conduct their own research. Certain universities have recently established nonthesis option programs for the master's student; other institutions have eliminated the thesis entirely, substituting a comprehensive written examination. It is now possible for students to complete a graduate program without ever becoming involved in independent research activity. And, if they do not become involved while in graduate school, it is highly unlikely that they will become involved after leaving the university setting. For this second group of students, an applied course in statistics or research is irrelevant to their immediate or long-range needs.

To meet the needs of this second group of students, new courses in statistics and research are being developed and offered at various universities. These new courses are based upon the premise that, although he may not conduct his own research, a student should be capable of understanding the results of research carried out by others. Moreover, he should be able to evaluate this research.

Courses in statistics and research with this perspective are well justified as additions to the graduate catalog. In fact, they are long overdue. Individuals must possess a specific set of skills before they can read and evaluate journal articles. Unfortunately, many students complete a graduate program without ever developing the ability needed to read the professional literature critically. When asked to read a journal article, most students attempt to locate a statement of the author's intent and his summary or conclusions, bypassing the explanation of the design and treatment of data. Similarly, tables presenting numerical information are given little, if any, attention.

It is exceedingly important for students to read and understand the material presented between the statement of the problem and the conclusion, because, although these sections are more difficult to read, they often contain the crucial content of the article. It is the purpose of the new statistics and research courses to teach the individual who is not himself a researcher to comprehend that content. Ideally, the student leaves such a course with the ability to read, to understand, and to evaluate the average research paper within his field of specialization. He should be able to decide, for any given journal article, whether a legitimate design was used, whether the data were subjected to an appropriate statistical analysis, and whether the conclusions drawn were justified in terms of the information provided.

Many texts have been written for applied courses in statistics and research. These books are designed to help students learn the correct procedures to be used by a research investigator and are invaluable for the individual who plans to engage in his own research. However, the traditional statistics and research texts are simply inappropriate for the individual who needs to learn how to read journal articles. At the present time, there are few, if any, books designed specifically for a "consumer" course in statistics or research. This book has been written to fill this unfortunate void in the literature.

The specific objectives of this book are to provide the reader with (1) an understanding of basic statistical terms, (2) an ability to decipher statistical tables, (3) a knowledge of what specific research questions can be answered by each of a variety of statistical procedures, (4) an understanding of the hypothesis testing procedure, (5) an ability to detect the misuse of statistics, (6) a facility for distinguishing between good and poor research designs, and (7) a feeling of confidence with regard to the reading of statistically oriented research reports. In the following paragraphs each of these objectives will be discussed in detail.

A highly specialized vocabulary of terms and symbols has been developed within the field of research methodology and statistics. The reader of professional journals frequently encounters terms such as reject, significant interaction, null hypothesis, t test, 2×3 analysis of variance, and quasi-experimental design. An individual who is not familiar with the precise meaning of these terms will be severely limited in his ability to understand what a researcher actually did. In other words, the reader of professional journals must have a working knowledge of the technical language and the jargon used in journal articles if he is to grasp fully what an author is trying to communicate.

To help students gain a proficiency in reading the professional literature in the behavioral sciences, the authors have included a variety of excerpts from actual journal articles. By bringing the topic under consideration, that is, journal articles, directly to the reader, the authors hope to facilitate the rate at which the student will be able to improve his skills in reading the professional literature.

The results of statistical analyses are frequently presented in tabular form. These tables are not included because the author had some extra space to fill in. They are included to aid the reader with a clear presentation of data. The information contained in a table may help to clarify the author's discussion in the body of the article or the table may provide additional information not specifically discussed in the text. A reader should not ignore these tables; he should use them as an aid to understanding and evaluate the data analysis and results.

A variety of statistical techniques are used by contemporary researchers. The "consumer" of professional journals should be able to specify the type of research questions that can be answered by each of these statistical procedures. For example, the reader of a journal article should be aware that the Spearman correlation is used to measure the degree of relationship between two rank-ordered variables, that Duncan's new multiple range test is used to locate significant differences following a significant F ratio, and so forth. This objective is especially important since researchers do not always use a correct statistical technique to answer their specific research questions.

Whenever a statistical test is used by a researcher, it means that he followed (sometimes unconsciously) the five basic steps of the hypothesis testing procedure. An individual with a basic understanding of these steps will be in a far better position to understand what a significant difference is, how the author decided whether to reject or accept the null hypothesis, the rationale behind the notation $p < .05$, and other related issues. It is not the purpose of this book to explain specific statistical tests. However, a delineation of the process of hypothesis testing should allow the reader of journal articles to have an awareness of the principles behind the researcher's statistical techniques.

As indicated earlier, researchers do not always use correct statistical procedures in their studies. An author may report having used a certain technique when a different one would have been more appropriate. If the reader is capable of detecting the misuse of statistics, he has the power to decide for himself whether the conclusions drawn by an author are justified. Throughout this book, attention will be directed to the appropriate use of statistical procedures.

In experimental studies it is possible for an author to use a legitimate statistical technique with an inappropriate research design. If a design is inadequate, even the most sophisticated statistical analyses cannot increase confidence in the conclusions of the research. The authors have placed a great deal of emphasis on Campbell and Stanley's classification of preexperimental, quasi-experimental, and true experimental designs. In addition, the various threats to the internal and external validity of experiments are thoroughly discussed.

The objectives discussed above all fall into the cognitive domain. The final goal of this book more appropriately falls into the affective domain. The terms statistics and research traditionally create a high level of anxiety among students. This anxiety is simply unnecessary. A concerted effort has been made by the authors to present the material in this book in such a manner as to relieve student anxiety and the fear of statistics. Students should be able to read journal articles with far more emotional confidence after studying this text.

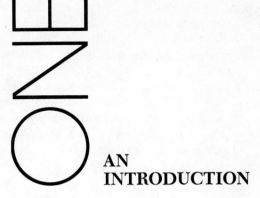

AN INTRODUCTION

There are three types of people in the world: people who plan to conduct their own research studies, people who do not engage in the research process but are interested in finding out about the results of others' investigations, and people who not only do not conduct research studies but are not even interested in the results obtained by other people. If you are a member of the third group, this book is not for you. Return it to the bookstore and get a refund on the purchase price.

If you are a member of the first group, this book may be helpful as a supplementary text, but it cannot serve as the only source for the things you need to learn. To carry out research projects, a person needs to learn how to set up researchable questions, how to choose between alternative experimental designs, and how to use various statistical techniques. A person seriously interested in conducting his own research will need to study one or more of the many excellent textbooks written to help a person develop skills in these and related areas.

If you are a member of the second group, that is, if you are interested in understanding other people's research, then this book is definitely for you! The authors have written this book for the express purpose of helping the reader to develop the skills that are needed to understand and evaluate published journal articles. To be more specific, the authors have designed the text to assist the reader (1) to gain an understanding of the terms and symbols used by the authors of journal articles, (2) to learn how to decipher statistical tables, and (3) to realize that the research process is not as mysterious as it might seem on the surface.

The authors have assumed that you are not a sophisticated mathematician or statistical whiz. In fact, if you know how to add, subtract, multiply, and divide, you possess all the prerequisite skills to understand the small amount of math included in the book. The authors have also assumed that you are not accustomed to reading professional journals. Accordingly, let us now turn to Chapter 1 and a discussion of the typical format used in most journal articles.

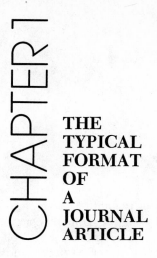

CHAPTER 1

THE TYPICAL FORMAT OF A JOURNAL ARTICLE

Almost all journal articles dealing with research studies are divided into different sections by means of headings and subheadings. Although there is some variation among authors with respect to the terms used as the headings or the order in which the different sections are arranged, there does appear to be a relatively standard format for published articles. A reader of the professional literature will find that he can get the most mileage out of the time he invests if he is familiar with the typical format of journal articles and the kind of information that is likely to be included in each section of the article.

We are now going to look at a particular journal article which does an excellent job of illustrating the basic format that most authors use as a guide when they are writing their articles. The different sections of our model article could be arranged in outline form as follows.

1. Abstract
2. Introduction
   a. Review of the literature
   b. Statement of purpose

3. Method
   a. Subjects
   b. Design
   c. Materials
   d. Procedure
4. Results
5. Discussion
6. References

Let us now examine each of these items.

ABSTRACT

An abstract is a summary of the entire journal article which appears at the beginning of the article. Although it normally contains no more than five or six sentences, the abstract usually provides the following information: (1) a statement of the purpose or objective of the study, (2) a description of the individuals who served as participants, (3) a brief explanation of what the participants did during the study, and (4) a summary of the important results.

Excerpt 1.1 is the abstract from our model article. As in most articles, it follows immediately after the title and authors' names and is printed in smaller type than the rest of the article.

EXCERPT 1.1 ABSTRACT

To determine what effects type and knowledge of the objectives of instruction have on learning two experimental variables (type and knowledge of objectives) were combined in a 2 × 3 factorial arrangement of treatments. One hundred and twelve subjects were randomly assigned to one of six experimental conditions or a learning control condition. The objectives used were either general or specific (type), and were provided either to teachers only, to teachers and students, or to students only (knowledge). Although significant learning occurred for all instructed conditions compared to a test-only control, neither knowledge of the objectives nor type of objectives differentially influenced performance on a criterion test.

SOURCE: J. R. Jenkins and S. L. Deno. Influence of knowledge and type of objectives on subject-matter learning. *Journal of Educational Psychology*, 1971, *62*, 67–70. Copyright 1971 by the American Psychological Association. Reprinted by permission.

The sole purpose of the abstract is to provide the reader with an overview of the material he will encounter in the remaining portions of the article. Although the four sentences in Excerpt 1.1 probably do not make too much sense to you right now, they will after you read a few more chapters of this book. Once you have gained an understanding of the vocabulary used by researchers, you will be able to read an abstract in about ninety seconds and then decide whether or not to read the whole article. On the basis of the abstract you may be able to decide that the article in front of you is exactly what you have been looking for, or that it is not at all related to your interests. Either way, the abstract serves a useful purpose.

INTRODUCTION

The introduction of an article usually contains two items—a review of the literature and a statement of purpose.

Review of the Literature

Most authors start their article with a discussion of previous research studies that have been conducted by them or by other individuals. In discussing previous research studies, an author is attempting to demonstrate how his research project is related to previous ones. Sometimes, of course, an author will publish a completely new idea that is not connected with previous articles that have been published. In the vast majority of articles, however, the author's work can be thought of as an extension of previous knowledge. His review of the literature provides the reader with background information or a framework within which to view the author's study.

The discussion of previous research can be as short as one or two sentences or as long as several pages. This section of the journal article does not typically fall under any heading. Excerpt 1.2 is the review of the literature from our model journal article.

EXCERPT 1.2 REVIEW OF THE LITERATURE

A great deal has been written on the importance of stating educational goals in terms of observable student performance. Among the arguments for using behaviorally stated objectives are (a) unless a teacher (or seahorse; Mager, 1962) knows exactly where he is going he may find himself somewhere else; (b) behaviorally stated objectives reduce wasted time in temporary diversions, ephemeral entertainment, or other irrelevancies (Popham, 1968); and (c) stating goals in behavioral terms permits a more objective measurement of goal attainment.

Implicit in most of the recommendations for using behavioral objectives is the assumptions that with such objectives instruction will be improved by decreasing the time required for learning a particular subject matter, or by increasing the amount of subject matter acquired under conditions of fixed study time. Presently, however, more verbiage than data has spilled on this assumption and the general topic of behavioral objectives. The logical arguments for using behavioral objectives, which are compelling, would be enhanced with some empirical data.

SOURCE: J. R. Jenkins and S. L. Deno. Influence of knowlede and type of objectives on subject-matter learning. *Journal of Educational Psychology,* 1971, *62,* 67–70. Copyright 1971 by the American Psychological Association. Reprinted by permission.

Statement of Purpose

After having reviewed the relevant literature so as to provide a rationale for his study, an author usually states the specific purpose or goal of his particular study. This statement of purpose is one of the most important parts of a journal article, since, in a sense, it explains what the author's "destination" was. It would be impossible for us to evaluate whether or not his "trip" was successful—in terms of his research findings, his conclusions from the findings, and his presentation of them—unless we know where he was headed.

In some journal articles the statement of purpose is found at the end of the final paragraph reviewing the literature. In other articles, a separate paragraph is devoted to the statement of purpose. In Excerpt 1.3 we see that the authors of our model article used a separate paragraph to point out the specific goal of their study.

EXCERPT 1.3 STATEMENT OF PURPOSE

The present study was designed to determine the empirical validity of the assumption that providing either teachers, students, or teachers and students with general or specific instructional objectives increases the amount learned during fixed instruction time. If knowledge of and type of objectives are significant instructional variables, then as students' and teachers' knowledge of specific objectives increases so should learning increase.

SOURCE: J. R. Jenkins and S. L. Deno. Influence of knowledge and type of objectives on subject-matter learning. *Journal of Educational Psychology*, 1971, *62*, 67–70. Copyright 1971 by the American Psychological Association. Reprinted by permission.

In most articles the review of the literature and the statement of purpose are not identified by separate headings nor are they usually found under a common heading. If a common heading were to be used, however, the word Introduction would probably be most appropriate since these two items set the stage for the substance of the article—an explanation of what was done and what the results were.

METHOD

In the method section of a journal article, an author explains in detail how he conducted his study. Ideally, his explanation should contain enough information so that a reader could duplicate the study. To accomplish this goal, the author usually addresses himself to four questions: (1) Who participated in the study? (2) What type of research design was used? (3) What type of materials were needed? (4) What were the participants required to do? The answer to each of these questions is generally found under an appropriately titled subheading of the method section.

Subjects

Each of the individuals who participates in a research study is considered to be a subject (abbreviated S). Within the subjects section of a journal article, an author usually indicates how many subjects he used, who the subjects were, and how they were selected.

A full description of the subjects is needed because the results of a study will often vary according to the nature of the Ss used. This means that conclusions of a study, in most cases, are valid only for individuals who are similar to the Ss used by the author. If a researcher compares two different types of counseling techniques and finds a difference in terms of the improvement of client anxiety, it is imperative that the author indicate whether his Ss were college students, adults, patients in a mental hospital, or whatever. What works for a counselor in a mental hospital may not work at all for a counselor in a normal high school.

It is also important for the author to indicate how he obtained his subjects. Were they randomly selected from a larger pool of potential Ss? Were they volunteers? Were any particular standards of selection used? Did the researcher simply use all of the members of a certain college or high school course? As we shall see in Chapter 12, certain procedures for selecting Ss allow the researcher to generalize his results far beyond the specific Ss that were used in the study, while other procedures for selecting Ss limit the valid range of generalization.

Excerpt 1.4 is the subjects section of our model journal article. The authors of this article did a good job of indicating how many Ss were used, who the Ss were, how the Ss were obtained, and what sort of reward the Ss were given in return for their participation. (The 16 students who served as teachers in this study are not considered to be Ss.)

EXCERPT 1.4 SUBJECTS

METHOD

Subjects

The 112 students who served as subjects were volunteers from a sophomore level educational psychology course. For their participation these students received points toward their final grade. In addition, 16 secondary education students from a senior level educational psychology course served as teachers. The teachers, by participating in the experiment, also earned points for their final grade. The teachers were seniors who had completed most of their course work except student teaching. They were familiar with the content of the instructional materials from previous course work in education and psychology.

SOURCE: J. R. Jenkins and S. L. Deno. Influence of knowlede and type of objectives on subject-matter learning. *Journal of Educational Psychology*, 1971, *62*, 67–70. Copyright 1971 by the American Psychological Association. Reprinted by permission.

Design

In most research studies the researcher attempts to acquire or confirm information about a particular question by making comparisons among two or more groups. The term experimental design is used to describe the way in which the groups are arranged prior to the statistical analysis of the data. There are several different types of designs that can be used, and Chapters

11 through 15 are devoted to a discussion of the more popular designs that are employed by contemporary researchers. After studying these chapters, you should be able to read an author's description of the design he used and then construct a mental picture of the arrangement of the comparison groups.

EXCERPT 1.5 DESIGN

Design

Two experimental variables were arranged in a 2×3 factorial design. The first variable, type of objective, consisted of (1) general (nonbehavioral) and (2) specific (behavioral) objectives. The second variable, knowledge of objective, consisted of (1) teacher knowledge (i.e., possession) of objectives prior to and during instruction, (2) teacher knowledge (as in Level 1) and student knowledge (i.e., possession) during instruction, and (3) student knowledge alone during instruction. It is well to note here that Level 3 (student knowledge alone) is con-

founded with the manner in which content was presented. In the third level, teachers were eliminated, and on the day of instruction students received both the objectives and a set of teaching materials from which to study. The teaching materials provided the students were identical to the materials given in advance to the teachers in the other conditions.

A seventh group, a learning control condition, was made up of subjects who were neither taught by teachers nor given study materials, but who took the criterion examination.

SOURCE: J. R. Jenkins and S. L. Deno. Influence of knowledge and type of objectives on subject-matter learning. *Journal of Educational Psychology*, 1971, *62*, 67–70. Copyright 1971 by the American Psychological Association. Reprinted by permission.

Excerpt 1.5 contains the design section of our model journal article, and in Figure 1.1 we have constructed a diagram to depict the 2×3 factorial design discussed in the first paragraph of the excerpt. Once you have the ability to construct diagrams (such as the one in Figure 1.1) to correspond with the

KNOWLEDGE OF OBJECTIVES

		TEACHER ONLY	TEACHER AND STUDENT	STUDENT ONLY
TYPE	GENERAL			
OF *OBJECTIVE*	SPECIFIC			

FIGURE 1.1. *A diagram to correspond to Excerpt 1.5.*

designs that researchers use, you will be in a far better position to understand how the study was conducted and what types of statistical results are likely to be presented later in the article.

Since the design of a research study is closely related to the statistical procedures for data analysis, this particular section of a journal article is sometimes labeled Data Analysis or Statistical Procedures. And since the design and statistical analysis are so closely related, some authors wait until the results section before indicating what particular design was used. However, no matter where the design is discussed and regardless of the subheading used to label this discussion, it is imperative that you have the ability to understand the type of design used by an author if you are going to understand completely the study which he conducted.

Materials

In this section of a journal article the author describes the materials used in the study. Often the only material involved is the measuring device (for example, the test or questionnaire) used to collect data from the subjects. If this is the situation, the author should indicate whether the measuring device is a new instrument constructed especially for his study or an older instrument developed by someone else. If a new instrument was constructed, the researcher should report whether it was found to be reliable and valid. The concept of reliability deals with whether or not the instrument can measure the same trait consistently upon repeated measurements, while validity deals with whether the instrument is truly measuring the specific trait that it is supposed to measure. When an established, reputable measuring device is used, there is less need for the author to discuss the reliability and validity of his instruments.

In some research studies special equipment is used. The author of an article should carefully describe any such materials that were involved in the conduct of the study. For example, if an experiment were to involve showing Ss pictures of different styles of clothing, the author should report how the pictures were presented, whether the pictures were in color, who wore the clothes in the pictures, how many pictures there were, and all other relevant details.

Sometimes the description of the equipment employed by the researcher is put in a special section labeled Apparatus, but this is usually done only in those studies in which either a lot of equipment or highly sophisticated equipment is involved.

In Excerpt 1.6, the materials section of our model, notice how the authors tell us where their materials came from and how they modified the original Baker criterion test.

EXCERPT 1.6 MATERIALS

Materials

The objectives, the instructional materials, and the criterion examination were taken from an instructional unit developed by Baker (1967). Examples of general objectives are:

This unit seeks to develop in students certain abilities which will make it easier for them to understand how knowledge in the social sciences is systematically accumulated. Broadly, the unit will help students to:

1. analyze written material critically
2. become familiar with some of the general procedures used in social science research
3. learn what common pitfalls in social science research must be avoided
4. know the prominent social scientists and their contribution.

The specific objectives were stated more precisely and each objective was accompanied by a sample test item. Some examples are:

1. From a set of alternative orders, the student should be able to select the proper ordering of the components of an experiment. These components are:

Hypothesis
Procedure
Selection of population and sample
Description of sample
Data collection
Data analysis
Statement of results
Interpretation of results

Sample Test Item. DIRECTIONS: CHOOSE THE BEST ORDER OF STEPS FOR CONDUCTING AN EXPERIMENT.

    a. analyze data, state results, form hypothesis
    b. state results, analyze data, form hypothesis
    c. form hypothesis, analyze data, state results
    d. analyze data, form hypothesis, state results

2. Given the name of one of the data collection procedures listed below, the student should be able to select from alternatives the major advantage or limitation of the technique.

Observations   Statistical records
Interviews   Mass communi-
Questionnaires   cation records
Projective tests   Experiments

Sample Test Item. DIRECTIONS: CHOOSE THE BEST ANSWER FOR THE FOLLOWING QUESTION. ON YOUR ANSWER SHEET FILL IN THE SPACE UNDER COLUMNS a, b, c, OR d NEXT TO THE NUMBER OF THE QUESTION YOU ARE ANSWERING.

One limitation of using projective tests is:
    a. the subjects easily find out the purpose of the test and do not behave naturally
    b. the subjects may not make the same responses on repeated tests
    c. the tests cannot be given to young children or other subjects who cannot read or write

3. Given a description of the key contribution of a social scientist, the student should be able to identify this social scientist from sets of four alternatives which include:

Thomas Carlyle   Bronislaw
Auguste Comte   Malinowski
Sigmund Freud   Karl Marx
Herodotus   Margaret Mead
Alexander von   Plato
  Humboldt   Ivan Pavlov
John Locke   Adam Smith

Sample Test Item. DIRECTIONS: CHOOSE THE SOCIAL SCIENTIST WHO FITS THE FOLLOWING DESCRIPTION.

This historian produced the frontier theory of history. He suggested that the ever-expanding frontier with its limitless free land in America could be an explanation for the differences between Americans and Europeans.
    a. F. J. Turner
    b. Thomas Carlyle
    c. Bronislaw Malinowski
    d. Alexander von Humboldt

The test was modified from the original by the addition of several test items which were generated out of the original objectives and the deletion of other items. In all, the test contained 56 items. The test was not available to any of the participants until after instruction was completed.

SOURCE: J. R. Jenkins and S. L. Deno. Influence of knowledge and type of objectives on subject-matter learning. *Journal of Educational Psychology,* 1971, *62,* 67–70. Copyright 1971 by the American Psychological Association. Reprinted by permission.

Procedure

How the study was conducted is explained within the procedure section of the journal article. In this part of the article the author describes what the Ss did or what was done to them during the research investigation. Sometimes an author will even include a verbatim account of the instructions given to the Ss.

It should be remembered that the method section is included so as to permit a reader to replicate the study. To accomplish this desirable goal, the author must outline his procedures clearly, providing answers to questions such as: Where was the study conducted? Who conducted the study? How long were the Ss allowed to work on examinations that were associated with the study? Did any of the Ss drop out of the study prior to its completion? (In Chapter 11, we shall see that subject drop-out can cause the results to be distorted.)

Excerpt 1.7 is the procedure section of the model article. As you read over this excerpt, ask yourself whether the authors have provided enough information to allow you to set up and conduct the same study. Of course, to do this you would need to go back and reconsider the other parts of the method section.

EXCERPT 1.7 PROCEDURE

Procedure

The teachers were individually contacted 2 weeks prior to the experiment and given the teaching materials. Each teacher was randomly assigned to one of four treatment combinations, general or specific objectives, and teacher or teacher-and-student knowledge of the objective, so that there were four teachers in each of the four conditions requiring teachers. When the teachers received their advance instructions they were provided with objectives, and in the appropriate conditions, with additional copies of the objectives that they would give to their students on the day of instruction.

Sixteen subjects were randomly assigned to each of the six treatment combinations and to the control. A teacher and four students were then assigned at random to one of 16 classrooms. In the two conditions where no teachers were present subjects were provided with objectives and instructional materials, and assigned to one of two classrooms where they could study. In the control conditions subjects were assigned to a classroom where they took the criterion test. In these latter three conditions a monitor was present while the subjects studied and were tested.

In all conditions, except the control, subjects either studied or were taught for 1¾ hours, after which they were administered the criterion tests. All treatment occurred simultaneously. Subjects had unlimited time to complete the criterion test although no subjects took longer than 40 minutes.

SOURCE: J. R. Jenkins and S. L. Deno. Influence of knowledge and type of objectives on subject-matter learning. *Journal of Educational Psychology*, 1971, *62*, 67–70. Copyright 1971 by the American Psychological Association. Reprinted by permission.

RESULTS

There are three ways in which an author can report the results of the statistical treatment of the data that were collected. First, he can describe the results within the text of the article. Second, he can summarize the results in one or more tables. Third, he can construct a graph (which is technically called a figure) to help the reader gain a more complete understanding of how the study turned out. In Excerpt 1.8 we see that the authors of our model article present their results by means of three paragraphs of text and one table.

EXCERPT 1.8 RESULTS

RESULTS

The means and standard deviation for the number correct on the test for the six treatment conditions and the control condition are presented in Table 1.

To ensure that learning had occurred performance of all experimental subjects was compared against the performance of learning control subjects (test only). The mean score for the experimental subjects reliably exceeded that for control subjects on the criterion test ($F = 21.2$, $df = 1/105$, $p < .001$).

A 2×3 analysis of variance was conducted to test the significance of the main effects—type of objective and knowledge of objectives—and their interaction. Neither the main effects nor interaction were statistically significant (type of objective, $F = 2.1$, $df = 1/90$, $p > .10$; knowledge of objectives, $F = 1.90$, $df = 2/90$, $p > .10$; interaction, $F = 1.3$, $df = 2.90$, $p > .10$).

TABLE 1 MEANS AND STANDARD DEVIATIONS OF THE NUMBER CORRECT ON THE CRITERION TEST FOR GROUPS DIFFERING IN KNOWLEDGE OF THE OBJECTIVES AND THE TYPE OF OBJECTIVES

Type of Objective	Knowledge of Objectives					
	TEACHERS ONLY		*TEACHERS AND STUDENTS*		*STUDENTS ONLY (NO TEACHERS)*	
	M	*SD*	*M*	*SD*	*M*	*SD*
General	44.5	3.7	42.9	2.9	46.2	3.0
Specific	42.5	5.1	43.6	4.4	43.9	4.2
Control	38.6	5.5				

SOURCE: J. R. Jenkins and S. L. Deno. Influence of knowledge and type of objectives on subject-matter learning. *Journal of Educational Psychology,* 1971, *62,* 67–70. Copyright 1971 by the American Psychological Association. Reprinted by permission.

Although the results section of a journal article contains some of the most crucial information about the study (if not *the* most crucial information), many readers of the professional literature totally disregard it. When required to read and evaluate research-oriented articles, the typical undergraduate and graduate student will simply skip over the author's presentation of the results.

It is essential that you, as a reader of journal articles, be able to read, to understand, and to evaluate the results provided by the author so that you are not forced into the unfortunate position of uncritical acceptance of the printed word. Researchers are human, they sometimes make mistakes, and, unfortunately, the reviewers who serve on the editorial boards do not catch all of these errors. Therefore, there will sometimes be an inconsistency between the results discussed in the text of the article and the results presented in tables. Or a researcher may use an inappropriate statistical test. Or the conclusions drawn may extend far beyond the realistic limits of the actual data that were collected.

The authors of this book believe that a person does not have to be a sophisticated mathematician before he can read and understand the results section of most journal articles. However, he must become familiar with the terminology, symbolism, and logic used by researchers, and the major part of this text is concerned with just this.

Look at Excerpt 1.8 once again. The concepts of means and standard deviations are discussed in Chapter 2, so after reading the next chapter you should be able to feel comfortable with Table 1 of the excerpt.

The second paragraph of Excerpt 1.8 presents the results of a one-way analysis of variance (although the authors do not state this explicitly). This particular statistical technique is discussed in Chapter 4, and the logic underlying this test (and other similar tests) is taken up in Chapter 3. The final paragraph of the excerpt concerns a factorial analysis of variance, and this topic is dealt with in Chapter 5. Thus, by the time you finish reading the next four chapters of this book, you should be able to return to Excerpt 1.8 and find it to be "easy reading." By the time you finish the entire book, you should be able to read comfortably the results section of the vast majority of journal articles.

DISCUSSION

The results section of a journal article normally contains a technical report of how the statistical analyses turned out, and the discussion section is usually devoted to a nontechnical interpretation of the results. In other words, the author will use the discussion section to explain what the results mean in regard to the central purpose of the study. The statement of purpose, which appears near the beginning of the article, usually contains an underlying or obvious research question, and the discussion section should provide a direct answer to that question.

In addition to telling us what the results mean, many authors use this section of the article to explain why they think the results turned out the way they did. Although such a discussion will occasionally be found in articles where the data support the researcher's hunches, authors are much more inclined to point out possible reasons for the obtained results when these results are inconsistent with their expectations.

Sometimes an author will use the discussion section to suggest ideas for further research studies. Even if the results do not turn out the way the researcher had hoped they would, the study may be quite worthwhile in that it might stimulate the researcher to identify new types of studies that need to be conducted. Although this form of discussion is more typically associated with unpublished master's theses and doctoral dissertations, you will occasionally encounter it in a journal article.

It should be noted that some authors use the term *conclusion* to label this part of the journal article rather than the term *discussion*. These two terms are used interchangeably, and it is unusual to find an article which contains both a conclusion and a discussion.

Excerpt 1.9 contains the discussion section that appeared in our model journal article. Notice that the first two sentences constitute a nontechnical explanation of the results, while the remainder of the first paragraph and the next four paragraphs are used by the authors to discuss possible reasons for the results.

EXCERPT 1.9 DISCUSSION

DISCUSSION

The results clearly do not support the proposition that type and knowledge of the objectives of instruction facilitate learning. It is also interesting to note that students learned at least as much when they studied without the aid of a teacher. The nonsignificant findings cannot be explained away by suggesting that subjects were unmotivated and therefore were inattentive to instruction since significant learning did occur. Further, since the mean number of errors committed by the experimental groups ranged from 9.8–13.5, the nonsignificant differences are probably not the result of a ceiling effect.

The possibility remains, however, that type and knowledge of objectives were insignificant variables because they received inadequate attention from both the teachers and the students. Since teachers and students rarely are exposed to the explicit objectives of instruction they might fail to use these objectives appropriately either because their value is not recognized or because one must learn how to use explicit objectives.

Perhaps a stronger test of the benefits potentially derivable from explicit objectives could be obtained if students or teachers received some incentive to use the objectives, or were given practice in their use.

At least one other possibility exists for why knowledge and type of objective failed to emerge as significant variables. When a unit is well structured, that is, designed to facilitate the attainment of particular objectives, explicitly stated objectives may be superfluous in that teachers and students are able to "read through" the materials to the objectives. If this is the case specific objectives will influence learning only indirectly through their influence on the design of the curricular materials.

An additional conclusion suggested by the data is that the argument which suggests that explicitly stating behavioral objectives produces improvement in learning is a difficult argument to support empirically.

SOURCE: J. R. Jenkins and S. L. Deno. Influence of knowledge and type of objectives on subject-matter learning. *Journal of Educational Psychology*, 1971, *62*, 67–70. Copyright 1971 by the American Psychological Association. Reprinted by permission.

REFERENCES

A journal article concludes with a list of the books or other journal articles that were referred to by the authors. Most of these references will probably come from the review of the literature. Excerpt 1.10 is the reference section of our model.

EXCERPT 1.10 REFERENCES

REFERENCES

Baker, E. L. *Social science research methods: An experimental instructional unit.* Los Angeles: University of California, Department of Education, 1967.

Mager, R. F. *Preparing instructional objectives.* Palo Alto: Fearon Publishers, 1962.

Popham, W. J. Probing the validity of arguments against behavioral goals. Paper presented at the annual convention of the American Educational Research Association, Chicago, February 1968.

SOURCE: J. R. Jenkins and S. L. Deno. Influence of knowledge and types of objectives on subject-matter learning. *Journal of Educational Psychology,* 1971, *62,* 67–70. Copyright 1971 by the American Psychological Association. Reprinted by permission.

The references can be very helpful to you if you want to know more about the particular study that you are reading. Journal articles are usually designed to cover one particular study or a narrowly defined area of a subject. Unlike more extended writings, long papers and books, they do not include background information and descriptions of related studies which would aid the reader's comprehension of the study. Reading books and articles listed in the references will provide some of this information and probably give a clearer understanding as to why and how the author conducted his particular study. Before hunting down any particular reference item, it is a good idea to look back into the article to reread the sentence or paragraph containing the original citation. This will give you an idea of what is in each reference item.

REVIEW TERMS

abstract	method
design	procedure
discussion	references
figure	results
materials	subjects

REVIEW QUESTIONS

1. Where is the abstract usually found in a journal article, and what type of information is normally contained in the abstract?
2. What does the author usually talk about immediately following the review of the literature?
3. What is the abbreviation for the term *subjects?*
4. Should an author take the time to explain how he obtained his subjects?
5. If an author has done a good job of writing the method section of the article, what should the reader of the article be able to do?

6. What are three forms in which an author can present the results of the statistical analyses?
7. Will a nontechnical interpretation of the results usually be found in the results section or the discussion section?
8. Why doesn't a summary usually appear at the end of a journal article?
9. What is the technical name for the bibliography which appears at the end of an article?
10. If an article is published, can we assume that it is free of mistakes?

STATISTICS

In Chapter 1 we saw that most research studies
are conducted in order to answer an underlying question. The research question is typically
presented in the statement of purpose of the
article, immediately following the review of
the literature. To answer the research question,
data are collected (from the Ss) and analyzed
with one or more statistical procedures. Thus,
the statistical analysis of the research study
can be thought of as a stepping stone which
the researcher uses in crossing a stream from
one bank (the question) to the other bank
(the answer).

Although there are hundreds of different
statistical techniques used to analyze data,
they all can be classified into one of two categories. Statistical procedures that do nothing
more than summarize large groups of numbers
are called *descriptive statistics*, since they are
designed solely to describe the characteristics
of a large group of numbers. In Chapter 2 we
will examine three types of descriptive techniques: measures of central tendency, measures of variability, and measures of relationship.

The second category of statistical techniques
involves procedures that are called *inferential
statistics*. By using these techniques, the re-

searcher can go beyond a simple description of the numbers he obtains to more generalized statements. The researcher obtains the numbers he uses from a group of subjects which is called the *sample*. The sample is thought of as having come from a larger group of Ss which is called the *population*. Although the researcher is interested in the characteristics of the population, he only has information (data) from the Ss in the sample. With inferential statistics, the researcher uses the sample data to make scientific guesses (i.e., inferences) about the population. Chapters 4 through 10 are devoted to a variety of inferential techniques.

In Chapter 3 we will discuss the hypothesis testing procedure. This is the five-step process that is followed by a researcher when he uses sample data to make inferences about populations. Chapters 4 through 7 are used to discuss inferential techniques that determine whether significant differences exist between group means. Specifically, t tests, the analysis of variance, and the analysis of covariance are covered in these chapters. In Chapters 8 and 9 we will examine three closely related inferential topics: multiple regression, discriminant function analysis, and multivariate analysis of variance. Finally, Chapter 10 is devoted to a discussion of nonparametric (distribution-free) inferential procedures.

It should be noted that all of the statistical techniques to be discussed in this section of the book can be used either in *descriptive research* studies or in *experimental research* investigations. In experimental research the investigator tries to determine whether a cause-and-effect relationship exists between the treatment variable and the data. For example, an experimental research study might be conducted to see whether different methods of counseling produce varying degrees of client satisfaction. In this hypothetical example, and in all experimental research studies, an attempt is made to identify "if-then" relationships: *If* a client has counseling method 1, *then* he will feel more satisfaction. The purpose of descriptive research is to describe things the way they are, rather than to investigate a cause-and-effect relationship. For example, a descriptive research study might be conducted to determine whether college males differ from college females in terms of vocabulary. The data would be used to establish the existence of a difference, not to point to a cause as to why males differ from females.

Regardless of whether a statistical technique is used within the context of descriptive or experimental research, we will use the term *dependent variable* to refer to the data that the researcher analyzes. In the above examples, therefore, the dependent variable would be vocabulary scores in the descriptive study and some numerical index of client satisfaction in the experimental study. The excerpts from journal articles in the next nine chapters will illustrate the statistical techniques that we will be discussing and show that both descriptive or inferential techniques are used in both descriptive and experimental research studies.

CHAPTER 2

DESCRIPTIVE STATISTICS

The two types of statistics considered in this book are descriptive and inferential statistics. The starting point for both types is usually a set of numbers obtained from measurements or observations. Descriptive statistics are methods used to derive from these raw data certain indices that characterize or summarize the entire set of data. Thus descriptive statistics transform large groups of numbers into more manageable form. Inferential statistics are methods that allow the researcher to generalize characteristics from his set of data to a larger population. While descriptive statistics are concerned only with characteristics of the set of data obtained by the researcher, inferential statistics are concerned with generalizations to a population larger than the set of data obtained by the researcher. This chapter will be devoted entirely to descriptive statistics. The topic of inferential statistics will be covered in later chapters.

FREQUENCY DISTRIBUTIONS

As stated above, descriptive statistics make large groups of data more manageable. For example, suppose a teacher has given a 100-point quiz to 25 pupils. The results of this test might resemble the data in Table 2.1. With

TABLE 2.1 SCORES OBTAINED ON A 100-POINT QUIZ

79	90	95	90	86
86	83	73	81	94
89	91	88	94	83
98	88	88	75	87
84	82	78	85	75

the data in this form, it is fairly difficult to describe the class's performance. By using a method of descriptive statistics, a frequency distribution, the data might look like that in Table 2.2. In this frequency distribution the scores have been arranged from the highest to the lowest. Instead of listing each specific score, however, the scores are combined into class intervals. The width of these intervals is specified in the table ($i = 4$). The frequency of scores that fall within each interval is given in the column labeled f. At the bottom of the table the total number of scores (N) is given.

TABLE 2.2 FREQUENCY DISTRIBUTION OF TEST SCORES

CLASS INTERVAL $i = 4$	f
96–99	1
92–95	3
88–91	7
84–87	5
80–83	4
76–79	2
72–75	3
	$N = 25$

In Excerpt 2.1 we see a table of frequency distributions that was included in an actual journal article. In this study the author was trying to see whether the Manifest Anxiety (MA) Scale could be used to distinguish among university students who (1) seek help from a counseling center regarding a vocational-educational (VE) problem, (2) seek help from the counseling center regarding a personal-emotional (PE) problem, or (3) do not seek help from the counseling center (NC, or nonclient). The total group of subjects was divided into a male group and a female group, and then each of these two groups was further subdivided into VE, PE, and NC groups to make six subgroups in the study. A separate frequency distribution was constructed for each of these groups and also for the total male group and the total female group.

EXCERPT 2.1 FREQUENCY DISTRIBUTION

TABLE 4 FREQUENCY DISTRIBUTION FOR MALE AND FEMALE GROUPS ON MA SCALE

| | Group[a] | | | | | | | |
| | MALES | | | | FEMALES | | | |
SCORE	VE	PE	NC	TOTAL	VE	PE	NC	TOTAL
38–39	0	1	0	1	0	0	0	0
36–37	1	0	1	2	0	0	2	2
34–35	0	1	2	3	1	0	0	1
32–33	0	1	3	4	0	0	0	0
30–31	1	2	4	7	1	0	2	3
28–29	3	1	7	11	3	0	3	6
26–27	3	0	7	10	2	1	3	6
24–25	9	1	9	19	4	6	2	12
22–23	13	1	14	28	5	1	3	9
20–21	13	3	16	32	1	2	6	9
18–19	23	1	29	53	8	1	17	26
16–17	14	3	26	43	10	4	20	34
14–15	19	1	25	45	11	3	15	29
12–13	23	1	39	63	11	1	24	36
10–11	19	2	48	69	14	2	17	33
8–9	24	1	61	86	19	1	18	38
6–7	20	2	60	82	17	4	17	38
4–5	18	2	39	59	11	1	14	26
2–3	11	3	38	52	6	2	8	16
0–1	3	0	7	10	0	0	2	2
Total	217	27	435	679	124	29	173	326

a VE = vocational educational, PE = personal emotional, NC = nonclient.
SOURCE: R. B. Simono. Anxiety and involvement in counseling. *Journal of Counseling Psychology,* 1968, *15,* 497–499. Copyright 1968 by the American Psychological Association. Reprinted by permission.

The frequency distribution can answer several important questions. We can see the most frequently occurring class of scores. We can identify any pattern in the distribution of scores. If most of the scores are concentrated in the middle of the frequency distribution with a few high and a few low scores, then the group of scores resembles what is called a normal distribution. On the other hand, if the bulk of scores are concentrated at either the high or the low end of the frequency distribution with a few scores spread out at the other end, then the distribution would be described as skewed. Normal and skewed distributions are discussed near the end of the next section.

MEASURES OF CENTRAL TENDENCY

In addition to the pattern of scores as shown by the frequency distribution, it may be important to know the most typical or representative score in the group. In this case a measure of central tendency is needed. A measure of central tendency gives a numerical index of the average score of the distribution. The three most frequently used measures of central tendency are the mode, the median, and the mean.

The *mode* is simply the most frequently occurring score. Given the scores 2, 2, 3, 4, 4, 5, 5, 5, 6, 6, 7, the mode would be 5. When reporting this measure, the word mode is usually given and no abbreviation is used. In Excerpt 2.2 the author reports the modal number of residence halls, residence assistants, and students per resident assistant according to institutional enrollment. In the top row of the table, we see that the modal number of residence halls on campuses where the institutional enrollment is 600–900 is 5; that is, the most frequent number of residence halls for schools in this enrollment category is 5.

EXCERPT 2.2 USE OF THE MODE AS
A MEASURE OF CENTRAL TENDENCY

TABLE 1 AVERAGE NUMBERS OF RESIDENCE HALLS, RESIDENT ASSISTANTS, AND STUDENTS PER RESIDENT ACCORDING TO INSTITUTIONAL ENROLLMENT

INSTITUTIONAL ENROLLMENT	RESIDENCE HALLS		RESIDENT ASSISTANTS PER CAMPUS		RESIDENT ASSISTANT-STUDENT RATIO	
(N = 244)	MEAN	MODE	MEAN	MODE	MEAN	MODE
600–900	6	5	28	18	1:34	1:32
901–1,200	6	4	40	23	1:32	1:28
1,201–1,500	7	5	39	23	1:34	1:28
1,501–1,800	9	7	73	38	1:30	1:18
1,801–2,100	7	5	51	38	1:36	1:31
2,101–2,500	11	7	66	25	1:28	1:20
Total	7	4	46	13	1:32	1:28

SOURCE: G. K. Dixon. Undergraduate resident assistant programs in small private colleges. *Journal of College Student Personnel*, 1970, *11*, 135–140. Copyright 1970 by the American Personnel and Guidance Association.

The *median* is the number which lies at the midpoint of the distribution; it divides the distribution of scores into two equal halves. For example, for the scores 6, 2, 7, 2, 5, 6, 3, the median is 5. Three scores are larger than 5; three scores are smaller than 5. Notice that only when there is an odd number of scores can the median be an actual number in the distribution. When there is an even number of scores, the median is a number halfway between the two middle scores. For example, if 7 were omitted from the set of scores given, the median would be 4, that is, the number halfway between 5 and 3.

In presenting the median in a journal article, authors generally either spell out the word median or abbreviate it as Mdn. Excerpt 2.3 includes a table in which four medians are presented. The first median, 1.57, indicates that half of the 82 individuals in the student enrollment-change group had cumulative grade point averages above 1.57, while the other half had cumulative GPA's below 1.57.

EXCERPT 2.3 USE OF THE MEDIAN
AS THE MEASURE OF CENTRAL TENDENCY

Table 2 presents data for this 82 student enrollment-change group. The first item shows the cumulative GPA to date of this group to be 1.57 or a D+ average. The second line shows that the initial enrollments would have necessitated a 2.41 median GPA in order to meet the minimal demands of the STEP Scale at the end of the spring quarter and, therefore, to be eligible to return to school the following quarter.

TABLE 2 MEDIAN GPA'S OF THE 82 STUDENT ENROLLMENT-CHANGE GROUP

	MEDIAN	RANGE
Cumulative GPA to Date	1.57	0.8–1.9
Term GPA Needed to Stay in School	2.41	1.2–5.0
Excess of Term GPA Over Cumulative GPA	0.97	0.0–4.0
Term Credit Load— Spring	14.0	9–28

SOURCE: A. E. Juola, J. W. Winburne, and A. Whitmore. Computer-assisted academic advising. *Personnel and Guidance Journal*, 1968, *47*, 146–150. Copyright 1968 by the American Personnel and Guidance Association.

The *arithmetic mean*[1] is the sum of all the scores divided by the number of scores. For example, if the sum of 7 scores is 28, then the mean is equal to 4. This mean is often called the average, but the word average can also be used to describe the other measures of central tendency. When presented in a journal article, the mean is usually indicated by the word mean or the symbols M or $\bar{X}$. On rare occasions, an author uses the symbols m or $\bar{x}$ to designate the mean.

In Excerpt 2.4 we see the results of a study that was conducted to find out how a group of graduate students in guidance and counseling felt about the courses that were included in their program. A total of 15 means are presented and each mean was found by adding up all of the ratings for a particular course and then dividing by the number (N) of ratings that were available for that particular course. (Note how the ratings for the statistics course turned out!)

To understand the particular sensitivities of each of these three measures of central tendency, a comparison can be made of the three measures. Given the scores 3, 2, 5, 6, 9, 7, 3, the mode is 3, the median is 5, the mean is

[1] There are also geometric and harmonic means, but since the arithmetic mean is the most important and popular mean, it will be the only one discussed here.

EXCERPT 2.4 USE OF THE MEAN AS A MEASURE OF CENTRAL TENDENCY

TABLE 3 RESPONDENTS' MEAN RATING OF INSTRUCTIONAL EFFECTIVENESS

COURSE TITLE	N	MEAN
Principles of guidance	51	2.02*
Occupational information	72	1.85
Statistics	80	1.96
Mental measurement	79	2.19
Counseling theory and techniques	71	2.10
Counseling practicum	75	1.71
Organization and administration	50	2.38
Seminar in guidance	49	2.02
Introd. to college stud. pers.	25	2.04
Seminar in college stud. pers.	14	2.14
American colleges and universities	20	2.25
Field practice in guidance	31	2.39
Group processes	28	3.28
Psychological foundations	42	3.05
Sociological foundations	34	3.41

* Note that in this table 1 = superior and 5 = inferior.
SOURCE: B. Shertzer and J. England. Follow-up data on counselor education graduates—relevant, self-revealing, or what? *Counselor Education and Supervision*, 1968, 7, 363–370. Copyright 1968 by the American Personnel and Guidance Association.

also 5. Suppose we change the score 9 to 16. The mode is still 3; the median is still 5. Yet, the mean has been affected by the change and is now equal to 6. When extreme scores are present in the data, the mode and median are relatively unaffected, but the mean is pulled toward the extreme scores.

To illustrate this differential sensitivity, let us look at the three measures of central tendency in normal and skewed distributions. In a normal distribution all three measures would be the same (Figure 2.1). In a skewed distribution the mean will be the closest measure to the tail of the distribution; the mode will be the farthest from the tail; the median will be between the mean and the mode. For example, in a positively skewed distribution (Figure 2.2A) the extreme scores at the tail of the distribution have increased the value of the mean. In a negatively skewed distribution (Figure 2.2B) with extreme

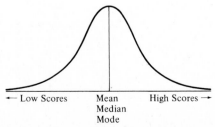

FIGURE 2.1 *Measures of central tendency in a normal distribution.*

FIGURE 2.2. *Measures of central tendency in skewed distributions.*

scores at the lower end of the distribution, the three measures of central tendency would be arranged in the opposite order.

The mean is usually the preferred measure of central tendency because it is based on all of the scores and the quantity of the scores. However, the researcher will, of course, report the measures which are most appropriate for his study. For example, if a number with a decimal would be illogical (for example, number of children per family), the researcher might use the mode for a measure of central tendency. If there are a few extreme scores in one direction, he might report the median as the most representative measure of central tendency.

MEASURES OF VARIABILITY

We have learned that a single numerical index (the mean, median, or mode) can be used to help describe a large group of scores. However, a measure of central tendency may not be enough; additional information may be needed for an accurate description of the larger group of scores. One type of additional information is a measure of variability.

The Meaning of Variability

Most groups of scores possess some degree of variability, that is, the scores differ (vary) from one another. A measure of variability simply indicates the degree of this *dispersion* among the set of scores. If the scores are very similar, there is little dispersion and little variability. If the scores are very dissimilar, there is a high degree of dispersion (variability). In short, a measure of variability does nothing more than indicate how spread out the scores are.

The term variability can also be used to pinpoint where a group of scores might fall on an imaginary homogeneous-heterogeneous continuum. If the scores are similar, they are homogeneous (and have low variability). If the scores are dissimilar, they are heterogeneous (and have high variability).

Even though a measure of central tendency provides a numerical index of the average score in a group, we need to know the variability of the scores to better understand the entire group of scores. For example, consider the following two groups of IQ scores.

GROUP I	GROUP II
102	128
99	78
103	93
96	101

In both groups the mean IQ is equal to 100. However, although the two groups have the same mean score, their variability is obviously different. While the scores in the first group are very homogeneous (low variability), the scores in the second group are far more heterogeneous (high variability).

The above example demonstrates the importance of knowing not only something about the central tendency of a group of scores, but also something about the degree of variability which is present. While a measure of central tendency provides some information, a measure of variability is also needed to gain a more accurate understanding of the entire set of scores. Since both measures are important, authors of journal articles usually include a measure of variability whenever they report a measure of central tendency.

Different Measures of Variability

The *range* is the simplest measure of variability. It is the difference between the lowest and the highest scores. For example, in Group I in the above example, the range is equal to 7 (103 − 96). The range is usually given as a single number, the difference between the extreme scores, but occasionally it is given by reporting the extreme scores themselves. Using the latter format, the range for Group II is equal to 78–128. For notation most authors write out the word range, although the symbol R is occasionally used to stand for this first measure of variability.

In Excerpt 2.5 examples of the range are presented. Within the group of 34 parents of psychotics (the top row of the table) the lowest parental IQ was 79 and the highest 134. Instead of recording the range of IQ scores for this group as 79–134, the authors could have entered the number 55 (134 − 79) in the same place. Using the table, we can compare the variability or range of different factors for different experimental groups. For example, the IQ scores for the parents of psychotics were more variable (range = 79–134) than the IQ scores for the parents of retardates (range = 80–125).

Although relatively simple to calculate, the range is not very stable (reliable) because it is based upon only two of the scores. Since the range does not indicate anything at all about the variability of those scores which lie between the highest and lowest scores, it is not considered to be a very sophisticated measure of variability.

EXCERPT 2.5 TABLE PRESENTING RANGES

TABLE 1 AGE, SOCIAL CLASS, YEARS OF EDUCATION, AND IQ OF THE THREE
PARENT GROUPS

EXPERIMENTAL GROUPS	PARENTS AGE	PARENTS SOCIAL CLASS	PARENTS YEARS OF EDUCA- TION	PARENTS IQ
Parents of psychotics ($N = 34$)				
Range	26–51	1–4	10–21	79–134
M	36	2	14	111
Parents of retardates ($N = 22$)				
Range	24–49	1–5	8–21	80–125
M	36	3	14	110
Parents of normals ($N = 42$)				
Range	24–48	1–3	12–25	102–131
M	34	1	18	121

SOURCE: Eric Schopler and Julie Loftin. Thinking disorders in parents of young psychotic children. *Journal of Abnormal Psycho ogy*, 1969, 74, 281–287. Copyright 1969 by the American Psychological Association. Reprinted by permission.

Two additional measures of variability, the *variance* and the *standard deviation,* are better indices of the dispersion among a group of scores than is the range. Both the variance and standard deviation are based upon all of the scores in the group. The variance is determined by (1) figuring out how much each score deviates from the mean and (2) putting these deviation scores into a computational formula. From this formula, the variance (often a decimal fraction) is derived. The standard deviation is found by taking the square root of the variance. Generally the researcher reports the variance and/or the standard deviation, but not the deviation scores or the formula.

In reporting the variance, most authors use the symbol σ^2 or s^2 or simply the word variance. In Excerpt 2.6 the author presents the mean and variance of the length of treatment in four clinics. The author uses the word variance, rather than σ^2 or s^2. The variance of treatment length for Clinic C is 19.35. Notice in the excerpt that the variance is the smallest for Clinic A and the largest for Clinic B.

EXCERPT 2.6 TABLE PRESENTING VARIANCES

TABLE 2 LENGTH OF TREATMENT IN MONTHS

CLINICS	A	B	C	D
Mean	1.56	20.83	8.90	4.90
Variance	4.88	149.44	19.35	16.197
N	22	18	20	19

SOURCE: Hershel Berkowitz. A preliminary assessment of the extent of interaction between child psychiatric clinics and public s hools. *Psychology in the Schools*, 1968, 5, 291–295.

In reporting a standard deviation, authors may use the abbreviation *SD*, the symbols *s* or *σ*, or write out the word sigma. A few record standard deviation in a plus and minus form, for example, 14.83 ± 2.51, where the first number (14.83) stands for the mean and the second number (2.51) stands for the standard deviation.

In Excerpt 2.7 the mean and standard deviation are presented for two groups, an experimental group and a control group. The author of this table used the notation *SD* to symbolize the standard deviation. For the full-scale IQ, notice how the variability of the experimental group was almost twice as great as the variability of the control group. In fact, on all three measures (full scale, verbal, and performance) the scores in the experimental group were more variable than those in the control group.

EXCERPT 2.7 TABLE PRESENTING MEANS AND STANDARD DEVIATIONS

TABLE 2 IQ, MEANS, AND STANDARD
DEVIATIONS FOR EXPERIMENTAL AND
CONTROL GROUPS

	EXPERIMENTAL	CONTROL
Full scale		
M	100.85	106.53
SD	11.22	6.62
Verbal		
M	99.70	105.85
SD	11.29	7.38
Performance		
M	101.93	106.15
SD	14.54	8.77

SOURCE: J. G. Lyle and J. Goyen. Performance of retarded readers on the WISC and educational tests. *Journal of Abnormal Psychology*, 1969, 74, 105–112. Copyright 1969 by the American Psychological Association. Reprinted by permission.

Excerpts 2.8–2.10 are examples of different ways of presenting standard deviation. The author of Excerpt 2.8 used the symbol *s* for the standard deviation; the authors of Excerpt 2.9 used the symbol *σ*. In Excerpt 2.9, look at the data for the white boys and white girls who were 6 years of age. Although these two groups of children had similar mean IQ scores, the variability of the boys' IQ scores was almost 2½ times as large as the variability of the girls' IQ scores (15.0 vs. 6.3). The authors of Excerpt 2.10 used the plus and minus format to present means and standard deviations. For the male group on the Manifest Anxiety scale, the mean was 13.18 and the standard deviation was 7.58.

Relationships exist among these first three measures of variability which may help in their interpretation. First, the standard deviation is always equal to the square root of the variance. Thus, if an author reports that the variance of a group of scores is equal to 17.68, we know that the standard deviation of the same set of scores would be equal to a little more than 4

EXCERPT 2.8 USING THE SYMBOL *s*
TO STAND FOR STANDARD DEVIATION

POSITIVENESS OF THE SELF-CONCEPT IN 1954 AND IN 1956

		1954			1956	
	N	*M*	*s*	*N*	*M*	*s*
Girls						
8th–10th grade	56	359.98	32.01	45	362.76	29.41
10th–12th grade	40	358.40	36.82	37	365.59	38.44
Boys						
8th–10th grade	48	351.29	34.68	45	352.25	37.15
10th–12th grade	27	360.81	23.03	24	369.75	25.72

SOURCE: Mary Engel. The stability of the self-concept in adolescence. *Journal of Abnormal and Social Psychology*, 1959, *58*, 211–215. Copyright 1959 by the American Psychological Association. Reprinted by permission.

EXCERPT 2.9 USING THE SYMBOL σ
TO DESIGNATE A STANDARD DEVIATION

MEANS AND STANDARD DEVIATIONS OF WISC FS IQs
FOR WHITE ($N = 141$) AND NEGRO ($N = 134$) CHILDREN AT FIVE AGE LEVELS

	AGE IN YEARS					
Ss	5	6	7	8	9	TOTAL
White						
Boys						
N	19	22	12	15	12	80
M	110.5	104.2	97.8	105.1	98.6	
σ	11.5	15.0	12.2	15.6	13.9	
Girls						
N	7	15	15	12	12	61
M	116.6	104.9	97.3	105.8	94.3	
σ	7.9	6.3	7.7	15.2	16.4	
Negro						
Boys						
N	13	14	12	20	12	71
M	83.5	92.6	90.7	98.3	89.6	
σ	13.1	11.3	8.1	9.9+	7.8	
Girls						
N	13	12	12	14	12	63
M	82.5	103.2	90.8	96.6	85.1	
σ	13.9+	7.2	12.3	11.2	11.3	

SOURCE: Ira J. Semler and I Iscoe. Structure of intelligence in Negro and White children. *Journal of Educational Psychology*, 1966, *57*, 326–336. Copyright 1966 by the American Psychological Association. Reprinted by permission.

EXCERPT 2.10 USING THE PLUS AND MINUS FORMAT
TO PRESENT STANDARD DEVIATIONS

MEANS AND STANDARD DEVIATIONS OF Ss ON PERSONALITY MEASURES

MEASURE	MALES ($N = 55$)	FEMALES ($N = 24$)	TOTAL ($N = 79$)
Manifest Anxiety scale (MA scale)	13.18 ± 7.58	14.96 ± 8.76	13.72 ± 7.94
Extraversion (MPI)	30.29 ± 9.37	28.96 ± 10.62	29.89 ± 9.72
Neuroticism (MPI)	20.51 ± 10.39	21.04 ± 10.95	20.67 ± 10.49

SOURCE: E. V. Piers and E. P. Kirchner. Eyelid conditioning and personality. *Journal of Abnormal Psychology*, 1969, *74*, 336–339. Copyright 1969 by the American Psychological Association. Reprinted by permission.

(since the square root of 16 is 4). Perhaps 4.2 would be a good approximation. Second, the range is usually (but not always) about 5 times as large as the standard deviation. Thus, if the standard deviation of a set of scores is equal to 4, we know that the range is probably equal to about 20.

The three measures of variability discussed above are popular and appear frequently in journal articles. In reporting the variability of a group of scores, almost all authors use one of these three measures. There are two additional measures of variability—the *semi-interquartile range* and the *average deviation*. Since these measures are rarely found in the professional literature, we will not discuss their meaning. However, the reader of journal articles should at least recognize that these two terms do signify measures of variability.

CORRELATION

Measures of central tendency and variability describe statistically only one variable, such as the mean height of all males or the range of scores on a test. A correlation describes a relationship between two variables. There are three general types of relationships: (1) positive correlations, (2) negative correlations and (3) zero correlations. Using various correlation techniques a researcher can determine the nature and degree of relationships between variables.

For example, if we measured the height and weight of 100 people and found that the tallest people generally weighed the most and that the shortest people generally weighed the least, we would say that a *positive correlation* exists between the variables of height and weight. When a positive correlation exists, high scores on one variable are paired with high scores on the other variable, and, conversely, low scores on one variable are paired with low scores on the other variable. Most teachers would probably like to think that there is a positive correlation between the amount of time spent studying and the grade obtained on a test. A positive correlation reflects what is called a *direct relationship* between the two variables.

Not all correlations are positive; a researcher might report a *negative correlation*. To illustrate a negative correlation, think of the relationship between a car's age and its monetary value. With the exception of classic and antique cars, as the age of a car increases, the value of the car decreases. Conversely, as the age of the car decreases, the value of the car increases. Thus, for a negative correlation, low values of one variable are paired with high values of the other variable; similarly, high values of one variable are paired with low values of the other variable. A negative correlation can be said to reflect an *inverse relationship* between the two variables.

Finally, there is a third possible correlation between two variables. If we compared the size of shoe worn with scores on a statistics test for any number of people, we would undoubtedly find that no systematic relationship exists between the two variables being measured. We then say that there is no correlation or that a *zero correlation* exists between the two variables.

The nature and size of the relationship between two variables is measured by a correlation coefficient (symbolized by r). The range of possible values is from -1.00 for a perfect negative correlation to $+1.00$ for a perfect positive correlation, so a correlation coefficient is expressed as a decimal fraction preceded by either a plus ($+$) or minus ($-$) sign. The sign indicates whether the correlation is positive or negative; the numbers indicate the strength of the correlation. The closer the coefficient is to either $+1.00$ or -1.00, the higher or stronger the correlation is; the closer the coefficient is to zero, the lower or weaker the correlation. Some examples of correlation coefficients are:

$+.95, +.85, +.93, +.87$	high positive correlation
$+.23, +.17, +.18, +.20$	low positive correlation
$+.02, +.01, \quad .00, -.03$	no systematic relationship
$-.21, -.22, -.17, -.19$	low negative correlation
$-.92, -.89, -.90, -.93$	high negative correlation

A correlation coefficient presented without a plus or a minus sign is interpreted as being positive ($r = .75$ is the same as $r = +.75$).

Correlation Techniques: Pearson Product-Moment Correlation and Spearman's rho

The two most common correlation techniques are the Pearson product-moment correlation coefficient and Spearman's rho. The Pearson correlation coefficient is a parametric technique[2] using continuous data, such as height and weight. Spearman's rho is a nonparametric technique using data in the form of ranks. For example, the rankings of class members on the midterm might be correlated with the rankings of the same class members on the final.

[2] The distinction between parametric and nonparametric techniques is discussed at the beginning of Chapter 10.

Since Spearman's correlation procedure involves ranks, it is often referred to as the rank-order correlation technique. Let us now look at some excerpts from journal articles in which these two correlational procedures were used.

In Excerpt 2.11 we see eight Pearson product-moment correlation coefficients. Each of the two measures of alienation (alienation from work and alienation from expressive relations) was correlated with (1) hierarchy of authority, (2) participation in decisions, (3) job codification, and (4) rule observation. Notice that there is a negative correlation between both measures of alienation and participation in decision. This negative correlation indicates that as participation in decisions increases, both alienation from work and alienation from expressive relations decrease.

EXCERPT 2.11 PEARSON PRODUCT-MOMENT CORRELATION

There is a correlation of 0.49 between the degree of hierarchy and alienation from work as shown in panel A of Table 2. The relationship between degree of hierarchy and alienation from expressive relations is approximately the same (0.45). . . . As shown in panel B of Table 2, there is a direct relationship between the degree of job codification and both alienation from work ($r = 0.51$) and alienation from expressive relations ($r = 0.23$). The former relationship is much stronger than the latter. . . . There is a strong direct relationship between the index of rule observation and both alienation from work ($r = 0.55$) and alienation from expressive relations ($r = 0.65$).

TABLE 2 PRODUCT-MOMENT CORRELATIONS BETWEEN MEASURES OF ALIENATION AND MEASURES OF CENTRALIZATION AND FORMALIZATION, FOR SIXTEEN WELFARE ORGANIZATIONS

MEASURES OF CENTRAL- IZATION AND FORMALIZATION	*MEASURES OF ALIENATION*	
	ALIEN- ATION FROM WORK	ALIENATION FROM EXPRESSIVE RELATIONS
A. Centralization		
Hierarchy of authority	.49	.45
Participation in decisions	−.59	−.17
B. Formalization		
Job codification	.51	.23
Rule observation	.55	.65

source: M. Aiken and J. Hage. Organizational alienation: a comparative analysis. *American Sociological Review*, 1966, *31*, 497–507.

In Excerpt 2.12 the authors present a table of six rank-order correlation coefficients. In this study there were two groups of adolescents, a "high IQ" group and a "high creativity" group. A questionnaire (called the Outstanding Traits Test) was administered to the members of both groups, and scores on three different parts of the questionnaire were correlated for the members of each group. Of the six correlations reported, five are positive and one is negative. The correlation of .81 indicates a strong relationship, the correlations of .67, .62, and .59 indicate moderate relationhips, and the correlations of −.25 and .10 indicate weak relationships.

EXCERPT 2.12 USE OF SPEARMAN'S RHO
TO CORRELATE TWO VARIABLES

Perhaps the most striking and sugges-
tive of the differences between the two
groups are observed in the relation of the
qualities they want for themselves to the
qualities they believe lead to adult success and the qualities they believe teachers favor.

For the high-I.Q. group, the rank-
order correlation between the qualities
they would like to have themselves and
the qualities making for adult success was
.81; for the high creativity group it was
.10. For the high-I.Q. group, the correlation between the qualities they would like
to have themselves and the qualities they
believe teachers favor was .67; for the
high-creativity group it was −.25. The
data are presented in Table 2.

TABLE 2 RANK-ORDER CORRELATIONS
AMONG SUBSECTIONS OF THE
OUTSTANDING TRAITS TEST

| | SUBJECTS | |
COMPONENTS OF CORRELATIONS	I.Q. (N = 28)	CREATIVE (N = 26)
"Personal traits be-lieved predictive of success" and "per-sonal traits believed favored by teachers"	.62	.59
"Personal traits pre-ferred for oneself" and "personal traits believed predictive of adult success"	.81	.10
"Personal traits pre-ferred for oneself" and "personal traits believed favored by teachers"	.67	−.25

SOURCE: J. W. Getzels and P. W. Jackson. The highly intelligent and the highly creative adoles-
cent: A summary of some research findings. In C. W. Taylor (Ed.), *The Third (1959) University
of Utah Research Conference on the Identification of Creative Scientific Talent.* Salt Lake City:
University of Utah Press, 1959.
   The above edition is out-of-print; however, this article has been reprinted in R. G. Kuhlen and
G. G. Thompson (Eds.), *Psychological studies of human development.* New York: Appleton-
Century-Crofts, 1963. Pp. 370–381.

The Correlation Matrix

A more specialized form of correlation table is the correlation matrix.
The correlation matrix usually presents intercorrelations of a list of variables,
that is, the matrix presents all the possible combinations of correlations between a certain number of variables. Excerpt 2.13 is an example of a correlation matrix. Notice that the numbers across the top of the table correspond
to the variables listed at the left of the table. For example, the number 8 in
the top row of numbers represents the variable of Latin grades. Each figure
in the table represents the correlation between the variables corresponding
to the row and column in which the figure is located. Thus, the correlation of
.50 in the table represents a correlation between IQ on the HSPT and language arts on the HSPT; .57 is the correlation between social studies grades
and English grades.

   In Excerpt 2.13 the intersection of row 1 with column 1 contains
a dash. The correlation of HSPT, IQ with itself would be equal to 1.00, and
the same is true for row 2 column 2, row 3 column 3, etc. Most researchers

EXCERPT 2.13 CORRELATION MATRIX

TABLE 1 INTERCORRELATIONS AMONG THE 11 VARIABLES

VARIABLES	1	2	3	4	5	6	7	8	9	10	11
1. HSPT, IQ	—	.63	.49	.50	.70	.36	.31	.26	.32	.38	.40
2. HSPT, Reading		—	.32	.42	.74	.36	.35	.20	.21	.37	.37
3. HSPT, Arithmetic			—	.42	.76	.37	.32	.38	.46	.34	.47
4. HSPT, Language Arts				—	.80	.38	.30	.32	.26	.33	.39
5. HSPT, Composite					—	.48	.42	.40	.41	.45	.53
6. English grades						—	.57	.63	.55	.60	.82
7. Social studies grades							—	.56	.48	.54	.78
8. Latin grades								—	.60	.56	.84
9. Algebra grades									—	.54	.79
10. General science grades										—	.80
11. Overall average grades											—

SOURCE: J. Impellitteri. Predicting academic achievement with the High School Placement Test. *Personnel and Guidance Journal*, 1967, *46*, 140–143. Copyright 1967 by the American Personnel and Guidance Association.

leave the upper-left to lower-right diagonal blank or insert dashes, as in Excerpt 2.13. A few authors, however, do put 1 or 1.00 into this diagonal. Excerpt 2.14 shows this alternative procedure for setting up a correlation matrix.

EXCERPT 2.14 CORRELATION MATRIX WITH 1.00 IN THE DIAGONAL

Finally, Table 5, which shows the Pearson product-moment correlations among the four types of drugs, demonstrates that students who use one type of drug are likely to use other types. This is particularly true of marijuana and the hallucinogenic and amphetamine drugs, where the correlations are relatively high ($r = $ .49 to .59). Of special interest is the fact that the amphetamines show the highest association ($r = .41$) with hard drugs.

TABLE 5 CORRELATIONAL MATRIX OF VARIOUS TYPES OF DRUG USE

DRUG USE TYPE	1	2	3	4
Marijuana (1)	1.00	.59	.49	.26
Hallucinogens (2)		1.00	.56	.34
Amphetamines (3)			1.00	.41
Hard drug (4)				1.00

SOURCE: D. L. Hager, A. M. Vener, and C. S. Stewart. Patterns of adolescent drug use in middle America. *Journal of Counseling Psychology*, 1971, *18*, 292–297. Copyright 1971 by the American Psychological Association. Reprinted by permission.

Usually, there are no numbers in the lower-left part of a correlation matrix. If they were included, the numbers would be a mirror image of the numbers in the upper-right portion of the table. To illustrate this point, look at Excerpt 2.15. Notice that the numbers in the lower-left half are the same numbers as those in the upper-right half.

EXCERPT 2.15 CORRELATION MATRIX WITH
NUMBERS IN BOTH HALVES OF THE TABLE

TABLE 7 INTERCORRELATION AMONG PREFERENCES FOR DIFFERENT
RESOURCES WHEN MONEY IS GIVEN

RESOURCE PREFERRED	LOVE	STATUS	INFOR- MATION	MONEY	GOODS	SERVICES
Love		56	40	−22	22	21
Status	56		38	−11	26	36
Information	40	38		−04	35	33
Money	−22	−11	−04		49	32
Goods	22	26	35	49		65
Services	21	36	33	32	65	

SOURCE: J. L. Turner, E. B. Foa, and U. G. Foa. Interpersonal reinforcers: classification, inter-relationship, and some differential properties. *Journal of Personality and Social Psychology,* 1971, *19,* 168–180. Copyright 1971 by the American Psychological Association. Reprinted by permission.

Correlation and Causality

When interpreting correlations it is absolutely essential to remember that a high correlation does not necessarily indicate that a causal relationship exists between the two variables. For example, a high positive correlation between the amount of time spent studying and score on a test does not prove that more studying will cause one's test scores to increase. The correlation might be due to a third factor, perhaps intelligence. In short, correlation coefficients, even when they are very strong, do not provide sufficient information to infer causality. In Chapters 12 and 13 we will discuss several experimental designs that can be, and are, used by researchers to demonstrate that one variable has a causal effect on a second variable. Also, in Chapter 14 we will present a design in which a researcher can infer causality by using several correlation coefficients.

Other Types of Correlation Procedures

Although the Pearson and Spearman correlation procedures are used with a high degree of frequency by researchers, other techniques are also occasionally used. For example, the *point-biserial procedure* is used to correlate a dichotomous variable (e.g., sex) with a continuous variable (e.g., IQ). The *biserial technique* also measures the relationship between a dichotomous variable and a continuous variable, but it is used when an artificial dichotomy is imposed upon a truly continuous variable. For example, if people were dichotomized on the basis of yearly income (really a continuous variable) into those who make $10,000 or more vs. those who make less than $10,000 and if the second variable were IQ, the biserial procedure would be used.

The *phi correlation* is used to correlate two truly dichotomous variables (e.g., sex and "handedness"). Still another technique, *tetrachoric correlation,* is used to correlate two artificially dichotomous variables.

The six correlation procedures mentioned in the previous paragraphs are designed for measuring the degree of relationship between two variables. In some studies a researcher will be interested in correlating (1) a set of two or more predictor variables with (2) a single criterion variable. For example, a director of admissions at a college or university might like to correlate high school GPA and SAT scores (predictor variables) with first-quarter college GPA (criterion variable). For problems such as this the procedure of *multiple correlation* (sometimes called *multiple regression*) must be used. (Since multiple correlation is used by many researchers, we shall examine it in greater detail in Chapter 8.) Finally, a complex procedure called *canonical correlation* is available to researchers who want to correlate two or more predictor variables with two or more criterion variables.

REVIEW TERMS

average	median
average deviation	mode
class interval	negatively skewed
correlation	normal distribution
direct relationship	Pearson product-moment correlation
dispersion	coefficient
frequency distribution	positively skewed
heterogeneous	range
homogeneous	semi-interquartile range
inverse relationship	Spearman's rho
mean	standard deviation
measure of central tendency	variance
measure of variability	

REVIEW QUESTIONS

1. What do each of the following symbols stand for: N, $\overline{X}$, SD, r, σ^2, f?
2. What is a good one-word synonym for the phrase measure of central tendency? What are the three most popular measures of central tendency?
3. Which of the three measures of central tendency is most affected by extreme scores?
4. If a researcher wanted to compare two groups to see which group was most homogeneous, would he compute a measure of central tendency or a measure of variability for each group?
5. Name three measures of variability that are frequently reported in journal articles. Which of these is the simplest to calculate? Which is the least accurate?

6. Can the numerical value of a measure of variability ever turn out to be negative?
7. If the variance of a certain set of scores is equal to 4, how large will the standard deviation be?
8. Is the average deviation a measure of central tendency or a measure of variability?
9. Correlation coefficients can assume numerical values between _____ and _____. Within this range of possible values, a low correlation would be reflected by a coefficient equal to _____.
10. If a researcher wants to correlate the rank order of finish in a foot race with the rank order of finish in a swimming race, would he be likely to use the Pearson or Spearman correlation procedure?
11. If a researcher computes a Pearson correlation for each possible pair of four variables and puts the results into a correlation matrix, what is the minimum number of correlation coefficients that will be inside the table?
12. Does a correlation of +1.00 between the amount of time spent studying for a test and the total number of test questions answered correctly prove that a person will do better on the test if he studies more?

CHAPTER 3

INTRODUCTION TO INFERENTIAL STATISTICS

In the previous chapter we learned that researchers may use measures of central tendency, variability, and correlation to summarize groups of data. These measures are descriptive statistics, that is, they describe the characteristics of the data collected by the researcher. Often researchers are not satisfied with only describing their data and would like to use the characteristics of their data to generalize to a larger group which they have not measured. *Inferential statistics* are the statistical techniques which are used by researchers to generalize from characteristics of a small group to a larger group unmeasured by the researcher, that is, inferential statistics allow the researcher to infer the characteristics of a larger group from the characteristics of a smaller group.

POPULATION AND SAMPLE

We have been discussing inferential statistics in terms of small and large groups; however, more precise terms are used to denote these groups. The larger group, to which the researcher is generalizing, is known as a *population*. A population is all the possible members of a group defined by the researcher. For example, a population might be all the third-graders in a school system, all the male students at a university, or all registered voters in a state. The smaller group from which the researcher generalizes is known as a *sample;* a sample is a subgroup of a population.

When a researcher is using inferential statistics, he selects a sample from a population, derives data from the sample, and then generalizes from the sample data to the population. For example, suppose a researcher would like to know the mean weight of all the females at a large university. He could measure each female and calculate the mean from the data. This approach would involve considerable time and expense. However, by applying the theory of inferential statistics, the researcher can select a sample of females from the population, calculate the mean weight of the females in the sample, and then make an inference about the mean weight of all the females at the university. This hypothetical example demonstrates the simplest application of inferential statistical theory.

Obviously, the accuracy of the researcher's inference depends on how representative the sample is of the population. To make the sample more representative, most researchers randomly select the sample from the population. A *random sample* means that all members of the population had an equal chance to be selected for the sample. Thus, in our above example, all females at the university have an equal chance (or equal probability) to be selected and weighed. In most cases, if the sample is not randomly selected, inferences and generalizations to the population should not be made. Randomly selected samples will be discussed in greater detail in Chapter 12.

Excerpts 3.1 and 3.2 illustrate how descriptions of random sampling can be reported in a journal article. In Excerpt 3.1 a sample of 40 is randomly selected from a population of 80. Excerpt 3.1 is simply a statement that a

---

EXCERPT 3.1 DESCRIPTION OF A RANDOM SAMPLE

Eighteen male and 22 female graduate and undergraduate resident dormitory counselors at the University of Arkansas were randomly selected from a population of 80 counselors.

SOURCE: J. Bozarth, S. Rubin, K. Mitchell, and J. Pelosi. Verbal protocol patterns of college dormitory counselors. *Counselor Education and Supervision*, 1970, *10*, 23–29. Copyright 1970 by the American Personnel and Guidance Association.

---

random sample was taken. Excerpt 3.2 gives more details about the process used to obtain the sample. The population was "all freshmen who had enrolled for the 1954 fall term and who had graduated from the Portland public schools." The sample of 228 freshmen was selected by using a table of random numbers. (Although researchers can, and sometimes do, obtain random samples by picking names out of a hat, a table of random numbers allows the researcher to accomplish this same goal more easily and more scientifically.)

EXCERPT 3.2 DESCRIPTION OF THE
PROCESS OF RANDOMLY SELECTING A SAMPLE

The study was based on a random sample of 228 freshman students at Portland State College. To obtain this sample, the names of all freshmen who enrolled for the 1954 fall term and who had graduated from the Portland public schools were put into an alphabetical list. The names on this list were then assigned consecutive numbers from 1 to 647. By the use of a randomized sample table the sample was selected.

SOURCE: L. R. Pierson. High school teacher prediction of college success. *Personnel and Guidance Journal*, 1958, 37, 142–145. Copyright 1958 by the American Personnel and Guidance Association.

HYPOTHESIS TESTING PROCEDURE

In the above section we discussed an example of the use of inferential statistics to estimate a population mean by using a sample mean. However, most applications of inferential statistics are more complex. Usually a researcher compares two or more groups to determine if the corresponding populations are similar. In making these comparisons and in most applications of inferential statistics, five steps are usually followed by researchers even though each step may not be reported in every journal article. These five steps constitute the hypothesis testing procedure.

1. Stating the Hypothesis

The initial step is a statement of the hypothesis. According to statistical theory, the hypothesis is always stated as a *null hypothesis*, that is, a statement that no difference exists between the populations being compared. Then the results of the statistical test are stated in terms of the probability that the null hypothesis is false. Excerpt 3.3 illustrates an author's statement of his null hypotheses.

EXCERPT 3.3 NULL HYPOTHESES

In this investigation of other-referred and self-referred students who used the University of Denver Counseling Service, the following hypotheses were formulated:

Hypothesis 1: There is no difference in college graduation rates between students who were referred to the counseling service and students who utilized the counseling service at their own initiative.

Hypothesis 2: There is no difference between the average rate of increase or decrease in grade point average between the self-referred and the other-referred students.

SOURCE: R. Redding. Self-referred students and other-referred students using college counseling services. *Journal of Counseling Psychology*, 1971, 18, 22–25. Copyright 1971 by the American Psychological Association. Reprinted by permission.

Frequently, in journal articles a researcher will state his hypothesis as an *alternative hypothesis* rather than as a null hypothesis. An alternative hypothesis is the logical opposite of a null hypothesis. If a null hypothesis states that there is no difference, an alternative hypothesis states that there is a difference. Excerpt 3.4 presents an author's alternative hypotheses. If hypothesis 1 in Excerpt 3.4 had been stated as a null hypothesis, it would have

---

EXCERPT 3.4 ALTERNATIVE HYPOTHESES

1. Teacher behavior changes as a function of changes in student behavior.
2. Teacher behavior is more positive (i.e., indirect) during periods of positive student behavior than during periods of negative student behavior.
3. Teacher behavior is more positive (indirect) during periods of positive student behavior than during periods of natural student behavior.
4. Teacher behavior is more positive (indirect) during periods of natural student behavior than during periods of negative student behavior.

SOURCE: S. Klein. Student influence on teacher behavior. *American Educational Research Journal*, 1971, 8, 403–421. Copyright 1971 by the American Educational Research Association, Washington, D.C.

---

read: There will be no teacher behavior changes as a function of changes in student behavior. In Excerpt 3.5 the author uses both null and alternative forms in presenting his hypotheses. The first hypothesis is stated in alternative form, the other two in null form.

EXCERPT 3.5 USE OF BOTH NULL AND ALTERNATIVE FORMS TO PRESENT HYPOTHESES

Hypotheses

The major hypothesis tested was that recommendations made by counselors who have participated in simulated counselor-teacher conferences will be rated more useful by teachers.

Two minor null hypotheses were also tested. The first was that years of teaching experience of the raters would not be related to their ratings of the counselors' recommendations. The second was that there would be no significant interaction between the teachers' ratings of counselors' recommendations and the teachers' years of teaching experience.

SOURCE: E. E. Panther. Simulated consulting experiences in counselor preparation. *Counselor Education and Supervision*, 1971, 11, 17–23. Copyright 1971 by the American Personnel and Guidance Association.

---

2. Selecting a Level of Significance

Having stated the hypothesis, the next step in the hypothesis testing procedure is the selection of a *level of significance*. The level of significance is a probability that defines how rare or unlikely the sample data must be before the researcher can reject the null hypothesis. The two most common levels are the .05 and the .01 level. Of these two, the .01 level requires a more rare and unlikely set of data than the .05 level. For example, if the researcher were testing for differences between group means, the means would have to be farther apart at the .01 level than at the .05 level before he could reject the null hypothesis. But on the other hand, a researcher can be more confident that the population means are actually different if he rejects the null hypothesis at the .01 level than at the .05 level.

Occasionally researchers choose levels of significance other than .01 and .05. Other significance levels occasionally used by researchers are the .001, .025, and .10. Levels of significance larger than .05 (e.g., .20, .30, etc.) are rarely used for hypothesis testing. It is important that the researcher choose his level of significance before he collects his data.

Levels of significance are sometimes called by other names in journal articles. In Excerpt 3.6 the authors use the term level of significance,[1] while in Excerpt 3.7 the term *level of probability* is used. Note that in Excerpt 3.6 the .05 level was used, while the .01 level was used in Excerpt 3.7. In Excerpt 3.8 the author uses the algebraic expression $\alpha = .05$ to denote the level of significance (α is the Greek letter alpha). Finally, the .95 *level of confidence* means the same thing as the .05 level of significance.

EXCERPT 3.6 USING THE TERM LEVEL OF SIGNIFICANCE

The .05 level of significance was used to evaluate the relevant F-ratios.

SOURCE: S. W. Huck and N. D. Bowers. Item difficulty level and sequence effects in multiple-choice achievement tests. *Journal of Educational Measurement*, 1972, 9, 105–111. Copyright 1972 by the National Council on Measurement in Education, East Lansing, Mich.

EXCERPT 3.7 USING THE TERM LEVEL OF PROBABILITY
INSTEAD OF LEVEL OF SIGNIFICANCE

In all of the hypothesis testing, the .01 level of probability was utilized.

SOURCE: D. J. Mickelson and R. R. Stevic. Differential effects of facilitative and nonfacilitative behavioral counselors. *Journal of Counseling Psychology*, 1971, 18, 314–319. Copyright 1971 by the American Psychological Association. Reprinted by permission.

EXCERPT 3.8 USING THE SYMBOL α TO
STAND FOR LEVEL OF SIGNIFICANCE

For $\alpha = .05$ . . . , none of the second and fifth grade means were significantly different from each other.

SOURCE: J. M. Kean. Second- and fifth-grade teachers' oral classroom language. *American Educational Research Journal*, 1968, 5, 599–615. Copyright 1968 by the American Educational Research Association, Washington, D.C.

---

3. Computing the Calculated Value

After the data are collected, a formula corresponding to the appropriate inferential statistical test is applied to the data to obtain the *calculated value*.

---

[1] The phrase relevant F ratios in Excerpt 3.6 refers to an inferential statistical test that we will study in later chapters.

Although the formula and data rarely appear, the calculated value is presented in most journal articles. For the purposes of this book, the reader will not have to compute a calculated value. In Chapter 4, we will illustrate the way the researcher uses the calculated value.

4. Obtaining the Critical Value

After computing the calculated value, the researcher must look up the *critical value* in the appropriate statistical table in the back of a statistics book. The critical value is a criterion (a scientific cut-off point) related to the level of significance chosen earlier by the researcher. In Excerpt 3.9 the author reports his calculated and critical value. The number 28.12 represents the calculated value computed from the data with the appropriate formula.[2] The number 5.30 represents the critical value obtained from a statistical table.

EXCERPT 3.9 CALCULATED AND CRITICAL VALUE

An over-all F test of differences in stability between self-concept groups resulted in an F ratio of 28.12, greatly exceeding the ratio of 5.30 needed for significance at $p = .05$.

SOURCE: M. Engel. The stability of the self-concept in adolescence. *Journal of Abnormal and Social Psychology*, 1959, 58, 211–215. Copyright 1959 by the American Psychological Association. Reprinted by permission.

---

5. Rejecting or Failing to Reject the Null Hypothesis

In the final step of the hypothesis testing procedure, the researcher must make a decision about the null hypothesis. If the critical value is larger than the calculated value, then the researcher *accepts* the null hypothesis.[3] When accepting the null hypothesis, the researcher often states that there were "no significant differences" between his samples. In Excerpt 3.10 the researcher failed to reject the null hypothesis.

EXCERPT 3.10 FAILING TO REJECT THE NULL HYPOTHESIS

The null hypothesis, therefore, is accepted, and the third hypothesis—that there are no significant differences between underachievers and achievers in the proportions of classroom time spent listening and watching, speaking to the teacher, and being socially friendly with peers—is substantiated.

SOURCE: H. V. Perkins. Classroom behavior and underachievement. *American Educational Research Journal*, 1965, 2, 1–12. Copyright 1965 by the American Educational Research Association, Washington, D.C.

---

[2] The inferential statistical test referred to as the F test or the F ratio is discussed in the next four chapters.

[3] Technically the researcher does not actually "accept" the null hypothesis; he "fails to reject" the null hypothesis.

On the other hand, if the calculated value is larger than the critical value, the researcher *rejects* the null hypothesis. After rejecting the null hypothesis, the researcher often states that *significant differences* exist between his samples. Excerpts 3.11 and 3.12 illustrate how some authors report rejection of the null hypothesis.

EXCERPTS 3.11–3.12 REJECTING THE NULL HYPOTHESIS

Generally speaking, the performance of the group trained in abstract thinking on the criterion tests of problem solving tended to lead toward rejection of the null hypothesis.

SOURCE: S. W. Lundsteen. Manipulating abstract thinking. *American Educational Research Journal*, 1970, 7, 373–396. Copyright 1970 by the American Educational Research Association, Washington, D.C.

The first hypothesis that there would be no difference in the number of trials needed to learn the two tasks was rejected ($p < .001$).
The second hypothesis that the good readers would learn lists of paired associates in fewer trials than the poor readers was supported as the difference was significant ($p < .001$).

SOURCE: J. W. Giebink and L. L. Goodsell. Reading ability and associative learning. *American Educational Research Journal*, 1968, 5, 412–420. Copyright 1968 by the American Educational Research Association, Washington, D.C.

Generally, the researcher will report the probability level at which he was able to reject the null hypothesis. For example, if the level of significance was .05, the researcher will reject the null hypothesis at a probability that is less than .05. Instead of writing this out in words, the researcher will use the equation $p < .05$ (p meaning probability, the symbol $<$ meaning "less than"). In other words, the symbolic statement $p < .05$ means that (1) if the null hypothesis is true, the chances are less than 5 out of 100 that the researcher would get the sample data he did and (2) since the sample is so different from what would be expected under the condition of a true null hypothesis, the researcher concludes that the null hypothesis was probably incorrect to begin with.

POSSIBLE ERRORS IN HYPOTHESIS TESTING

There is a possibility that the researcher will make the wrong decision concerning the null hypothesis, in either accepting or rejecting it. The decision he makes is based on his measurement of the samples and not on the populations. Even though the samples may have been randomly selected from their populations, there is always a possibility (though often small) that the samples are not representative of their corresponding populations. One of these errors is possible regardless of the decision made about the null hypothesis.

A *Type I Error* is rejecting the null hypothesis when it should be accepted. The probability of making a Type I Error is equal to the level of significance. For example, if the significance level is set at .05, then there is a probability of .05 (i.e., 5 chances out of 100) that the sample data will be extreme enough for the researcher to reject the null hypothesis when it is actually true. Obviously, if the researcher is primarily interested in avoiding a Type I Error he should use the .01 or .001 level of significance.

A *Type II Error* is accepting the null hypothesis when it should be rejected. Unlike a Type I Error, there is no simple relationship between the level of significance and the probability of making a Type II Error. However, if the researcher wants to decrease the probability of making a Type II Error, he could lower his level of significance from .05 to .10, or even lower.

While researchers and authors of journal articles rarely discuss Type I or Type II Errors, in Excerpt 3.13 the author points out the possibility of a Type II Error. Notice that this possibility is associated with the researcher's accepting the null hypothesis.

EXCERPT 3.13 TYPE II ERROR

Since no significant differences among the treatment means were found in either experiment one or two, the null hypothesis was accepted. Consequently, there is the possibility of a Type II Error. . . .

SOURCE: S. J. Samuels. The effect of letter-name knowledge on learning to read. *American Educational Research Journal,* 1972, 9, 65–74. Copyright 1972 by the American Educational Research Association, Washington, D.C.

Finally, it is impossible for both errors to be made at the same time. If the researcher rejects the null hypothesis, there is a possibility of making a Type I Error (but not a Type II Error). If the researcher accepts the null hypothesis, there is a possibility of making a Type II Error (but not a Type I Error). Furthermore, since inferential statistics is concerned with inferences from samples and since population characteristics are rarely known, the researcher seldom knows if a Type I or a Type II Error has occurred.

ONE- AND TWO-TAILED TESTS

When using some tests of significance, the researcher must also decide whether the test will be a one-tailed test or a two-tailed test. A *two-tailed test* is sensitive to significant differences in either direction (i.e., greater *and* less); the *one-tailed test* is sensitive to differences in only one direction (i.e., greater *or* less). Suppose, for example, that a researcher compares achievement test scores of a group of students exposed to a new method of instruction to the scores of another group instructed by the traditional method. If a two-tailed test is used to compare the scores of both groups, the researcher can answer two questions: (1) Do students under the new method score significantly higher? (2) Do students instructed with the traditional method score higher? On the other hand, if a one-tailed test is used, the researcher can answer only one question: (1) Do students under the new method score significantly higher?

Also, if differences are found to be significant at a certain level of significance with a one-tailed test, the same difference with a two-tailed test would be significant at a level of significance twice as large as that used with a one-tailed test. For example, if the researcher found a significant difference at the .025 level with a one-tailed test, the same data used with a two-tailed test would be significant only at the .05 level.

When the direction of the difference between the populations is unknown, then the researcher will use a two-tailed test. A researcher should use a one-tailed test only if he is certain of the direction of difference between the populations or if he is concerned only about a difference in a certain direction. For example, in testing a new drug against a placebo, the re-

---

EXCERPTS 3.14–3.16 DESCRIPTIONS OF THE USE OF ONE-TAILED AND TWO-TAILED TESTS

The comparison of the mean shifts of the two experimental conditions yielded a significant difference ($t = 1.94$, $p < .05$, one-tailed).

SOURCE: A. Vinokur. Distribution of initial risk levels and group decisions involving risk. *Journal of Personality and Social Psychology*, 1969, 13, 207–214. Copyright 1969 by the American Psychological Association. Reprinted by permission.

MEAN PHOTO RATINGS FOR THE +5 AND −5 EXPECTANCY GROUPS BY EVALUATION APPREHENSION LEVEL

EVALUATION APPRE- HENSION	EXPECTANCY +5	−5	DIFFERENCE[a]
High	+.16	−1.06	+1.22*
Low	−.78	−.59	−.19

[a] The signs (+ and −) in the difference column indicate whether the difference was or was not in the predicted direction.
* $p < .01$ (one-tailed).

SOURCE: M. W. Minor. Experimenter-expectancy effect as a function of evaluation apprehension. *Journal of Personality and Social Psychology*, 1970, 15, 326–332. Copyright 1970 by the American Psychological Association. Reprinted by permission.

TABLE 2 SHOCKS TAKEN BEFORE ANNOUNCING PAIN AND TOLERANCE, STUDY I

		Pain					Tolerance				
		NO. SHOCKS			DIF- FERENCE III−I[*]	% IMPROVE- MENT III−I/II−I[***]	NO SHOCKS			DIF- FERENCE III−I[***]	% IMPROVE- MENT III−I/II−I[**]
CONDITION	N	I	II	III			I	II	III		
Placebo	13	6.69	10.31	8.46	1.77	75%	13.08	24.08	14.23	1.15	10%
Drug	11	7.82	13.00	7.82	0	−4%	11.73	19.91	11.00	−.73	−26%

* $p = .12$, two-tailed Placebo versus Drug.
** $p = .06$, two-tailed Placebo versus Drug.
*** $p < .05$, two-tailed Placebo versus Drug.
SOURCE: G. C. Davison and S. Valins. Maintenance of self-attributed and drug-attributed behavior change. *Journal of Personality and Social Psychology*, 1969, 11, 25–33. Copyright 1969 by the American Psychological Association. Reprinted by permission.

searcher would be unconcerned with determining that the new drug is significantly less effective than the placebo. Therefore, the researcher might use a one-tailed test to determine if the drug was significantly more effective than the placebo. Excerpts 3.14–3.16 provide illustrations of one-tailed and two-tailed tests that may be reported in journal articles.

REVIEW TERMS

accept	one-tailed test
alternative hypothesis	population
calculated value	random selection
critical value	reject
descriptive statistics	sample
fail to reject	significant difference
hypothesis testing procedure	two-tailed test
inferential statistics	Type I Error
level of significance	Type II Error
null hypothesis	

REVIEW QUESTIONS

1. When using inferential statistics, does a researcher make inferences about a sample on the basis of his knowledge of the population, or does he make inferences about a population on the basis of his knowledge of the sample?
2. To insure that a sample is representative of the population, how should the researcher select individuals to be in his sample?
3. Can the size of a sample ever be larger than the size of the population?
4. Suppose a researcher conducts a study to see whether eighth-grade boys have a more extensive vocabulary than eighth-grade girls. A sample of 50 boys is compared with a sample of 50 girls in terms of the average number of words correctly defined from a list of 100 terms. What is the null hypothesis in this study?
5. What are the two most common levels of significance used by researchers?
6. What symbol is sometimes used in journal articles to represent level of significance? What is the name for this symbol?
7. Assuming that other things are held constant, is it easier for the researcher to reject the null hypothesis at the .001 level of significance or the .05 level?
8. In using an inferential statistical procedure, the researcher's data are put into a formula. What is the technical name for the numerical value that is derived from that formula after all of the calculations are completed?

9. Where does the researcher get his "critical value"?
10. If the researcher's calculated value is larger than the critical value, what will the researcher conclude about the null hypothesis?
11. Does the notation "$p < .05$" indicate that the null hypothesis was rejected or not rejected?
12. If a researcher finds a significant difference between two groups when using the .0001 level of significance, can he be completely confident that the null hypothesis is false?
13. Suppose a researcher's sample data cause him to make the mistake of rejecting a true null hypothesis. Does this mistake constitute a Type I Error or a Type II Error?
14. Which researcher does the better job of protecting against a Type I Error, the one who uses the .05 level of significance or the one who uses the .001 level of significance?
15. In using a single statistical test, is it possible for a researcher to make, at the same time, both a Type I Error and a Type II Error?

CHAPTER 4

t TESTS, ONE-WAY ANALYSIS OF VARIANCE, AND MULTIPLE COMPARISON PROCEDURES

In this chapter attention is directed to three of the more basic and popular statistical procedures: t tests, one-way analysis of variance, and multiple comparisons. These three procedures are similar in that they are used to compare two or more groups to see whether the differences between group means are large enough to assume that the corresponding population means are different.

It should be noted that certain statistical techniques exist which compare groups in terms of characteristics other than the mean. For example, there are tests which could be employed to determine whether Republicans differ from Democrats in terms of their median yearly income or percent unemployed. These and other similar tests will be discussed in Chapter 10. However, most educational and psychological research involves a comparison of means. There are many ways to compare groups in terms of their means, and this and the following four chapters are devoted to a discussion of these procedures. Let us now turn to the most basic of these techniques, the t test.

THE t TEST

Although the t test can be used for several different purposes (e.g., to see whether a correlation coefficient is significantly different from zero), researchers use the t test most often to compare the means of two groups. If the two sample means are far enough apart, the t test will yield a significant difference, thus permitting the researcher to conclude that the two populations probably do not have the same mean.

Suppose a researcher wants to see whether students in a Ph.D. program at his university have a higher average IQ than students in a Master's program. He could answer his research question by administering the Wechsler Adult Intelligence Scale (WAIS) to all graduate students attending his institution. However, this would be time consuming and expensive, since the WAIS can be administered to only one person at a time by a trained examiner. Thus, our hypothetical researcher might want to select a random sample of 15 students from each of the two populations. After obtaining IQ scores for each of the 30 students, the raw data might look like this.

GROUP 1 (MASTER'S STUDENTS)	GROUP 2 (PH.D. STUDENTS)
108	106
124	132
112	119
.	.
.	.
.	.
118	122
$\bar{X}_1 = 117.5$	$\bar{X}_2 = 125.6$

At first glance, it looks as though the Ph.D. students have a higher mean IQ than the Master's students. However, we must remember that our researcher collected data on only a small segment of the population. Although the two sample means provide the best estimate of the two population means, it is highly unlikely that the sample means would be identical to the population means. Some degree of error is inevitable since the means of 117.5 and 125.6 correspond to samples (and not populations). Thus, the statistical question becomes: Are the two sample means enough different to allow the researcher to conclude, with a high degree of confidence, that the population means are different from one another (even though he will never see the actual population means)? The t test will provide an answer to this question.

The Calculated *t* Value and the Critical Value

In using the *t* test, a researcher "plugs" his data—two sample means, the group variances, and the respective sample sizes—into a formula in order to obtain a number that can be called the calculated *t* value. In most cases the calculated *t* value will be a three- or four-digit number with two decimal places, such as 4.72, 12.43, 0.95, 16.18. (Sometimes a researcher will obtain a minus *t* value, such as −5.71. However, it is the magnitude of the numbers and not the plus or minus sign that is important.)

To decide whether the sample means do differ enough, the researcher must compare his calculated *t* value with an appropriate critical value. The critical value is found in a *t* table, located in the back of almost all statistics books. The *t* table of critical values contains information similar to that presented in Table 4.1. Two factors are used to find the correct critical value:

TABLE 4.1 CRITICAL VALUES FOR THE *t* TEST

df	.05	.01	df	.05	.01
2	4.40	10.03	22	2.17	2.92
4	2.88	4.70	24	2.16	2.90
6	2.55	3.81	26	2.16	2.88
8	2.41	3.46	28	2.15	2.86
10	2.33	3.27	30	2.14	2.85
12	2.28	3.16	40	2.12	2.80
14	2.25	3.08	50	2.11	2.78
16	2.22	3.02	60	2.10	2.76
18	2.20	2.98	80	2.09	2.74
20	2.19	2.95	100	2.08	2.73

Note: The critical values in this table are fictitious; they do, however, closely resemble the actual critical values for *t*.

(1) the degrees of freedom (*df*) associated with his calculated *t* value and (2) the level of significance. To find the *df*, the researcher simply subtracts 2 from the total number of data observations upon which his *t* test is based. In the hypothetical example concerning Master's and Ph.D. IQ scores, the *df* would be equal to 30 minus 2, or 28. And, as indicated in Chapter 3, the level of significance will usually be equal to .05 or .01. Looking at Table 4.1 we find that with 28 degrees of freedom at the .05 level of significance, the critical value is equal to 2.15.

After locating the critical value in the *t* table, the researcher compares his calculated *t* value against it. If the calculated *t* value, disregarding the plus or minus sign, is larger than the critical value, the researcher is then able to say that a significant difference exists between the two sample means. On the other hand, if the calculated t value is less than the critical value, the researcher will be forced to conclude that the two sample means are *not* significantly different from one another.

When reading a journal article, you will never have to use a t table to find the critical value; nor will you have to compare the calculated t value with the critical value to determine whether a significant difference exists. The author will do these things, presenting his conclusions both in the text of the article and in tables near the results section of the article. Usually, an asterisk next to the researcher's calculated t value in a table indicates that a significant difference has been found. As a reader of journal articles, however, it is helpful to know that the degrees of freedom associated with the critical value are equal to 2 less than the number of data observations. If the author indicates the number of degrees of freedom associated with his t test, it is possible to determine how many subjects were involved in the study by simply adding 2 to the degrees of freedom.

The Null Hypothesis

The basic research question, we remember, deals with the two population means, not the sample means. As we have seen, the null hypothesis is an assumption that no difference exists between the two population means. If the researcher finds a significant difference between his sample means, he will sometimes report that he has rejected the null hypothesis. If the t test yields a nonsignificant difference between the two sample means, the researcher will fail to reject the null hypothesis (that is, he will "accept" it).

In his journal article an author will usually say something about the sample means (whether there is a significant difference) or something about the null hypothesis (whether it was rejected or accepted). In most situations the author will not say both things, for one implies the other.

Two Forms of the t Test

There are two forms of the t test for comparing group means, one for independent samples and one for correlated samples. The *independent samples* t *test* is used in situations in which the scores in one group have absolutely no logical relationship with the scores in the other group. In the IQ example none of the Master's level IQ's could be logically related (paired) with any of the Ph.D. level IQ's. More generally, the two samples will be independent whenever (1) a large group of subjects are randomly assigned to two subgroups (possibly an experimental group and a control group) or (2) the subjects in the groups are selected at random from larger populations (as was the case with the hypothetical IQ study).

The second form of the t test is the *correlated samples* t *test*, which is also referred to as the matched t test, the correlated t test, and the paired t test. This t test is appropriate for three situations in which each of the data observations in the first group is logically tied to one of the scores in the second group. The first of these is the research situation in which a single group of subjects is measured twice, for example, measured under two different treatment conditions or before and after a common experience. Each score in the first group is logically tied to a specific score in the second group because it is obtained from the same person. In Excerpt 4.1 a correlated samples *t* test was called for because of the before-and-after nature of the study.

EXCERPT 4.1 USE OF A CORRELATED
SAMPLES *t* TEST FOR A PRE-POST SITUATION

Means and standard deviations of pre- and post-training were computed; differ- ences between means were analyzed by use of a *t* test for correlated samples.

SOURCE: R. F. Haase and D. J. DiMattia. The application of the microcounseling paradigm to the training of support personnel in counseling. *Counselor Education and Supervision*, 1970, *10*, 16–22. Copyright 1970 by the American Personnel and Guidance Association.

A second application of the *t* test for correlated samples is the two-group study in which subjects are matched. For example, suppose a researcher conducts a study involving the comparison of males and females in regard to their ability to improve their reading skills through a speed reading program. In an attempt to control for extraneous variables that could contaminate the study, the researcher might match each male with a female on the basis of IQ, initial reading ability, age, and so forth. By so doing, he would hope to eliminate all possible variables except one (i.e., sex) that could be used to explain a significant difference between the two group means. Because of the matching process, each score in the male group would be logically tied to a specific score in the female group.

A third situation which calls for a correlated *t* test is the comparison of two groups which are formed by splitting up pairs of twins. For example, suppose a group of 20 pairs of twins is available for a study dealing with a comparison of two types of diets. If one member of each pair were assigned to each of the two diets, once again there would be a logical connection between each score in the first group and one of the scores in the second group.

Although the formula for an independent samples *t* test is different from the formula for a correlated samples *t* test, the reader of a journal article does not have to worry about which one to use—this decision must be made by the researcher. Furthermore, interpretation of the calculated *t* value (i.e., the procedure for comparing it against the critical value) is the same for both *t* tests. The only difference that is important to the reader of the article concerns the degrees of freedom. For the independent samples *t* test, the *df* is equal to the total number of observations (in both groups) minus 2. For the correlated samples t test, however, the df is equal to the number of pairs of scores minus 1. This difference in figuring out the df must be noted if the reader of the article wants to be able to determine the number of subjects involved in the study solely on the basis of the df.

A simple rule-of-thumb may help in figuring out whether a researcher used an independent samples t test or a correlated samples t test. An independent samples t test must, of necessity, be used if there are an unequal number of scores in the two groups, because with an unequal number it is impossible to have logical connection between each score in the first group and one of the scores in the second group. However, if there are an equal number of people in each of the two groups, then either of the two types of t test could be appropriate, and the text of the article will usually state which one was used.

A Look at Some Actual t Tables

The format of the table in Excerpt 4.2 is typical of many statistical tables found in journal articles in that it contains (1) a title which tells what the dependent variable is, (2) a body of numbers, some of which have asterisks, and (3) a symbolic footnote which explains the asterisk. This table presents the means and standard deviations for each of two training groups. The number in the right-hand column (3.2) represents the researcher's calculated t value based upon a comparison of the two means (4.7 and 6.3). Since the subjects were randomly assigned to the two training methods, the t test was for independent samples, not correlated samples. The explanatory footnote beneath the table indicates that a significant difference was found to exist between the two means at the .01 level of significance, that is, the chances of the null hypothesis being true are less than 1 out of 100.

EXCERPT 4.2 TABLE PRESENTING RESULTS OF A t TEST

The children were randomly divided into two groups. The experimental group was to be given practice in identifying sounds in nonsense words; the control group was to be given practice in identifying the same sounds in meaningful terms.

TABLE 1 ACHIEVEMENT IN IDENTIFYING SELECTED SOUNDS IN SPOKEN WORDS FOLLOWING TWO TRAINING METHODS

TRAINING METHOD	M	SD	t
Meaningful-term method	4.7	2.2	
			3.2*
Nonsense-term method	6.3	1.6	

* $p < .01$.

SOURCE: J. D. McNeil and J. Stone. Note on teaching children to hear separate sounds in spoken words. *Journal of Educational Psychology*, 1965, *56*, 13–15. Copyright 1965 by the American Psychological Association. Reprinted by permission.

The table in Excerpt 4.3 is somewhat more complicated. This research study involved three feedback groups, the names of which are listed in the first column. Each of these groups was rated in terms of Accurate Empathy (AE) on two occasions, pre and post (that is, before and after they did something). The second column presents the pre mean score for each group, while the third column presents the corresponding standard deviation for each group. The fourth and fifth columns present the means and standard deviations for the three groups at the post time period. Finally, the author's calculated *t* values are presented in the last column.

EXCERPT 4.3 TABLE WITH THREE *t* TESTS

TABLE 1 *t* TEST FOR PRE AND POST ACCURATE
EMPATHY (AE) RATINGS OF FEEDBACK GROUPS

FEEDBACK GROUPS	M (PRE-AE MEASURE)	SD	M (POST-AE MEASURE)	SD	*t*
Immediate	1.18	.226	2.06	.803	4.02*
Delayed	1.14	.239	1.55	.364	4.30*
Control	1.13	.286	1.28	.285	1.80

* $p < .01$.
SOURCE: W. B. Reddy. Effects of immediate and delayed feedback on the learning of empathy. *Journal of Counseling Psychology,* 1969, *16,* 59–62. Copyright 1969 by the American Psychological Association. Reprinted by permission.

Since the *t* test compares means (not standard deviations), each *t* value in the last column resulted from a comparison of the two mean scores on the same row. For example, the *t* of 4.02 involved a comparison of 1.18 and 2.06. Since the same subjects were used to obtain the pre and post means compared by each *t* test, the researcher used a correlated samples *t* test. By means of the two asterisks and the footnote beneath the table, the author tells us that a significant difference was found between the pre and post means for the immediate group and the delayed group; however, no significant difference was found between the two means of the control group. To summarize, the table seems to indicate that either immediate or delayed feedback caused subjects to receive higher AE scores than a control group which received no feedback.

The table in Excerpt 4.4 reports the results of 12 *t* tests. The researcher had a total of four groups—two experimental groups and two control groups. The *t* test was used to compare Experimental Group 1 to Experimental Group 2, and also to compare Control Group 1 to Control Group 2. There were a total of six dependent variables involved in the study, and these are listed in the column at the left. Thus, all of the numbers in the top row of the table (beginning with 65.75) correspond to the first dependent variable, age.

EXCERPT 4.4 TABLE WITH SEVERAL t TESTS

TABLE 1 COMPARISON OF TWO EXPERIMENTAL AND TWO CONTROL GROUPS IN
TERMS OF AGE, WECHSLER WEIGHTED SCORES, AND PAIRED-ASSOCIATE LEARNING
TEST SCORES ASSESSED BY MEANS OF STUDENT'S t

| | Experimental Groups | | | | | | Control Groups | | | | | |
| | 1 ($N = 8$) | | 2 ($N = 10$) | | | | 1 ($N = 8$) | | 2 ($N = 10$) | | | |
VARIABLE	m	σ	m	σ	t	p	m	σ	m	σ	t	p
Age	65.75	8.19	70.80	5.67	.49	NS	67.25	7.17	71.50	5.36	.46	NS
Wechsler Weighted Scores												
Verbal scale[a]	33.88	8.90	33.20	9.07	.05	NS	30.50	8.42	34.70	5.62	.40	NS
Performance scale	13.50	13.77	14.10	10.58	.03	NS	26.50	9.75	28.10	6.77	.13	NS
Verbal minus performance	20.38	13.53	19.10	6.28	.08	NS	4.00	9.97	6.60	5.08	.23	NS
Full scale	47.38	18.84	47.30	18.67	.003	NS	57.00	15.25	62.80	11.36	.30	NS
Paired-associates (acquisition score)	59.25	23.60	54.80	28.07	.11	NS	14.13	4.55	10.70	5.16	.47	NS

a In each case the Verbal scale weighted score is based on five subtests; i.e., exclusive of the Information subtest.
SOURCE: J. Inglis. Learning, retention, and conceptual usage in elderly patients with memory
disorder. *Journal of Abnormal Psychology*, 1959, 59, 210–215. Copyright 1959 by the American
Psychological Association. Reprinted by permission.

The numbers under the heading m are means; the numbers under the
heading σ are standard deviations. For each of the six dependent variables, a
t test was used to compare the two experimental group means (the calculated
t values being presented in the first t column) and the two control group
means (the calculated t values being presented in the t column near the right
side of the table). Knowing this, the reader can easily tell that the t value of
.003 resulted from a comparison of the two experimental group means
(47.38 and 47.30) on the full scale of the Wechsler.

To the right of each column of t values, there is a column labeled p. Instead of interpreting his calculated t values in a footnote (as was done in
Excerpt 4.3), this author gives the interpretation right next to his t values. In
each case the NS stands for "not significant," and we thus know that each of
the calculated t values was smaller than the tabled critical value. If one of
the t values had been larger than the critical value, the author would have
replaced the NS with a symbolic statement such as $< .05$ or $< .01$. Since no
significant differences were found, the table indicates that the two experimental groups are not significantly different on any of the six variables, nor
are the two control groups different on any of the dependent variables.

An author will often report the results of his t test(s) within the text
of his article rather than in a table. Excerpt 4.5 shows how this is usually
done. Since each of the two t tests involved a pretest-posttest comparison,
correlated samples were involved. With this form of t test, the degrees of
freedom are found by subtracting 1 from the number of pairs of scores.
Knowing this, it is easy for us, the readers of this excerpt, to determine how
many Ss were in each group: 48 in the *New Content* group and 24 in the no
training group.

EXCERPT 4.5 RESULTS OF A *t* TEST PRESENTED WITHOUT A TABLE

There was a significant increase from pretest to posttest for the *New Content* group (*t* = 8.59, *df* = 47, *p* < .001), whereas a group of similar subjects who were given no experimental training showed no increase from pretest to posttest (*t* = 1.40, *df* = 23).

SOURCE: J. P. Williams and E. I. Levy. Retention of introductory and review programs as a function of response mode. *American Educational Research Journal*, 1964, *1*, 211–218. Copyright 1964 by the American Educational Research Association, Washington, D.C.

The Assumption of Equal Variances

The *t* test assumes that the scores in one group have about the same degree of variability as the scores in the second group. Sometimes a researcher will test this assumption to see if the variances are homogeneous. When this is done, the researcher hopes that the null hypothesis (of equal variability) will not be rejected. In other words, he wants the sample data to support the requirement of approximately equal variances. The null hypothesis would have to be rejected if the variance of the first sample was quite a bit different from the variance of the second sample. In this situation, the regular *t* test would be inappropriate and a special form of the *t* test would be needed.

If the two groups have an equal number of scores, a violation of the assumption of equal variances does not affect the validity of the *t* test. Since there are usually an equal number of people in each group, most authors do not test or discuss this assumption. However, if the two groups do not contain the same number of scores, then the researcher should check whether the

---

EXCERPT 4.6 TEST FOR THE ASSUMPTION OF HOMOGENEITY OF VARIANCE

TABLE 1 MEAN, VARIANCE, AND RESULTS OF TESTS OF SIGNIFICANCE BETWEEN MEANS AND VARIANCES OF LORGE–THORNDIKE INTELLIGENCE QUOTIENTS FOR HIGH- AND LOW-CURIOSITY GROUPS OF BOYS AND GIRLS

	HIGH CURIOSITY			LOW CURIOSITY			F°	$t\dagger$
	N	MEAN	VARIANCE	N	MEAN	VARIANCE		
Lorge-Thorndike Intelligence Quotients (Boys)								
I	16	119.50	252.40	12	110.16	237.60	1.06	1.56
II	37	116.35	257.90	30	114.93	305.02	1.18	.34
III	50	109.54	386.53	41	107.17	271.49	1.42	.62
Lorge-Thorndike Intelligence Quotients (Girls)								
I	15	116.20	283.17	14	108.78	255.10	1.11	1.21
II	50	110.42	252.98	40	111.87	295.54	1.16	−.41
III	60	109.00	235.76	49	110.53	279.29	1.18	−.49

$^{\circ}$ *F* test—tests the homogeneity of the variance.
$\dagger$ *t* test—tests significance of difference between means.
SOURCE: W. H. Maw and E. W. Maw. Self-appraisal of curiosity. *Journal of Educational Research*, 1968, *61*, 462–465.

sample data support the assumption of homogeneous variances. A failure to test this assumption indicates that the author has not done as thorough a job as he should have.

An author sometimes simply states within the text of his article that the assumption of equal variances was tested. Some authors indicate through a table that this assumption was tested. Excerpt 4.6 demonstrates the latter technique. As indicated by the first footnote beneath the table, the column near the right labeled F corresponds to a test for the homogeneity of variance. Each number in this column resulted from a comparison of the two variances in that same row. Since none of these six F tests was significant, the researcher was able to proceed with the t test, the results of which are presented in the last column on the right.

ONE-WAY ANALYSIS OF VARIANCE

A one-way analysis of variance (abbreviated ANOVA) is an inferential statistical procedure which has the same general purpose as the t test: to compare groups in terms of the mean scores. The difference between the two procedures lies in the number of groups that can be compared. Whereas the t test is designed for comparing two groups, a one-way ANOVA can be used to compare two or more groups. Both procedures yield identical results in a two-group comparison, but the one-way ANOVA is more versatile because it can also be used to compare three or more groups. The one-way ANOVA is, in effect, an extension of the t test to a greater number of groups compared.

The null hypothesis in a t test states that there is no difference between the two population means. It is the same in the one-way ANOVA procedure, except that there may be more than two population means assumed to be equal. The procedure for assessing the validity of the null hypothesis is also similar: (1) the original raw data are put into a formula in order to obtain a calculated value, (2) the resulting calculated value is compared against a critical value, and (3) the null hypothesis is rejected if the calculated value is larger than the tabled critical value, or accepted if the calculated value is less than the critical value.

The ANOVA Summary Table

The results of a one-way ANOVA are usually presented by means of a summary table. As we shall see later, authors do not always follow the same format for their summary tables. Nonetheless, an understanding of a model summary table will allow the reader of journal articles to decipher most, if not all, of the variations found in the literature.

To help understand how the summary table works, imagine that a researcher has conducted an experiment to compare three different ways of arranging the items on a test. In one test form the items were arranged in ascending order of difficulty (from the easiest to the most difficult item); in

TABLE 4.2 ANOVA SUMMARY TABLE FOR EXPERIMENT COMPARING DIFFERENT ITEM ARRANGEMENTS

SOURCE	*df*	*SS*	*MS*	*F*
Test forms	2	16	8	4*
Within-groups	15	30	2	
Total	17	46		

* $p < .05$

the second test form the items appeared in descending order of difficulty (from the most difficult to the easiest item); and in the third test form the items were arranged in a random order. The researcher had a pool of 18 subjects available for his study, and he randomly assigned 6 subjects to each of the three test forms. The dependent variable was the total number of items answered correctly by each subject.

Table 4.2 presents an ANOVA summary table to correspond with this hypothetical experiment. The ANOVA summary table has five columns: source, *df*, SS (for sum of squares), MS (for mean square), and *F*. Under the first column are the names for the three rows. The first row is identified by the name of the independent variable, that is, the variable that distinguishes the groups from one another. In the sample summary table, test forms in the first row indicates that the three groups differed in terms of the test form which they received. Another popular name for the first row is simply between-groups. The second row is usually labeled either within-groups (as in Table 4.2) or error. The third row is always labeled total.

The numbers in the column labeled *df* are easy to compute. To find the *df* for the first row, the researcher simply subtracts 1 from the number of groups involved in his study. There were 3 groups in the hypothetical study; therefore, the test forms source has 2 degrees of freedom. To find the *df* for the total row, the researcher simply subtracts 1 from the total number of subjects. Since there were 18 subjects involved in the study, the total *df* is equal to 17. There are two different procedures for finding the within-groups *df*. One procedure is to subtract 1 from the number of subjects in each group and then add up the resulting figures. In the hypothetical experiment there were 6 people in each of the 3 groups. Thus, there are 5 degrees of freedom in each of these groups, for a total of 5 × 3 or 15 degrees of freedom within-groups. The second procedure for getting the within-groups *df* is easier: the *df* in the first row are simply subtracted from the total *df*. In Table 4.2, 2 subtracted from 17 is 15, the same number of degrees of freedom for within-groups as we obtained when using the first procedure. As shown here, the two methods for getting the within-groups *df* always result in exactly the same answer.

The various SSs (sums of squares) are determined by the researcher. He uses a relatively complicated formula to obtain each SS, and it is impossible for the reader of a journal article to check the accuracy of the first two numbers in the SS column. However, the bottom number in this column, that is, the total SS, should always be equal to the sum of the other two SSs. In Table 4.2 the numbers for the various SSs are fictitious; they were chosen to make the math of the summary table easy to follow. In reality, the numbers for the three SSs will almost always contain a decimal point.

Each value in the column labeled MS (mean square) is found by dividing the SS by the df found in the same row. Thus, the test forms MS of 8 was obtained by dividing 16 by 2. Note that there are only two values in the MS column; the summary table does not contain a value for total MS. If an author puts a number in this particular place in the summary table, he has made a mistake. In our demonstration summary table there is one number in the F column. This number is found by dividing the MS for test forms (8) by the within-groups MS (2). As indicated in Table 4.2, no F value for within-groups or total is found in the summary table.

The most important number in the entire summary table is the single value in the F column. All the remaining eight numbers in the summary table are used to calculate the important F value. The F value allows the researcher to decide whether there are significant differences between (among) the means of the groups being compared.

The Critical Value

To interpret his calculated F value, the researcher must first find the critical value in the F table which is located in the back of most statistics books. The F table of critical values contains information similar to that presented in Table 4.3. Although the F table is even more complicated than the t table, three factors allow the researcher to locate the critical value without any difficulty at all. First, the number of degrees of freedom associated with the test forms (or between-groups) row in the summary table tells the researcher which column of the F table to use. Since the test forms source had a df of 2, the researcher would disregard everything in the F table except the column labeled 2. Second, the number of degrees of freedom associated with the within-groups row in the summary table tells the researcher which row of the F table to use. Since the within-groups source had a df of 15, the researcher disregards all of the numbers in the second column except those in the row labeled 15. Third, two numbers are usually given together, in any combination of one column and one row, the first for the .05 level of significance and the second for the .01 level. Looking at the 2 column and the 15 row of Table 4.3, we find two numbers, 3.78 and 6.46. If the researcher is using the .05 level of significance, 3.78 is the critical value for F.

If the researcher's calculated F value is larger than the tabled critical value, then there is significant difference between the sample means, and the

TABLE 4.3 CRITICAL VALUES FOR F°

df WITHIN-GROUPS FOR	*df* for Between-Groups					
	2	4	6	8	10	20
5	5.89 (13.4)	5.29 (11.5)	5.05 (10.8)	4.92 (10.4)	4.87 (10.2)	4.66 (9.65)
10	4.20 (7.66)	3.58 (6.09)	3.32 (5.49)	3.17 (5.16)	3.07 (4.95)	2.87 (4.51)
15	3.78 (6.46)	3.16 (4.99)	2.89 (4.42)	2.74 (4.10)	2.65 (3.90)	2.43 (3.46)
20	3.59 (5.95)	2.97 (4.53)	2.70 (3.97)	2.55 (3.66)	2.45 (3.47)	2.22 (3.04)
40	3.33 (5.28)	2.71 (3.93)	2.44 (3.39)	2.28 (3.09)	2.17 (2.90)	1.94 (2.47)
120	3.17 (4.89)	2.55 (3.58)	2.28 (3.06)	2.12 (2.76)	2.00 (2.57)	1.75 (2.13)

Note: The critical values in this table are fictitious; they do, however, closely resemble the actual critical values for *F*.
° Critical values at .05 are given first; the critical values at .01 are given in parentheses.

null hypothesis (of equal population means) will be rejected. On the other hand, if the critical value is the larger of the two, then the sample means do not differ significantly from one another, and the null hypothesis will be accepted. If there is a significant difference between the sample means, the author will usually place one or two asterisks next to his calculated *F* value in the summary table, and he will indicate beneath the table the precise level of significance by a statement similar to $p < .05$.

A Few Summary Tables from Actual Articles

As indicated earlier, authors do not always use the same format for presenting their summary tables. In fact, the model shown in Table 4.2 is probably used less often than are variations of it. Nevertheless, an understanding of the model summary table should allow a person to interpret most of the variations. Let us now turn to five summary tables which have been taken directly from actual journal articles.

The table in Excerpt 4.7 is similar to the model summary table in many ways. However, there are a few differences. First, the first column of the table is labeled "Source of Variation" rather than simply "Source." Second, the phrases sum of squares and mean square are written out rather than abbreviated as SS and MS. Third, the column labeled df is placed between the SS and MS columns rather than before them. Finally, the items listed in the first column are labeled "Between treatment" and "Within treatment" rather than "Between-groups" and "Within-groups."

EXCERPT 4.7 ANOVA SUMMARY TABLE
SIMILAR TO THE MODEL SUMMARY TABLE

TABLE 1 ANALYSIS OF VARIANCE OF SAI
DIFFIDENCE-EGOISM SCORES FOR THE
LEADER-STRUCTURED, GROUP-
STRUCTURED AND CONTROL GROUPS

SOURCE OF VARIATION	SUM OF SQUARES	df	MEAN SQUARE	F
Between treatment	109.00	2	54.50	3.116*
Within treatment	1364.96	78	17.49	
Total	1473.96	80		

* Significant at the .05 level.

SOURCE: S. H. Gilbreath. Group counseling with male underachieving college volunteers. *Personnel and Guidance Journal*, 1967, 45, 469–476. Copyright 1967 by the American Personnel and Guidance Association.

In spite of minor differences in terminology and arrangement, the mathematics of this summary table is exactly the same as that in the model. The between and within SS add up to the total SS, as do the degrees of freedom. Each MS is found by dividing the SS by the appropriate df, and the F of 3.116 is found by dividing the between-treatment MS of 54.50 by the within-treatment MS of 17.49.

Through this table, the author tells us a great deal about the study. The title indicates what the dependent variable was and how the groups differed (in terms of structure). The between-treatment df of 2 indicates that there were three groups involved in the study, while the total degrees of freedom

---

EXCERPT 4.8 SUMMARY TABLE WITHOUT A COLUMN FOR SS

TABLE 1 SUMMARY OF ANALYSIS OF
VARIANCE ON NUMBER OF INSTANCES
REQUIRED FOR SOLUTION

SOURCE	df	MS	F
Between	3	7.15	1.96
Within	147	3.65	
Total	150		

SOURCE: L. Sechrest and J. Wallace. Assimilation and utilization of information in concept attainment under varying conditions of information presentation. *Journal of Educational Psychology*, 1962, 53, 157–164. Copyright 1962 by the American Psychological Association. Reprinted by permission.

of 80 indicates that there were 81 subjects distributed among the three groups. Finally, the asterisk next to the *F* of 3.116, in conjunction with the footnote beneath the table, indicates that the author's calculated *F* was bigger than the tabled critical value, implying that the three group means are far enough apart to permit the researcher to conclude that the chances are less than 5 out of 100 that the three population means are alike. Thus, the null hypothesis is rejected and a significant difference is said to exist among the three sample means.

The summary table in Excerpt 4.8 is similar to the previous table, except that there is no SS column. We could calculate what the figures in this column would be since we know that SS divided by *df* equals *MS* (Be-

---

EXCERPT 4.9 SEVERAL ONE-WAY ANOVA'S COMBINED IN ONE SUMMARY TABLE

TABLE 5 ANALYSIS OF VARIANCE SUMMARIES

SCALE	SS	*df*	*MS*	*F*
Hs				
Between Ss	2,840	2	1,420	6.44**
Within Ss	16,542	75	221	
D				
Between Ss	1,399	2	699	1.98
Within Ss	26,263	75	350	
Hy				
Between Ss	1,124	2	562	3.85*
Within Ss	10,956	75	146	
Pd				
Between Ss	139	2	69	0.45
Within Ss	11,667	75	156	
Mf				
Between Ss	19	2	9	0.07
Within Ss	9,590	75	128	
Pa				
Between Ss	230	2	115	0.67
Within Ss	12,816	75	171	
Pt				
Between Ss	1,474	2	737	2.55
Within Ss	21,692	75	289	
Sc				
Between Ss	1,559	2	779	2.29
Within Ss	25,510	75	340	
Ma				
Between Ss	51	2	26	0.14
Within Ss	13,936	75	186	
Si				
Between Ss	1,313	2	657	4.52*
Within Ss	10,892	75	145	

* $p < .05$.
** $p < .01$.

SOURCE: N. N. Markel. Relationship between voice-quality profiles and MMPI profiles in psychiatric patients. *Journal of Abnormal Psychology*, 1969, 74, 61–66. Copyright 1969 by the American Psychological Association. Reprinted by permission.

tween SS would be 21.45). Again we see that the table conveys a great deal of information: that the dependent variable involved the number of instances required for solution, that there were four groups, that there were 151 subjects, and that no significant difference was found among the means of the four groups.

In Excerpt 4.9 the results of 10 one-way ANOVA's are combined into one summary table. Each of the 10 analyses is related to a different dependent variable, and anyone who has had a course in testing should recognize the letters in italics as being the scales of the MMPI. Since the *df* for between Ss was always equal to 2, we know that there were three groups compared in the study. And since the total degrees of freedom would have been equal to 77 if the author had included it, we know that there were 78 subjects distributed among the three groups. Three *F* values were found to be significant, indicated by the asterisks next to the *F*s and by the explanatory footnotes. For two of the MMPI scales, *Hy* and *Si,* the sample means were such that the chances are less than 5 out of 100 that the population means are the same. For one scale, *Hs,* the sample means were even more divergent, allowing the researcher to conclude that the probability is less than 1 out of 100 that the population means are the same.

In Excerpt 4.10 we find a table that is quite a bit different from the model summary table. Here the author combined the results of seven one-way ANOVA's into one table. A separate analysis was conducted for each of

---

EXCERPT 4.10 ALTERNATIVE COMBINATION OF SEVERAL ONE-WAY ANOVA'S IN ONE SUMMARY TABLE

ANALYSES OF VARIANCE FOR THE VARIABLES

SOCIOMETRIC VARIABLE	MEAN SQUARE BETWEEN GROUPS[a]	MEAN SQUARE WITHIN GROUPS[b]	F	p
Positive status	457.7	66.1	6.92	.005
Negative status	266.1	179.7	1.48	.25
Net status	1052.2	317.0	3.31	.01
Total activity	610.4	281.7	2.16	.10
Asymmetrical reciprocity	19.7	7.5	2.63	.025
Symmetrical reciprocity	11.1	4.8	2.29	.05
Index of attraction	2824.5	809.4	3.49	.005

[a] $df = 6.$
[b] $df = 58.$

SOURCE: M. Cohen et al. Sociometric study of the sex offender. *Journal of Abnormal Psychology,* 1969, *74,* 249–255. Copyright 1969 by the American Psychological Association. Reprinted by permission.

the sociometric variables listed in the first column. Instead of a complete summary table for each of these analyses, excerpts from each table were presented, with each row of the table corresponding to a different analysis. The footnotes beneath the table indicate that a *df* of 6 was associated with the between-groups mean square and a *df* of 58 was associated with the within-groups mean square. Thus, there were seven groups with a total of 65 subjects.

The *F* of 6.92 for the variable positive status was found by dividing the between-groups mean square of 457.7 by the within-groups mean square of 66.1. Each of the remaining six *F* ratios was found in a similar manner. The *p* column at the right gives the interpretation of each *F* value. This column is similar in purpose to the two *p* columns found in Excerpt 4.4. The smaller the value for *p*, the greater the difference between the seven means. And since it is customary to use .05 as a cut-off point, a significant difference can be said to exist for all of the variables except negative status and total activity. If the author had wanted to eliminate the column of *p* values, he could have put asterisks next to his *F* values with explanatory footnotes (*p* < .05, **p* < .01) beneath the table as was done in Excerpt 4.9.

In Excerpts 4.8 and 4.10 we saw that an author will sometimes omit information regarding the SS, presenting only the *df*, *MS*, and *F* values in the table. An author will sometimes present only the *F* value, as in Excerpt 4.11 in which one-way ANOVA's were used to compare three diagnostic groups in terms of age, vocabulary, and word count. The table indicates that the three groups did not differ from one another on any of the variables.

EXCERPT 4.11 TABLE CONTAINING ONLY THE *F* VALUES FROM THREE ONE-WAY ANOVA'S

AGE, VOCABULARY, AND WORD COUNT SCORES BY DIAGNOSIS

	PARANOID SCHIZOPHRENICS	NONPARANOID SCHIZOPHRENICS	CONTROLS
N	23	23	23
Age			
M	34.52	32.65	33.22
σ	10.20	11.34	10.24
F = .19			
Vocabulary			
M	27.61	27.39	28.04
σ	7.34	7.38	5.30
F = .05			
Word count			
M	11.28	10.71	11.29
σ	3.23	3.46	3.83
F = .84			

SOURCE: R. H. Goldstein and C. F. Salzman. Proverb word count as a measure of overinclusiveness in delusional schizophrenics. *Journal of Abnormal Psychology*, 1965, *70*, 244–245. Copyright 1965 by the American Psychological Association. Reprinted by permission.

Excerpt 4.12 shows how an author can report the results of a one-way ANOVA within the text of his article instead of presenting the results in a table. Within the first set of parentheses in the second paragraph, the author presents (1) his calculated F ratio, (2) the between-groups df and the within-groups df, separated by a slashed line, and (3) a probability statement indicating that the F value was found to be significant at the .01 level of confidence. Note that the first df value allows us to figure out how many groups there were $(2 + 1 = 3)$, and that the sum of the two degrees of freedom permits us to determine how many Ss were involved in the study $(df_{tot} + 1 = N; 45 + 1 = 46)$.

EXCERPT 4.12 RESULTS OF A ONE-WAY ANOVA PRESENTED WITHOUT A SUMMARY TABLE

At the end of the spring and fall semester (1966–67) students completing individual testing or practicum in school psychology were asked to help the instructor score some short forms of the WISC. As the students came to the instructor's office they were randomly assigned to one of three scoring conditions: 16 students scored the Comp., Sim., and Voc. subtests under conditions where they were led to believe that the WISC was obtained from a slow learner; 15 students scored the same responses under conditions suggesting that the child was of above-average intellectual ability; and 15 students scored the same responses under conditions where no prior information was given.

A one-way analysis of variance . . . was used in analyzing the data for each subtest. The raw scores were used in the statistical analysis and a significance level of .05 selected. The results showed significant differences between scoring groups on the Comp. $(F = 7.73, df = 2/43, p < .01)$ and Sim. $(F = 6.42, df = 2/43, p < .01)$. The mean differences between scoring groups on the Voc. subtest failed to reach significance $(F = 1.83, df = 2/43)$.

SOURCE: B. Egeland. Examiner expectancy: effects on the scoring of the WISC. *Psychology in the Schools*, 1969, *6*, 313–315.

The Assumption of Equal Variances

Like the t test, a one-way analysis of variance is based upon the assumption that the scores in each of the various groups have approximately the same variance. If there are an equal number of scores in each of the various groups, the researcher does not have to test this assumption. Previous experiments have shown that the F test is valid when group variances are dissimilar, as long as the sample sizes are constant; that is, the F test is *robust* to violations of the homogeneity of variance assumption provided that the number of scores in the groups is the same.

If the various groups do not contain the same number of subjects, the researcher is obligated to test the assumption of equal variances. Several different procedures have been developed to make that test, with the four most popular tests being Bartlett's chi-square, Hartley's F-max test, Cochran's C test, and Levene's test. If a researcher is using a one-way ANOVA with groups that vary in size, then he should report that he tested this assumption

---

EXCERPTS 4.13–4.15 REPORTING OF
TESTS FOR HOMOGENEITY OF VARIANCE

Stimulus and response data were analyzed separately by means of independent one-way analyses of variance. When initial inspection of the response data indicated the possibility of non-homogeneity of cell variances, a test of this factor was made using Levene's technique, as suggested by Glass (1966). However, the degree of heterogeneity was not significant ($F = 2.52$, $df = 3/28$, $p > .05$).

SOURCE: R. H. Bruning and L. J. Lantinga. Effects of encounter with verbal units in prose contexts on subsequent associative learning. *Journal of Educational Psychology*, 1971, 62, 308–314. Copyright 1971 by the American Psychological Association. Reprinted by permission.

A conventional Bartlett's test on the within-cell variances yielded a χ^2 of 9.8455 ($p < .30$) indicating homogeneity among variances.

SOURCE: H. E. Anderson et al. Generalized effects of praise and reproof. *Journal of Educational Psychology*, 1966, 57, 169–173. Copyright 1966 by the American Psychological Association. Reprinted by permission.

A Hartley test indicated extreme non-homogeneity of variance ($F_{max (12,5)} = 276.5$, $p < .001$).

SOURCE: J. L. Wolff. Effects of verbalization and pretraining on concept attainment by children in two mediation categories. *Journal of Educational Psychology*, 1967, 58, 19–26. Copyright 1967 by the American Psychological Association. Reprinted by permission.

---

with one of these four procedures. Excerpts 4.13–4.15 show how different authors report having tested for homogeneity of variance. If the researcher fails to reject the null hypothesis of equal variances when using one of these tests, he may then go ahead and use the regular ANOVA procedure. If he rejects the null hypothesis, then he cannot analyze his data with the traditional method, because the results would not be valid. (When the variances are shown to be unequal, researchers can solve the problem by transforming their raw data and then use the transformed scores in the ANOVA formulas.[1]) As should be obvious, a researcher hopes that his data supports the assumption of equal variances, for he may then proceed with the analysis by testing for differences between the means.

MULTIPLE COMPARISON PROCEDURES

Suppose a researcher is using a one-way ANOVA to compare three or more groups in terms of the group means. If the calculated F value turns out to be larger than the critical value found in the F table, the researcher will conclude that he has obtained a significant F ratio. Thus, a significant F ratio simply means that the sample data are such as to allow the researcher to reject the null hypothesis (of equal population means). However, the significant F ratio by itself does not tell the researcher which of the group means are significantly different from the others.

---

[1] Four different types of transformations are used by researchers. These are the square root, the logarithmic, the arc sine, and the inverse (or reciprocal) transformations.

Let us return for a moment to the hypothetical experiment involving different ways of arranging test items. Table 4.2 shows that the F value for this study was significant. However, this significant F ratio tells us only that at least one of the group means is significantly different from one of the other group means. But there are many different ways in which there could be one or more significant differences. If we use the letters A, D, and R to stand for the three different test forms (ascending order of difficulty, descending order of difficulty, and random arrangement), it would be possible for: (1) A to be significantly different from both D and R, with D and R not significantly different; or (2) D to be significantly different from A and R, with A and R not significantly different; or (3) R to be significantly different from A and D, with A and D not significantly different; or (4) all three test forms to be significantly different from one another. In general, with three groups involved, there will always be four possible meanings of a significant F ratio; with four groups, there are 14 possible meanings; with five groups, 56 possible meanings; etc. Therefore, a researcher cannot stop his analysis after getting a significant F; he must locate the cause of the significant F. To do this, he must perform a follow-up analysis.

Specific Multiple Comparison Tests

Several procedures of follow-up analyses, referred to as multiple comparisons or post hoc comparisons, have been developed to help the researcher find out exactly where the significant differences lie after a significant F ratio has been obtained. These techniques, all named after the people who developed them, include: Fisher's LSD, Duncan's new multiple range test, Newman-Keuls, Tukey's HSD, and Scheffé's test. To locate the significant differences, these statistical procedures analyze each possible pair of means to determine if the two means are significantly different from one another.

The various multiple comparison procedures may be likened to several t tests applied to the data following a significant F ratio. In fact, you may be wondering why the researcher isn't able to apply a regular t test to each possible combination of means. The answer involves the level of significance. Suppose the researcher wanted to use the .05 level of significance in the hypothetical experiment of test item arrangements. If he made three t test comparisons of two means at a time after getting a significant F, his t tests would not really be at the .05 level of significance. Although the researcher would be making each individual t test at the .05 level, the three tests, as a group, would be at the .14 level. As the number of groups and t tests increases, the probability of getting a significant result by chance alone increases. For example, if a one-way ANOVA were used to compare 7 groups and if a significant F were obtained, there would be a total of 21 t test comparisons of different combinations of the 7 means. If all t tests were conducted at the .05 level of significance, one of them would be expected to result in a significant difference by chance alone. If more than one of the *t* tests resulted in a significant difference, the researcher would be in a dilemma, for he would not know which of the significant *t*s were due to chance and which were indicative of a true difference in population means.

The five multiple comparison procedures mentioned above adjust the level of significance to reduce the influence of chance due to having more than just one comparison. The five procedures differ from one another in the degree of the adjustment; some procedures make it easier for the researcher to find a significant difference between two means than others which require the two sample means to be farther apart before a significant difference can be said to exist. A *liberal procedure* will find a significant difference between two means that are relatively close together, whereas a *conservative procedure* will indicate that two means are significantly different only when the means are far apart.

Of the five multiple comparison tests, Fisher's LSD is the most liberal and Scheffé's the most conservative. Table 4.4 shows the relative placement of the various multiple comparison procedures on an imaginary liberal-conservative continuum.

TABLE 4.4 THE FIVE BASIC MULTIPLE COMPARISON PROCEDURES ON A LIBERAL-CONSERVATIVE CONTINUUM

Liberal ▲	Fisher's LSD
	Duncan's new multiple range test
	Newman-Keuls
	Tukey's HSD
Conservative ▼	Scheffé's test

A sixth multiple comparison procedure, Dunnett's test, is somewhat different from the five basic ones discussed above. It is used when a researcher wants to compare all of his experimental (treatment) groups with a control group. There is no comparison between the various experimental groups. Thus, if there are a total of five groups involved in the original one-way ANOVA, Dunnett's test would permit just four comparisons—the control group with each of the other four groups. The number of comparisons made by Dunnett's test will always be equal to one less than the total number of groups involved in the study. It would be inappropriate for a researcher to apply Dunnett's test if he does not have a control group or if he wants to make comparisons among his experimental groups, and since researchers usually want to compare the experimental groups against one another, Dunnett's test is not as popular as the other procedures mentioned.

Reporting Results

There are two main procedures for reporting the results of a multiple comparison investigation. With the first, the *underlining method,* the various groups are arranged in order on the basis of the size of the means. A line is then drawn beneath those groups which do not differ significantly from one another. Thus, any groups which are not underlined by the same line are significantly different from each other.

Excerpt 4.16 demonstrates the use of the underlining method. The title of the table tells us that the multiple range test (Duncan's procedure) was used to identify which specific age groups were significantly different from one another. The table shows that there were six age groups, labeled A_1–A_6. Sample sizes are presented in the middle of the table. Near the bottom, the six means are presented. (Note that the means are arranged in order of magnitude from the largest to the smallest.)

EXCERPT 4.16 RESULTS OF A MULTIPLE COMPARISON
TEST REPORTED BY THE UNDERLINING METHOD

TABLE 4 RESULTS OF THE MULTIPLE RANGE
FOR THE MAIN EFFECT AGE: MOTIVATOR SCALE

Age	A_3 (30–34)	A_5 (40–49)	A_6 (50+)	A_4 (35–39)	A_2 (25–29)	A_1 (20–24)
	$N = 18$	$N = 32$	$N = 25$	$N = 21$	$N = 27$	$N = 25$
Mean	4.21	4.19	4.11	3.96	3.90	3.62

source: J. C. Bledsoe and I. D. Brown. Role perceptions of secondary teachers as related to pupils' perceptions of teacher behavioral characteristics. *Journal of Educational Research,* 1968, *61,* 422–429.

Groups 3, 5, 6, 4, and 2 are underlined by the top line; therefore, these groups did not differ significantly from one another. The lower line indicates that groups 6, 4, 2, and 1 are also not significantly different. Thus, two significant differences are revealed by the table, the first between groups 3 and 1 and the second between groups 5 and 1. Neither of these pairs is underlined by the same line.

For want of a better term, the second method for reporting the results of a multiple comparison test can be referred to as the *triangular table method.* Excerpt 4.17 demonstrates the use of this method. Three groups were involved in this study, and the group means are arranged in order along the top of the table and also in the first column of the table. Each of the numbers within the body of the table (i.e., 3.71, 4.88, and 1.17) is the calculated value resulting from a comparison of the mean at the left of that row and at the top of that column. Thus, 3.71 was the calculated value obtained by the author when the means for group A (4.60) and group B (5.39) were compared. An asterisk next to a calculated value signifies that the two means involved in the comparison are significantly different from one another. In Excerpt 4.17, in the three comparisons made, only one significant difference was found—between the means for group A (4.60) and group C (5.64).

EXCERPT 4.17 RESULTS OF A MULTIPLE COMPARISON TEST REPORTED BY THE TRIANGULAR TABLE METHOD

In order to determine how the three groups differ, an extension of Duncan's New Multiple Range Test to group means with unequal numbers was used (3). This test is used to determine which of the differences between group means are significant and which are not. The results of Kramer's extension of Duncan's New Multiple Range Test are summarized in Table 5.

An examination of the means in Table 5 indicates that the mean for the group counseling group is greater than the mean for the group-individual counseling group and the latter is greater than the mean for the control group. The difference between the group counseling group and the control group is large enough to be significant at the .05 level of confidence. No other differences are significant.

TABLE 5 KRAMER'S EXTENSION OF DUNCAN'S NEW MULTIPLE RANGE TEST OF THE FOURTH TERM GPA MEANS OF THE THREE EXPERIMENTAL GROUPS

	A CONTROL	B INDIVIDUAL	C GROUP	SHORTEST SIGNIFICANT
Means	4.60	5.39	5.64	RANGES
A—4.60		3.71	4.88*	$R_2 = 4.64$
B—5.39			1.17	$R_3 = 4.85$
C—5.64				

* Significant at the .05 level.
SOURCE: J. Mezzano. Group counseling with low-motivated high school students—comparative effects of two uses of counselor time. *Journal of Educational Research*, 1968, *61*, 222–224.

The derivation of the term triangular table method becomes obvious when one observes that numbers are found only in the upper-right part of the table, forming a triangle. Numbers placed in the lower-left part of the table would be repetitive, simply a mirror image of the numbers in the upper right. Furthermore, numbers never appear on the upper-left to lower-right diagonal of the table because such numbers would represent a meaningless comparison of a group mean against itself.

In the table in Excerpt 4.17 two numbers are listed under the heading Shortest Significant Ranges. These are simply the critical values used by the author in deciding whether a pair of means are significantly different from one another. The first shortest significant range (R_2) is the critical value to be used when comparing two means that are right next to one another in terms of ordering by size. Thus, the calculated values of 3.71 and 1.17 were compared against the critical value of 4.64, and since neither of the calculated values was larger than the critical values, no significant difference can be said to exist between A and B or between B and C. The other shortest significant range (R_3) is used when comparing two means that have a third mean (which is irrelevant to the comparison) between them. Thus, the calculated value of 4.88 was compared against the critical value of 4.85, and since the former is larger, the author is able to say that the mean for group A is significantly different from the mean for group C. If four groups had been involved in the study, R_4 would have been the critical value for two means when a total of four means are spanned by the comparison.

Instead of using either of the two table methods described above, many authors simply report the results of their multiple comparison tests within the text of their article. This was the case in Excerpt 4.18.

EXCERPT 4.18 RESULTS OF A MULTIPLE COMPARISON TEST REPORTED WITHIN THE TEXT OF THE ARTICLE

Duncan's New Multiple Range Test (Edwards, 1964, pp. 136–138) was used to determine the means between which significant differences existed. It was found that primary and intermediate regular-class teachers used a significantly greater proportion of indirect behavior to stimulate students to talk than did special-class teachers. The multiple-range test results revealed that the two groups of regular-class teachers did not differ significantly from one another.

SOURCE: G. B. Stuck and M. D. Wyne. Study of verbal behavior in special and regular elementary school classrooms. *American Journal of Mental Deficiency*, 1971, 75, 463–469.

REVIEW TERMS

ANOVA	multiple comparison
Bartlett's chi-square	*MS*
Cochran's C	Newman-Keuls
conservative	one-way analysis of variance
correlated samples	paired *t* test
Duncan's new multiple range test	post hoc comparison
	robust
Dunnett's test	Scheffé's test
Fisher's LSD	*SS*
Hartley's F-max	summary table
homogeneity of variance	*t* test
independent samples	triangular table
Levene's test	Tukey's HSD
liberal	underlining method
matched *t* test	

REVIEW QUESTIONS

1. How many groups are usually compared by a *t* test? On what basis does the *t* test compare the groups?
2. Suppose a researcher measures a single group of people before they receive some type of training and then a second time after the training. Should he use an independent samples *t* test or a correlated samples *t* test to compare the pretest and posttest scores?
3. If a researcher uses an independent samples *t* test and if the *df* for the *t* test is equal to 25, we know that _____ Ss were in the study.
4. Suppose a researcher is using an independent samples *t* test. Under what condition can he bypass testing the assumption of equal variances and under what condition should he test this assumption?
5. A one-way analysis of variance can be used to compare _____ or more groups.
6. How many *MS* values usually appear in a one-way ANOVA summary table? How does the researcher obtain these values? How are these *MS* values used to obtain the *F* value?
7. Suppose a researcher uses a one-way ANOVA to compare four groups. Each group contains 10 Ss. In the ANOVA summary table, how many degrees of freedom will there be for (a) between-groups, (b) within-groups, and (c) total?
8. Under what condition is the analysis of variance considered to be robust?
9. What is the purpose of Hartley's *F*-max test? Name three other tests sometimes used to accomplish the same thing.
10. What is the purpose of a post hoc multiple comparison test?
11. Suppose a researcher uses a one-way ANOVA and obtains a significant *F*. Under what condition would it be appropriate for him to neglect using a multiple comparison test?
12. How does Dunnett's test differ from other multiple comparison tests?
13. Suppose you encounter a table in a journal article in which there are four means lined up in a horizontal row. If the two middle means are underlined, how many significant differences are there between pairs of means?
14. Suppose a researcher compares five groups with a one-way ANOVA. If a significant *F* is obtained, the researcher might choose to use Tukey's multiple comparison test. If the results of the Tukey investigation were to be presented in a triangular table, how many numbers (i.e., calculated values) would there be within the table?
15. As compared with a liberal multiple comparison test, will a conservative test tend to yield more or fewer significant differences between means?

CHAPTER 5

FACTORIAL
ANALYSES
OF
VARIANCE

In Chapter 4, we learned that a one-way analysis of variance can be used to compare two or more group means. This form of ANOVA is considered to be one-way because the comparison groups differ from each other along just *one* dimension. For example, if a one-way ANOVA is used to compare three different approaches to counseling (directive vs. nondirective vs. rational), the single dimension in the study would be method of counseling. Stated differently, in a one-way ANOVA there is always one and only one independent variable.

The analysis of variance can also be used to compare groups which differ along two or more dimensions. In other words, the analysis of variance is appropriate for studies in which there are several independent variables. Within this chapter attention will be directed toward two types of ANOVA—one which involves two independent variables (two-way ANOVA) and one which involves three independent variables (three-way ANOVA).

TWO-WAY ANOVA

A two-way ANOVA is used to compare groups which differ from one another along two dimensions. A hypothetical example may help to show how this is possible. Suppose a researcher sets up an experiment to investigate the effects of (1) item arrangement and (2) time limit on the dependent variable of student performance on a 100-item multiple-choice final exam. For the first independent variable (item arrangement), our researcher might be interested in comparing three ways of ordering the items—one in which the items are arranged in accordance with the course syllabus (i.e., with questions covering topics discussed early in the academic term at the beginning of the test), a second form in which the items are arranged in the opposite order (i.e., with questions covering the most recent topics coming first), and a third form in which the items are not arranged in any systematic order at all. For easy reference, these three test forms, each containing the same questions, might be called the syllabus form, the backwards form, and the random form. With regard to the second independent variable (time limit), the researcher might want to compare how students do when there is a restrictive time limit which forces them to hurry through the test in order to finish versus how they do when they can take as long as they like to finish.

The two independent variables of this hypothetical experiment combine to form six treatment conditions: (1) the syllabus test form administered under a time limit, (2) the backwards test form administered under a time limit, (3) the random test form administered under a time limit, (4) the syllabus test form administered with no time limit, (5) the backwards test form administered with no time limit, and (6) the random test form administered with no time limit. If a pool of 60 students were available to the researcher, he would randomly assign 10 students to each treatment condition. The dependent variable would be the number of items answered correctly by each student.

Factors and Levels

In the field of statistics and research design, the term *factor* means the same thing as the term independent variable. Therefore, the term *factorial analysis of variance* has come to denote any ANOVA in which there are two or more factors. The many different types of factorial ANOVA's are distinguished from each other by the number of factors involved, and the terms two-way factorial ANOVA, two-factor ANOVA, two-way ANOVA, and simple factorial ANOVA are all different ways of referring to an ANOVA involving only two factors.

Each of the dimensions (independent variables) in a two-way ANOVA is called a factor; the categories or subgroups of each factor are called *levels*. In the hypothetical study described above, there are three levels of the first factor (item arrangement) and two levels of the second factor (time limit). By definition, there are always two factors in a two-way ANOVA; however, the number of levels in either factor can vary from two on up. If the hypothetical experiment were to be set up so as to compare four different item arrangements and five different time limits, it would still be a two-way

ANOVA, but in this case there would be four levels of the first factor and five levels of the second factor.

To help indicate the precise nature of the two-way ANOVA used, authors frequently specify the number of levels in each factor. For example, an author might state that he used a 2×4 factorial analysis of variance (where the 2 and the 4 correspond to the number of levels in each factor). Excerpt 5.1 gives such an example. Using this notational system, our hypothet-

---

EXCERPT 5.1 STATEMENT OF THE
NUMBER OF LEVELS IN A TWO-WAY ANOVA

A 2×2 analysis-of-variance factorial design was used to analyze the data obtained from the study.

SOURCE: R. H. Coop and L. D. Brown. Effects of cognitive style and teaching method on categories of achievement. *Journal of Educational Psychology*, 1970, *61*, 400–405. Copyright 1970 by the American Psychological Association. Reprinted by permission.

---

ical researcher would refer to his experiment as a 2×3 ANOVA. Sometimes, as in Excerpt 5.2, an author may be even more explicit by specifying the nature of each level in both factors.

EXCERPT 5.2 EXPLICIT DESCRIPTION OF THE NUMBER
AND NATURE OF LEVELS IN A TWO-WAY ANOVA

Two experimental variables were arranged in a 2×3 factorial design. The first variable, type of objective, consisted of (1) general (nonbehavioral) and (2) specific (behavioral) objectives. The second variable, knowledge of objective, consisted of (1) teacher knowledge (i.e., possession) of objectives prior to and during instruction, (2) teacher knowledge (as in Level 1) and student knowledge (i.e., possession) during instruction, and (3) student knowledge alone during instruction.

SOURCE: J. R. Jenkins and S. L. Deno. Influence of knowledge and type of objectives on subject-matter learning. *Journal of Educational Psychology*, 1971, *62*, 67–70. Copyright 1971 by the American Psychological Association. Reprinted by permission.

Two Main Effects and One Interaction

Whereas a one-way ANOVA answers just one research question, a two-way ANOVA answers three research questions. To gain a clear understanding of these three questions, let us return once again to the hypothetical experiment involving the 100-item final exam. A diagram corresponding to this experiment is presented in Table 5.1. From this diagram we see that (1) the two factors (independent variables) are item order and time limit and (2) there are three levels of the first factor and two levels in the second factor.

In addition, the diagram contains various means ($\overline{X}$'s). (The numbers for these means are, of course, fictitious; nonetheless, they will assist greatly

TABLE 5.1 A DIAGRAM TO CORRESPOND TO
THE HYPOTHETICAL EXPERIMENT

TIME LIMIT

		YES	NO	
I **T** **E** **M**	SYLLABUS	$\bar{X} = 69$	$\bar{X} = 75$	$\bar{X} = 72$
O **R** **D**	BACKWARDS	$\bar{X} = 68$	$\bar{X} = 74$	$\bar{X} = 71$
	RANDOM	$\bar{X} = 61$	$\bar{X} = 73$	$\bar{X} = 67$
E **R**		$\bar{X} = 66$	$\bar{X} = 74$	

in our discussion of the three research questions of a two-way ANOVA.)
Each of the 6 means within the various cells of the diagram corresponds to
the 10 Ss who were assigned to that particular treatment combination. For
example, the mean of 69 in the upper-left *cell* represents the average score
for the 10 Ss who had the syllabus test form administered with a restrictive
time limit. Each of the 3 means at the extreme right of the diagram is an
average of the 2 cell means in the same row. Thus, the mean of 72 was found
by averaging 69 and 75, and it represents the average score for all the Ss who
had the syllabus test form (disregarding whether or not a time limit was
involved). Each of the 2 means at the bottom of the diagram is an average
of the 3 cell means in the same column. Thus, the mean of 66 was found by
averaging 69, 68, and 61, and it represents the average score for all Ss who
took the final exam under a restrictive time limit (disregarding which specific
form of the test they had).

The first two research questions of a two-way ANOVA involve, respectively, the overall row means and the overall column means. The third research question deals with the individual cell means within the diagram. To
answer the first research question, a two-way ANOVA compares the row
means to see if there are any significant differences present. In our hypothetical experiment comparisons would be made between the overall mean for
students taking the syllabus test form ($\bar{X} = 72$), the overall mean for the
students taking the backwards test form ($\bar{X} = 71$), and the overall mean for
the students taking the random test form ($\bar{X} = 67$). If significant differences
are found to exist between these means, our researcher will report that there
is a significant *main effect* of item order.

The second research question involves a comparison of the column
means to see whether there are any significant differences present. In our

experiment a comparison would be made between the overall mean for students taking the test with a time limit ($\overline{X} = 66$) and the overall mean for students without a time limit ($\overline{X} = 74$). If these means are found to be significantly different from one another, our researcher would say that the main effect of time limit is significant.

Thus, the first two research questions of a two-way ANOVA concern the main effects of the two factors. As you may have been thinking, the researcher could have used a one-way ANOVA to test for a main effect of item order and then another one-way ANOVA (or t test) to investigate the main effect of time limit. However, a single two-way ANOVA is a better procedure than two separate one-way ANOVA's for three reasons. First, the two-way ANOVA is more *parsimonious*, that is, it answers the same questions more quickly and with less computation. Second, the two-way ANOVA is more *powerful*, that is, it is more sensitive to differences among the groups that are being compared. Thus, a two-way ANOVA might pick up significant differences which would remain undetected by the two separate one-way ANOVA's (even though both procedures are applied to the same data). Third, a two-way ANOVA can answer an additional research question that cannot be answered at all with the two separate one-way ANOVA's. This third research question concerns the possible existence of an interaction between the two independent variables.

In a two-way ANOVA an *interaction* is the effect on the dependent variable of the two independent variables operating together, as distinguished from the main effect of each independent variable. An interaction effect will show up in the data in this manner: the differential effectiveness of the levels of one factor will change according to how these levels (of the first factor) are combined with the levels of the second factor. Applying this definition of interaction to our hypothetical experiment, there would be no interaction if the disadvantage of having a time limit was the same for each of the three test forms. On the other hand, there would be an interaction if the lack of a time limit proved to be more beneficial for some test forms than for others.

Let us return to Table 5.1 for a moment to see whether or not an interaction does exist in the data. To answer this question, we must look at the individual cell means and ask, "Do the differences among the levels of one factor remain constant as we move from one level to another level of the second factor?" At the first level of item order (syllabus), the difference between the time limit group ($\overline{X} = 69$) and the no time limit group ($\overline{X} = 75$) is 6 points. Moving to the next level of item order (backwards), the difference between the time limit group ($\overline{X} = 68$) and the no time limit group ($\overline{X} = 74$) is again 6 points. So far, no interaction. However, when we move to the third level of item order (random), we find that the difference between the means is 12 points. Thus, the differential effect of having a time limit as opposed to not having a time limit does not remain constant for all three test forms. For this reason, an interaction may be said to exist between item order and time limit.

In an attempt to see whether an interaction existed in Table 5.1, we calculated the difference between the means in each row of the diagram, and then we compared these differences across the three rows. We could have calculated the differences among the means within each column, and then compared these differences across columns. Looking at the left column of means, the difference between the top mean and the middle mean is 1, and the difference between the middle mean and the bottom mean is 7. If these same differences were also found in the right column, no interaction would be present. However, when we move over to the right column, we find that the difference between the top and middle means is 1, and the difference between the middle and bottom means is also 1. The differences of 1 and 7 (from the left column) do not correspond with the differences of 1 and 1 (from the right column). Thus, no matter how we look at the cell means, we come up with the same conclusion—there is some degree of interaction present in the data.

Many authors will include in their articles a table of means similar to Table 5.1. For example, in Excerpt 5.3 we find a table of means which corresponds to a 2×2 ANOVA. When you see such a table, you do not have to compare the overall row and column means to see if there are significant main effects or the individual cell means to see if there is a significant interaction. In fact, it would be impossible for you to do this. Although you could determine whether there is some degree of a main effect or interaction present, you could not tell whether it is a significant main effect or a significant interaction. Only the statistical analysis can answer these questions.

EXCERPT 5.3 TABLE OF MEANS FOR A TWO-WAY ANOVA

TABLE 3 MEAN ERRORS TO CRITERION
AS A FUNCTION OF PROMPTED TRAINING
AND VERBAL TRAINING

VERBAL TRAINING	PROMPTED TRAINING		
	NO PROMPT	PROMPT	M
Verbalization	61.80	49.10	55.45
No verbalization	34.00	73.30	53.15
M	42.90	60.70	

SOURCE: J. K. Davis and H. J. Klausmeier. Cognitive style and concept identification as a function of complexity and training procedures. *Journal of Educational Psychology*, 1970, *61*, 423–430. Copyright 1970 by the American Psychological Association. Reprinted by permission.

Now that we understand what the three research questions of a two-way ANOVA are getting at, let us turn our attention to the various ways in which authors present the results of their data analyses.

Reporting the Results of a Two-Way ANOVA

The Model Summary Table. The results of a two-way ANOVA are usually presented by means of a two-way ANOVA summary table. As with the model summary table for a one-way ANOVA, all authors do not use the same format, but an understanding of the model table should enable a person to interpret the variations found in the literature. Table 5.2 shows how the model summary table would be constructed to present the results of our hypothetical experiment.

TABLE 5.2 MODEL SUMMARY TABLE FOR
THE HYPOTHETICAL TWO-WAY ANOVA EXPERIMENT

SOURCE	df	SS	MS	F
Item order	2	10	5	10**
Time limit	1	8	8	16**
Interaction	2	4	2	4*
Within (error)	54	27	5	
Total	59	49		

* $p < .05.$
** $p < .01.$

Like a one-way ANOVA summary table, a two-way summary table has five columns: source (or source of variation), df (degree of freedom), SS (sum of squares), MS (mean square), and F. While the one-way summary table had just three rows, a two-way summary table has five rows. The first row corresponds to the main effect for the first factor, the second row corresponds to the main effect for the second factor, the third row corresponds to the interaction between the two factors, the fourth row is the within-group data (or error), and the bottom row is the total.

As for the one-way summary table, the various degrees of freedom in the summary table for a two-way ANOVA are easy to compute. For each of the two main effects, the appropriate degrees of freedom are found by subtracting 1 from the number of levels in the factor. Since there are three levels of the first factor (item order) in our hypothetical experiment, there are 2 degrees of freedom for the first row of the table. Since there are two levels of the second factor (time limit), there is 1 df for the second row of the table. The df for the interaction is found simply by multiplying the degrees of freedom for the two main effects. Dropping down to the bottom row, the df for Total is found by subtracting 1 from the total number of subjects in the study. Since there were 60 students involved in the experiment, the total df is equal to 59. Finally, the df for the within (or error) source is found by subtracting the top three degrees of freedom from the total df. Thus, the within df is equal to $59 - (2 + 1 + 2) = 59 - 5 = 54$.

The various SS (sum of squares) are determined by the researcher; he uses a relatively complicated formula to obtain each SS, and it is impossible for the reader of a journal article to check the accuracy of the first four numbers in the SS column. However, the bottom number in this column, the total SS, should always be equal to the sum of the other four SS. (In our model table the numbers for the various SS are fictitious in order to make the math of the table easy to follow; in reality, however, the numbers for the SS will almost always contain decimal points.)

Each value in the MS (mean square) column is found by dividing the appropriate SS by the corresponding df on the same row of the table. For example, the MS value of 5 for the main effect of item order was found by dividing 10 by 2. Note that the summary table does not contain a value for total MS. If an author puts a number in this position, he has simply made a mistake.

There are three different numbers in the F column. Each of these numbers was found by dividing the appropriate value for MS by the within (error) MS. For the main effect of item order, the F value of 10 was found by dividing 5 (the MS for item order) by .5 (the within MS). For the interaction, an F of 4 was found by dividing 2 (the interaction MS) by .5 (the within MS). Note that the within MS is always divided into another MS to obtain an F value.[1] Also, note that no value for F is computed for either of the two bottom rows of the summary table.

The most important numbers in the summary table are the three values in the F column. Again, as in the one-way ANOVA summary table, the remaining numbers are simply the figures needed to calculate the important Fs. The three F values allow the researcher to determine whether there are significant main effects or a significant interaction. The researcher makes this decision by comparing each of his F values with a critical value in an F table. (As in a one-way ANOVA, the two df values associated with each calculated F allow the researcher to find the critical values. The first of these df values is located on the same row as the F value in question; the second df value corresponds to the within-groups [error] source.) The author usually indicates a significant difference by placing one or two asterisks next to the value for F and giving the precise level of significance in a note to the table by means of a statement such as $p < .05$.

Different Formats for the Two-Way ANOVA Summary Table. It would be impossible to present and discuss all of the different formats used by

---

[1] Although infrequently seen in the literature, there are exceptions to this general rule. If the researcher uses a mixed or random model for his ANOVA (rather than the more common fixed model), he may divide by something other than the within (error) MS to get his F values. We will discuss these three models in Chapter 13.

authors for their summary tables, but the following tables represent the most popular variations on the model table.

The summary table in Excerpt 5.4 is similar to our model, but it differs in three respects: (1) the *df* column appears between the *SS* and *MS* columns instead of preceding the *SS* column, (2) the phrase A × B has been used instead of interaction to identify the third row of the table, and (3) the term remainder has been used instead of within (or error) to identify the fourth row of the table. In spite of these differences, the mathematics of both tables is the same: the first four numbers in the *SS* and *df* columns add up to the bottom figure, the interaction *df* is equal to the product of the degrees of freedom of the two main effects, each *MS* value is found by dividing the *SS* on that row by the corresponding *df*, and each *F* value is found by dividing the *MS* on that row by the remainder (within/error) *MS*.

EXCERPT 5.4 TWO-WAY ANOVA SUMMARY TABLE
SIMILAR TO THE MODEL SUMMARY TABLE

TABLE 7 ANALYSIS OR VARIANCE: SPELLING TEST

SOURCE	SS	df	MS	F
Between groups (A)	6,611.34	1	6,611.34	137.74°
Between grades (B)	7,287.94	5	1,457.59	30.02°
A × B	254.82	5	50.97	1.05
Remainder	4,661.78	96	48.56	
Total	18,815.88	107		

° $p < .001$.
SOURCE: J. G. Lyle and J. Goyen. Performance of retarded readers on the WISC and educational tests. *Journal of Abnormal Psychology*, 1969, 74, 105–112. Copyright 1969 by the American Psychological Association. Reprinted by permission.

Through this table, the authors tell us a great deal about their study. Without reading a single word in the text of the article, we know that the 2 factors were groups and grades, that there were 2 levels in the first factor and 6 levels in the second factor (thus making it a 2 × 6 factorial ANOVA), that there were 108 subjects in the study, that both main effects were significant, that the interaction between groups and grades was not significant, and that the .001 level of significance was used.

In Excerpt 5.5 we see a summary table for a 2 × 3 ANOVA in which the authors have omitted the column for sum of squares (SS). However, since the SS values are used only to calculate the MS values and the MS values are included, the table gives us the same amount of information as the model summary table. This table also differs from the model with respect to the F column. The authors do not present an F value for the first main effect or the interaction (A × SAT) because they result in numbers that are smaller than 1.0, and only Fs that are larger than 1.0 have a chance of being significant. If the authors had put in all three Fs, the first one would have been equal to .79 (.243 divided by .306), and the third one would have been equal to .21 (.066 divided by .306).

EXCERPT 5.5 SUMMARY TABLE WITHOUT THE COLUMN FOR SS

TABLE 1 ANALYSIS OF VARIANCE OF
GRADE-POINT AVERAGE SCORES

SOURCE	df	MS	F
Forced-choice anxiety (A)	1	.243	—
Aptitude (SAT)	2	1.166	3.81°
A × SAT	2	.066	—
Within Ss	89	.306	
Total	94		

° $p < .05$.

SOURCE: O. Desiderato and P. Koskinen. Anxiety, study habits, and academic achievement. *Journal of Counseling Psychology*, 1969, *16*, 162–165. Copyright 1969 by the American Psychological Association. Reprinted by permission.

An author occasionally deletes not only the entire SS column, but the total row as well (see Excerpt 5.6). Even though this information is missing, the table still allows us to figure out that a 2 × 3 ANOVA was used, that there were a total of 24 subjects involved (by adding the degrees of freedom to get the total and then adding 1), and that the only significant finding dealt with the main effect of reinforcement.

EXCERPT 5.6 SUMMARY TABLE WITHOUT THE
COLUMN FOR SS AND THE ROW FOR TOTAL

TABLE 2 ANALYSIS OF VARIANCE SUMMARY FOR ERRORS IN
WORD RECOGNITION TASK IN MIDDLE CLASS POPULATION

SOURCE OF VARIATION	DEGREES OF FREEDOM	MEAN SQUARES	F
Sex	1	308.16	.528
Reinforcement	2	3251.29	5.57°°
Sex × Reinforcement	2	1094.29	1.87
Error	18	583.78	

°° $p < .025$.
SOURCE: J. Pukulski. Effects of reinforcement on word recognition. *The Reading Teacher*, 1970, *23*, 516–522. Copyright 1970 by the International Reading Association. Reprinted by permission.

In Excerpt 5.7 we find a table in which the authors present the results of 12 separate two-way ANOVA's. Each of these analyses corresponds to one of the scales listed in the left column. The other column headings indicate that the main effects in each analysis were sex and citizen status, and the interaction is referred to by the notation S × C. There are two rows of numbers for each scale. The first row contains the *MS* values from the analysis and the second row contains the three *F* values. Each of the three *F* values was found by dividing the *MS* values by the within *MS*. For example, the *F*

EXCERPT 5.7 TWELVE SEPARATE TWO-WAY ANOVA'S COMBINED INTO ONE TABLE

TABLE 2 ANALYSIS OF VARIANCE OF CPI SCALES

SCALE	SEX (s)	CITIZEN STATUS (c)	s × c	WITHIN
Social presence				
MS	9.67	733.33	6.30	107.94
F	.10	6.79*	.06	
Responsibility				
MS	100.77	561.69	50.34	83.32
F	1.21	6.74*	.60	
Socialization				
MS	555.05	50.93	21.82	87.18
F	6.37*	.58	.25	
Tolerance				
MS	53.03	920.18	68.59	109.51
F	.48	8.40**	.63	
Sense of well-being				
MS	470.77	841.79	26.01	149.71
F	3.15	5.62*	.17	
Good impression				
MS	526.74	.23	122.95	75.04
F	7.02*	.00	1.64	
Communality				
MS	94.57	582.11	83.59	98.13
F	.96	5.93*	.85	
Achievement via conformance				
MS	220.74	497.91	58.63	77.75
F	2.84	6.40*	.75	
Achievement via independence				
MS	10.47	777.50	95.52	104.42
F	.10	7.45**	.92	
Intellectual efficiency				
MS	63.34	499.26	30.11	107.95
F	.59	4.63*	.28	
Flexibility				
MS	296.42	601.17	23.37	121.49
F	2.44	4.95*	.19	
Femininity				
MS	35.15	342.51	460.11	85.33
F	.41	4.01*	5.39*	

Note. $df = 1$ for sex, citizen status, and S × C; $df = 82$ for within.
 * $p < .05$.
 ** $p < .01$.

SOURCE: S. L. M. Fong and H. Peskin. Sex-role strain and personality adjustment of China-born students in America: a pilot study. *Journal of Abnormal Psychology*, 1969, *74*, 563–567. Copyright 1969 by the American Psychological Association. Reprinted by permission.

of .10 for the main effect of sex on the first scale (social presence) was found by dividing 9.67 by 107.94. The note to the table indicates the *df* associated with each source of variation, thus allowing us to figure out that each analysis was a 2 × 2 ANOVA with 86 subjects.

Presenting the Results of a Two-Way ANOVA Without a Summary Table. As we mentioned earlier, the most important numbers in the summary table are the three *F* values. Because of the comparative unimportance of the other numbers, an author will sometimes take the three *F* values out of the summary table and put them into a table of means. Excerpt 5.8 shows how this can be done.

EXCERPT 5.8 THREE *F* VALUES FROM A TWO-WAY ANOVA IN A TABLE OF MEANS

TABLE 3 MEAN OVERACHIEVEMENT SCORES FOR MALES

LEVEL OF ANXIETY	LEVEL OF REPRESSION			ROW MEAN
	HIGH ($\geq$ 17)	MEDIUM (13–16)	LOW ($\leq$ 12)	
High ($\geq$ 20)	53.90 ($N = 21$)	49.73 ($N = 22$)	49.50 ($N = 16$)	51.15
Medium (12–19)	48.53 ($N = 19$)	47.89 ($N = 29$)	50.47 ($N = 32$)	49.08
Low ($\leq$ 11)	53.27 ($N = 22$)	46.28 ($N = 18$)	48.15 ($N = 26$)	49.35
Column mean	52.03	48.06	49.45	

F ratios: A, 1.77; R, 5.49*; A × R, 2.10

* $p < .01$ since $F_{.99}$ (2,196) = 4.71.
SOURCE: D. L. Stix. Discrepant achievement in college as a function of anxiety and repression. *Personnel and Guidance Journal*, 1967, 45, 804–807. Copyright 1967 by the American Personnel and Guidance Association.

The first of the three *F* ratios, 1.77, is for the anxiety main effect. It dealt with a comparison of the three row means (51.15, 49.08, 49.35). The second *F* ratio, 5.49, is for the main effect of repression, and it dealt with a comparison of the three column means (52.03, 48.06, 49.45). The third *F* ratio, 2.10, is for the anxiety-by-repression interaction, and it dealt with the individual cell means.[2]

Quite frequently, an author will state the results of his two-way ANOVA in the text of his article, not in a table. For example, consider Excerpt 5.9. Within the parentheses at the end of the second sentence, the authors first present the calculated *F* value, 53. Next, the degrees of freedom for this *F* are given. The first *df* value, 1, corresponds to the *df* for the main

---

[2] If the number of Ss in the cells of a factorial ANOVA varies from cell to cell (as in Excerpt 5.8), a *solution for unequal n* must be applied by the researcher. Often the author of an article will not take the time to mention having used a solution for unequal *n*. If he does say something, he will probably indicate that an *unweighted means solution* or a *least squares solution* was employed.

effect being tested (i.e., teaching method); the second *df* value, 76, corresponds to the within (error) *df*. Finally, the notation $p < .01$ indicates that the *F* of 53 was found to be significant at the .01 level.

EXCERPT 5.9 RESULTS OF A TWO-WAY
ANOVA PRESENTED WITHOUT A TABLE

The present study employed a 2×2 factorial design with teaching method (independent-problem-solving and teacher-structured-presentation) as one variable and cognitive style (analytic versus nonanalytic) as the other. The results of the analysis of variance to test the factual content criterion disclosed that there was a significant difference between the two teaching methods, with students in the teacher-structured-presentation method demonstrating superior achievement ($F = 53$, $df = 1/76$, $p < .01$). There was no significant difference between analytic students and nonanalytic students on the criterion nor was there a significant interaction between cognitive style and teaching method.

SOURCE: R. H. Coop and L. D. Brown. Effects of cognitive style and teaching method on categories of achievement. *Journal of Educational Psychology*, 1970, *61*, 400–405. Copyright 1970 by the American Psychological Association. Reprinted by permission.

Graphing the Interaction. If a researcher obtains a significant interaction from his two-way ANOVA, he may include a graph of the interaction in his article to help explain what caused it to be significant. In Excerpt 5.10 we see a table of means and a graph of the interaction from a 2×3 ANOVA. In this study the authors investigated how the dependent variable of test performance was influenced by (1) true achievement level and (2) false information given to students concerning their aptitude (potential) for doing well in the course. There were two levels for the first factor—high GPA and low GPA. There were three levels for the second factor—positive information, that is, telling the student he had a high aptitude for the course material; negative information, that is, telling him he had a low aptitude for the course; and no information at all.

As was mentioned earlier in the chapter, the interaction in a two-way ANOVA deals with the individual cell means, not the overall row or column means. Therefore, the graph of the interaction was constructed by placing each of the six cell means at the appropriate point on the abscissa (horizontal axis) representing treatment groups and the ordinate (vertical axis) representing mean scores. Finally, a dotted line was used to connect the three means for the high GPA students, and a solid line was used to connect the three means for the low GPA students. The graph does not give us any information beyond that already contained in the table of means, but it does illustrate in a straightforward manner that the effect of the treatments depends upon whether they are administered to students having high or low GPA's.

In our previous discussion of interaction, we said that an interaction exists if the difference between the levels of the first factor does not remain

EXCERPT 5.10 GRAPH OF A 2 × 3 INTERACTION

TABLE 1 MEAN SCORES OF VARIOUS TREATMENT
GROUPS ON COURSE CONTENT TEST

GROUP	POSITIVE INFORMATION	NEGATIVE INFORMATION	NO INFORMATION	COMBINED
High GPA	42.00	47.75	45.92	45.22
Low GPA	39.58	35.00	37.50	37.36
Combined	40.79	41.37	41.71	41.29

Note. $N = 72$ for the combined groups, 36 for high GPA, 36 for low GPA.

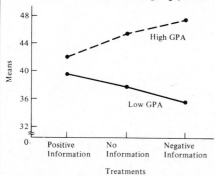

FIGURE 1. *Effect of aptitude information and achievement level on a course content test.*

SOURCE: R. S. Means and G. H. Means. Achievement as a function of the presence of prior information concerning aptitude. *Journal of Educational Psychology,* 1971, *62,* 185–187. Copyright 1971 by the American Psychological Association. Reprinted by permission.

constant as we move from one level to another level of the second factor. The graph clearly shows us that the difference between the high and low GPA students (i.e., the two lines in the graph) does not stay the same for each of the three treatment positions. If the difference between the GPA groups had been constant across the three treatment conditions, the two lines in the graph would have been parallel to one another. In fact, some authors of statistics books define interaction as a departure from *parallelism* when the cell means from the two-way ANOVA are graphed as in Excerpt 5.10.

When the interaction from a two-way ANOVA is graphed, two or more of the lines in the graph may cross. If this is the case, the interaction can be technically referred to as *disordinal.* On the other hand, if none of the lines cross (as in Excerpt 5.10), the word *ordinal* can be used to characterize the interaction.

Additional Analyses

If the F corresponding to a main effect is significant, a researcher will know that there are significant differences among the overall means for the levels making up the factor. However, he will not know which specific levels are significantly different from one another. To answer this question, a researcher will need to apply a follow-up test. Any of the multiple comparison procedures discussed in Chapter 4 could be used to find out which particular means are significantly different from one another.

In the hypothetical experiment dealing with the three test forms, a significant main effect for the first factor (item order) would indicate that significant differences exist somewhere among the three row means of Table 5.1. To locate the significant differences, our researcher might choose to employ the Newman-Keuls procedure. Results of this follow-up analysis might indicate that the mean for the syllabus test form ($\overline{X} = 72$) is not significantly different from the mean for the backwards test form ($\overline{X} = 71$), but that both of these means are significantly different from the mean for the random test form ($\overline{X} = 67$). If the results of the two-way ANOVA also indicated a significant main effect for the second factor (time limit), would a follow-up test be needed? The answer is "no" because there are only two levels of this factor, and we would automatically know that the mean for students having a restrictive time limit ($\overline{X} = 66$) is significantly lower than the mean for students who could take as much time as they wanted ($\overline{X} = 74$).

If the F value for the interaction is significant, a researcher will know that the difference between the levels of one factor change from one level to another level of the second factor. Such an interaction is a tip off to the researcher that direct interpretation of the two main effects may be misleading.

Consider a new hypothetical experiment in which the two factors are teaching method (lecture vs. discussion) and class size (small classes vs. large classes). The results of this 2×2 ANOVA might reveal a significant interaction—with the lecture method working better with large classes and the discussion method working better with small classes. However, both main effects might very well yield nonsignificant F values. If we were to interpret the main effects directly, we might be misled into thinking that there is no difference between the two teaching methods, when there really is if we simply take into consideration the variable of class size.

When a significant interaction is present in a two-way ANOVA, comparisons of the main effects are inappropriate. Instead of comparing the overall row or column means, a *comparison of simple main effects* should be made. This procedure involves comparing the various levels of one factor at each separate level of the other factor. Thus, a significant interaction would signal our researcher to make a comparison of the two teaching methods with regard to their relative effectiveness with small classes, and then make a second comparison with regard to their effectiveness with large classes.

Imagine a table of means for a two-way ANOVA. An investigation of simple main effects involves comparing the means that appear in the first row, then the means that appear in the second row, etc. (Or, the means in each separate column could be compared.) These comparisons are analogous to several t tests or several one-way ANOVA's. If only two levels of the factor are being compared (as would be the case if we compared the discussion method to the lecture method at each separate level of class size), then a t test could be used to make the simple main effect comparisons. However, if three or more levels are being compared, then a one-way ANOVA must be used. Thus, if four teaching methods had been used rather than just two, two one-way ANOVA's would be needed to compare the different levels of teaching method at each of the levels of class size. And if a significant F were found for either of the one-way ANOVA's, a multiple comparison procedure would be needed to identify precisely where the significant differences lie.

Excerpt 5.11 comes from an article in which the authors used a 3×4

EXCERPT 5.11 USE OF A SIMPLE MAIN EFFECTS INVESTIGATION TO HELP INTERPRET A SIGNIFICANT INTERACTION

TABLE 1 ANALYSIS OF VARIANCE SUMMARY FOR PERFORMANCE SCORES AS A FUNCTION OF ACHIEVEMENT-ANXIETY TYPE AND ITEM-DIFFICULTY SEQUENCE

SOURCE OF VARIATION	df	MS	F
Item-difficulty sequences (A)	2	107.50	1.05
Achievement-anxiety types (B)	3	442.41	4.32*
B for A_1	3	582.50	5.36*
B for A_2	3	409.37	4.00*
B for A_3	3	108.63	1.06
A × B	6	329.04	3.22*
Error	108	102.32	
Total	119		

Note. Abbreviated: A_1 = random sequence, A_2 = easy-hard sequence, A_3 = hard-easy sequence.
* $p < .01$.

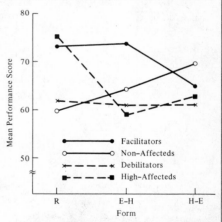

FIGURE 1. *Mean performance scores for achievement-anxiety types (facilitators, debilitators, high-affecteds, and nonaffecteds) on item-difficulty arrangements consisting of random (R), easy to hard (E-H), and hard to easy (H-E).*

There was a statistically significant interaction among the three item-difficulty orders and the four achievement-anxiety types ($F = 3.22$, $p < .01$); hence, Hypothesis 3 was supported. A simple main effects analysis indicated that within the achievement-anxiety factor the R arrangement and E-H arrangements were significant ($F = 5.36$, $p < .01$; $F = 4$, $p < .01$, respectively). Examining these two item-difficulty orders with the NKT revealed that (a) on the R form, facilitators and high-affecteds scored significantly higher than the debilitators ($p < .05$) and nonaffecteds, ($p < .01$), and (b) on the E-H form, facilitators scored significantly higher than the other three anxiety types ($p < .05$; see Figure 1).

SOURCE: D. C. Munz and A. D. Smouse. Interaction effects of item-difficulty sequence and achievement-anxiety reaction on academic performance. *Journal of Educational Psychology,* 1968, 59, 370–374. Copyright 1968 by the American Psychological Association. Reprinted by permission.

factorial ANOVA. The first factor was item sequence on the final exam; the three levels of the factor were an easy-to-hard sequence (E-H), a hard-to-easy sequence (H-E), and a random sequence (R). The second factor, achievement-anxiety, had four levels; the levels were four types of anxiety that students are known to have in test-taking situations. At the time of a final examination, the researchers randomly assigned students from within each anxiety classification to one of the three test forms. The dependent variable was the score earned on the final exam by each of the 120 students.

As shown in the summary table, the interaction between the two factors (A × B) was significant at the .01 level. (Note that the lines in the graph are not even close to being parallel, as we said would be the case when an interaction is present.) In making tests of simple main effects, the authors compared the four levels of the achievement-anxiety types factor at each separate level of the item-difficulty sequence factor. In conducting the analysis of simple main effects, the researchers did something that would be analogous to using a one-way ANOVA three times, once to compare the four points in the graph above R, once to compare the four points in the graph above E–H, and once to compare the four points in the graph above H–E. The results of these three analyses appear in the summary table immediately above the A × B interaction. Two of the comparisons for simple main effects yielded a significant F, and the Newman-Keuls (NKT) multiple comparison procedure was employed to identify exactly which means were significantly different from one another.

THREE-WAY ANOVA

As we mentioned at the beginning of this chapter, a three-way ANOVA always involves three factors (independent variables). As with a two-way ANOVA, the number of levels in each factor can vary from two up. Thus, the notation 2 × 3 × 4 ANOVA designates a three-way analysis of variance in which there are two levels of the first factor, three levels of the second factor, and four levels of the third factor. As we shall see, the results of a three-way ANOVA allow the researcher to see, first, whether there are overall differences among the levels of each factor and, second, whether the factors combine in such a way as to have a unique effect on the dependent variable. These two groups of research questions deal with the issues of main effects and interactions, respectively.

An Example from the Literature

To assist in our discussion of the various research questions dealt with by a three-way ANOVA, let us turn our attention to an actual study from the literature. In this study the researchers were primarily interested in comparing the way in which three types of reinforcement affected their ability to condition children to use the word *they* when making up sentences. Subjects were assigned to three groups: (1) Ss in the material reinforcement condition received an M & M candy immediately after using the word they at the beginning of a sentence; (2) Ss assigned to the praise reinforcement condition were reinforced by the experimenter's saying "good"; and (3) Ss in the symbolic reinforcement condition were simply given a plus mark. In addition to comparing these three different types of reinforcement, the researchers were also interested in seeing whether the different reinforcement conditions worked the same way for middle-class children and lower-class children, and for second-graders and sixth-graders. To answer these questions, a 3 × 2 × 2 factorial ANOVA was used, with the three factors being type of reinforcement, social class, and grade.

Whenever a three-way ANOVA is used, a minimum of eight different groups will be compared. (The number of groups is determined by multiplying the number of levels in the three factors together and the minimum must therefore be 2 × 2 × 2.) In this experiment, dealing with the three types of reinforcement, two social classes, and two grade levels, there were a total of 12 groups. The table of means and standard deviations in Excerpt 5.12 shows how each of these comparison groups can be defined by a unique combination of one level from each factor. For example, the mean in the upper left-hand corner of the table (5.66) corresponds to the group of middle-class children in grade 2 who were conditioned with material reinforcement. In this study, each of the 12 groups contained 12 children.

EXCERPT 5.12 TABLE OF MEANS AND STANDARD DEVIATIONS FOR A 3 × 2 × 2 FACTORIAL ANOVA

TABLE 1 MEANS AND STANDARD DEVIATIONS OF DIFFERENCE SCORES BY SOCIAL CLASS, GRADE LEVEL, AND TYPE OF REINFORCEMENT

| | Grade Level | | | | | |
| | GRADE 2 | | | GRADE 6 | | |
SOCIAL CLASS	MATERIAL	PRAISE	SYMBOLIC	MATERIAL	PRAISE	SYMBOLIC
Middle						
M	5.66	6.64	6.58	5.75	8.25	9.66
SD	2.32	3.28	2.98	1.63	4.12	3.37
Lower						
M	8.41	5.41	5.25	6.75	7.00	6.33
SD	4.23	3.63	2.74	4.20	4.16	3.22

SOURCE: J. D. Cradler and D. L. Goodwin. Conditioning of verbal behavior as a function of age, social class, and type of reinforcement. *Journal of Educational Psychology*, 1971, *62*, 279–284. Copyright 1971 by the American Psychological Association. Reprinted by permission.

Three Main Effects and Four Interactions

The main effects in a three-way ANOVA have the same meaning as the main effects in a two-way ANOVA, except we have three main effects rather than two. In other words, the main effect for any particular factor in a three-way ANOVA involves a comparison of the overall means for the levels making up that factor. With respect to the study of reinforcement, the main effect of grade level would involve a comparison of two overall means, one for all Ss in grade 2 and one for all Ss in grade 6 (disregarding social class and type of reinforcement). These overall means for the main effect of grade level are not shown in the table of Excerpt 5.12; however, the first one could be found by averaging the six means on the left side of the table, while the second one could be found by averaging the six means on the right side of the table.

For the main effect of social class, the statistical analysis compares the mean for all middle-class children with the mean for all lower-class children (disregarding grade level and reinforcement). To find these two means, we would find the average of the six means in the top row of the table and the average of the six means in the third row. The three overall means for the main effect of reinforcement could be found, respectively, by averaging the four means in the first and fourth columns, the four means in the second and fifth columns, and the four means in the third and sixth columns.

Of the four interactions associated with a three-way ANOVA, the first three deal with interactions between pairs of factors (i.e., each possible combination of two factors). If the interaction between reinforcement and social class turned out to be significant, this would indicate that the comparative effectiveness of the three reinforcement conditions changes as we move from middle-class children to lower-class children. In the same manner, an interaction between reinforcement and grade level would indicate that the comparative effectiveness of the three reinforcement conditions is not the same for sixth-grade children as for second-grade children. Finally, an interaction between social class and grade level would indicate that the relative performance of second-graders and sixth-graders (averaged across the three reinforcement conditions) changes as we move from middle-class children to lower-class children. As a group, these interactions are referred to as two-way or *first-order interactions*.

The fourth interaction in a three-way ANOVA involves all three factors. It is sometimes referred to as the three-way or *second-order interaction*. With respect to the study under consideration, a significant interaction between all three factors would indicate that the comparative effectiveness of the three reinforcement conditions changes as a function of both social class and grade level. If this were the case, the researchers would not be able to make a definitive statement about how the three reinforcements compared against one another without qualifying the statement as to a particular grade level and a particular social class.

Reporting the Results of a Three-Way ANOVA

In Excerpt 5.13 we see the ANOVA summary table for the study involving the three different types of reinforcement. The primary difference between a two-way and a three-way ANOVA summary table is the number of F values that are calculated. Whereas a two-way ANOVA will always provide three Fs (two for main effects and one for the interaction), a three-way ANOVA will always provide seven Fs (three for main effects, three for first-order interactions, and one for the second-order interaction).

EXCERPT 5.13 THREE-WAY ANOVA SUMMARY TABLE FOR THE DATA IN EXCERPT 5.12

TABLE 2 ANALYSIS OF VARIANCE FOR SOCIAL CLASS TYPE OF REINFORCEMENT AND GRADE LEVEL

SOURCE	df	MS	F
Social class (A)	1	12.250	.957
Grade level (B)	1	25.000	1.954
Type of reinforcement (C)	2	1.465	.114
A × B	1	11.111	.868
A × C	2	41.646	3.255*
B × C	2	47.396	3.704*
A × B × C	2	5.298	.414
Error	132	12.792	

* $p < .05$.

SOURCE: J. D. Cradler and D. L. Goodwin. Conditioning of verbal behavior as a function of age, social class, and type of reinforcement. *Journal of Educational Psychology*, 1971, 62, 279–284. Copyright 1971 by the American Psychological Association. Reprinted by permission.

Although there are more F values in the three-way ANOVA summary table, the mechanics of the table are the same as in the two-way ANOVA summary table. In particular, note that the main effect and interaction MSs were divided by the error MS in order to obtain the seven Fs. Also, note that the df for each main effect was found by subtracting 1 from the number of levels making up that factor, and that the df for each interaction can be found by multiplying together the main effect degrees of freedom for the factors involved in the interaction. As in the previous summary tables, the df for total (if it had been presented) would be equal to 1 less than the total number of Ss in the study.

As you might suspect, different authors use different formats for their three-way ANOVA summary tables. Whereas some authors delete the various sums of squares (SS), others include a column for these values. Whereas some authors delete the total row at the bottom of the table, others include it. Whereas some authors present all seven F values no matter what their size, others present only those Fs that are larger than 1.0, and a few authors present only those Fs that are significant. And sometimes, as shown in Excerpt 5.14, we may encounter a single summary table which contains the results of two or more separate analyses. (Note that the footnotes beneath the table have the triangular symbol pointing in the wrong direction. Also, the number in the total row of the Frequency MS column should not be there—this space should have been left blank.)

EXCERPT 5.14 TABLE CONTAINING THE RESULTS OF TWO SEPARATE THREE-WAY ANOVA'S

TABLE 1 ANALYSIS OF VARIANCE FOR FREQUENCY AND VARIETY OF INFORMATION-SEEKING BEHAVIOR IN THREE EXPERIMENTAL SCHOOLS

SOURCE OF VARIATION	df	FRE-QUENCY MS	F	VARIETY MS	F
School (A)	2	53.08	4.08*	7.13	1.81
Sex (B)	1	164.69	12.61**	50.17	12.76**
Treatment (C)	5	155.81	11.93	60.70	18.44**
A × B	2	28.77	2.20	5.67	1.44
A × C	10	8.65	.66	1.70	.43
B × C	5	7.14	.54	2.60	.66
A × B × C	10	4.87	.37	3.08	.78
Within groups	108	13.06		3.93	
Total	143	18.81			

* $p > .05$.
** $p > .01$.

SOURCE: J. B. Meyer, R. Strowig, and R. Hosford. Behavioral-reinforcement counseling with rural high school youth. *Journal of Counseling Psychology*, 1970, 17, 127–132. Copyright 1970 by the American Psychological Association. Reprinted by permission.

Authors may discuss the results of their three-way ANOVA's without presenting them in a table. As Excerpt 5.15 indicates, this technique for reporting results saves space in the article without reducing the amount of information provided.

EXCERPT 5.15 RESULTS OF A THREE-WAY ANOVA PRESENTED WITHOUT A TABLE

The trials to criterion on the concept-formation task were analyzed using a $4 \times 2 \times 2$ analysis of variance which assessed the effects of Instructional Treatments I-IV, the two schools, and the two levels of student IQ. A significant main effect was obtained for the instructional treatments ($F = 9.80$, $df = 3/48$, $p < .001$), and the interaction of treatments and student IQ was also significant ($F = 2.95$, $df = 3/48$, $p < .05$).

SOURCE: J. T. Guthrie and T. L. Baldwin. Effects of discrimination, grammatical rules, and application of rules on the acquisition of grammatical concepts. *Journal of Educational Psychology*, 1970, 61, 358–364. Copyright 1970 by the American Psychological Association. Reprinted by permission.

Graphing Interactions from a Three-Way ANOVA

To assist the reader in understanding the nature of a significant two-way (first-order) interaction, authors will sometimes present a graph similar to that shown in Excerpt 5.16. This particular graph and the accompanying table of means correspond to the results discussed in Excerpt 5.15. Since the interaction was between treatment conditions and IQ, the data points on the graph were found by averaging across the third factor, school. For example, the point at the right end of the solid line (i.e., the point corresponding to the high IQ Ss in treatment condition IV) represents the average of 33.25 and 54.50. In short, this graph shows us that the differential performance of high and low IQ students changes as a function of treatment conditions.

EXCERPT 5.16 GRAPH OF A TWO-WAY
INTERACTION FROM A THREE-WAY ANOVA

TABLE 2 MEANS AND STANDARD DEVIATIONS OF TRIALS
TO CRITERION ON THE CONCEPT-FORMATION TASK

TREATMENT CONDITION	School A				School B			
	HIGH IQ		*LOW IQ*		*HIGH IQ*		*LOW IQ*	
	$\bar{X}$	*SD*	$\bar{X}$	*SD*	$\bar{X}$	*SD*	$\bar{X}$	*SD*
I	4.75	3.90	4.50	4.75	4.00	3.00	1.00	0.00
II	2.50	2.60	8.50	6.34	19.50	15.63	6.75	2.17
III	18.25	21.51	61.00	29.44	25.75	30.90	53.50	31.50
IV	33.25	10.91	21.00	32.92	54.50	31.66	44.25	34.42

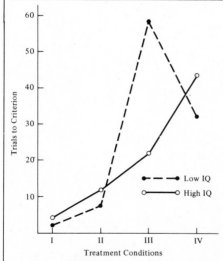

FIGURE 1. *Trials to criterion on the concept-formation task as a function of treatment conditions and IQ.*

A graph for a three-way interaction from a $2 \times 2 \times 2$ ANOVA is shown in Excerpt 5.17. Each of the points in the graph represents the mean score for one of the eight comparison groups involved in the analysis. The factors and levels involved are: (1) race, black and white; (2) DS, high and low; (3) condition, experimental and control. This graph can be thought of as containing two simple interactions. The two lines on the left side of the graph represent the race-by-DS interaction for the first (experimental) level of the third factor; the two lines on the right represent the race-by-DS interaction for the second (control) level of the third factor. The dissimilar patterns formed by the two "simple" interactions indicates a significant three-way interaction, just as nonparallel lines indicate significance in a graph for a two-way interaction. If the pattern of the simple interaction on the left had been the same as that on the right, then no three-way interaction would be present. Thus, this particular graph tells us that when exposed to the experimental treatment, black children performed better when in the high DS group and white children performed better when in the low DS group. For the Ss exposed to the control treatment, just the reverse was true: blacks performed better when in the low DS group, and whites performed better when in the high DS group.

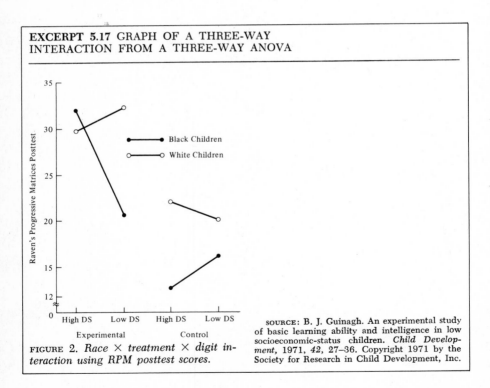

EXCERPT 5.17 GRAPH OF A THREE-WAY INTERACTION FROM A THREE-WAY ANOVA

●——● Black Children
○——○ White Children

Raven's Progressive Matrices Posttest

High DS   Low DS   High DS   Low DS
Experimental   Control

FIGURE 2. *Race × treatment × digit interaction using RPM posttest scores.*

SOURCE: B. J. Guinagh. An experimental study of basic learning ability and intelligence in low socioeconomic-status children. *Child Development*, 1971, *42*, 27–36. Copyright 1971 by the Society for Research in Child Development, Inc.

Additional Analyses

If one of the three main effects turns out to be significant and if there are three or more levels of the factor, the researcher will need to use one of the multiple comparison procedures to find out exactly where the significant differences lie. Excerpt 5.18 comes from a study in which a $3 \times 3 \times 2$ ANOVA was used for each of several dependent variables. One aspect of the study dealt with the effects of credibility, punishment, and sex on compliance to threats. The researchers used Duncan's new multiple range test to compare the overall means for the three levels of the credibility factor (10 percent, 50 percent, and 90 percent) and the overall means for the three levels of the punishment factor (5 points, 10 points, and 20 points). The main effect for the sex factor was not significant, but had it been so, a multiple comparison procedure would not have been applied since there are only two levels of this factor.

EXCERPT 5.18 USE OF A MULTIPLE COMPARISON PROCEDURE TO INVESTIGATE A SIGNIFICANT MAIN EFFECT IN A THREE-WAY ANOVA

The analysis of variance which is presented in Table 1 reveals significant effects of credibility ($F = 11.423$, $df = 2/72$, $p < .001$) and punishment magnitude ($F = 8.905$, $df = 2/72$, $p < .001$) on the measure of compliance. There were no sex differences nor did any of the interaction terms reach significance. A subsequent Duncan range test for multiple comparisons applied to the means for compliance associated with credibility revealed that the 10% and 50% ($p < .05$), the 50% and 90% ($p < .05$), and 10% and 90% ($p < .001$) conditions differed significantly from each other. The punishment factor showed that although the 5- and 10-point conditions did not significantly differ ($p < .10$), the 10- and 20-

($p < .01$), and the 5- and 20- ($p < .001$) point levels did.

TABLE 1 SUMMARY OF ANALYSIS OF VARIANCE FOR COMPLIANCE TO THREATS

SOURCE	df	MS	F
Credibility (C)	2	1074.234	11.423*
Punishment (P)	2	837.433	8.905*
Sex (S)	1	139.378	1.482
C × P	4	88.467	.941
C × S	2	138.078	1.468
P × S	2	101.144	1.076
C × P × S	4	16.944	.180
Error	72	94.039	

* $p < .001$.

SOURCE: J. Horai and J. T. Tedeschi. Effects of credibility and magnitude of punishment on compliance to threats. *Journal of Personality and Social Psychology*, 1969, 12, 164–196. Copyright 1969 by the American Psychological Association. Reprinted by permission.

If a two-way (first-order) interaction is significant, researchers will often not pay much attention to the results for the main effects. Instead, they will usually conduct an investigation of simple effects. This follow-up analysis involves nothing more than a comparison of the levels of one of the factors (which is involved in the interaction) at each separate level of the other factor (which is also involved in the interaction). Excerpt 5.19 shows how this is done.

As shown in the ANOVA summary table, the A × C (F level × Communication) interaction was significant. The accompanying table presents the means associated with this interaction. Each of these means has been averaged across the two levels of the third factor, source. (These particular

---

EXCERPT 5.19 USE OF A SIMPLE EFFECTS ANALYSIS TO INVESTIGATE A
SIGNIFICANT TWO-WAY INTERACTION IN A THREE-WAY ANOVA

The F × Communication Interaction
(Table 3) indicated that for plausible
communication, the high F category
showed significantly ($p < .001$) more attitude change than each of the other three

F levels. These latter three categories did
not differ from one another. There were
no significant differences due to F level
for either the unsubstantiated or the implausible communication.

TABLE 2 UNWEIGHTED MEANS
ANALYSIS OF VARIANCE

SOURCE	df	MS	F
F level (A)	3	358.58	2.84*
Source (B)	1	4,741.44	37.52**
Communica-tion (C)	2	14,713.75	116.42**
A × B	3	40.32	—
A × C	6	413.73	3.27**
B × C	2	805.35	6.37**
A × B × C	6	64.33	—
Within	444	126.38	

* $p < .05.$
** $p < .005.$

TABLE 3 MEAN ATTITUDE CHANGE AS A
FUNCTION OF AUTHORITARIAN LEVEL
AND TYPE OF COMMUNICATION

	AUTHORITARIAN LEVEL			
COMMUNI-CATION	LOW	LOW-MIDDLE	MIDDLE-HIGH	HIGH
	F	F	F	F
Plausible	16.69	16.26	16.44	27.02
Unsub-stantiated	3.82	4.59	2.48	3.08
Implaus-ible	.77	3.00	−1.06	.41

SOURCE: H. H. Johnson and R. R. Izzett. Relationship between authoritarianism and attitude change
as a function of source credibility and type of communication. *Journal of Personality and Social Psychology*, 1969, *13*, 317–321. Copyright 1969 by the American Psychological Association. Reprinted
by permission.

---

means, if plotted, would give us the graph of the A × C interaction.) In
conducting the investigation of simple effects, the researchers compared the
four means in each row of the table (using a procedure that is analogous to
a one-way ANOVA). After finding a significant difference among the means
in the top row, a multiple comparison procedure was then used to find out
which of the four means was significantly different from the others. The
results of this simple effects analysis seem to indicate that the cause of the
interaction was the large mean for the Ss in the high *F* and plausible communication group.

When the three-way (second-order) interaction is significant, the
researcher will often turn his attention once again to tests of simple effects
(rather than to the overall main effects). But this time he will be concerned
with the basic cell means from his analysis, that is, he will not average
across levels of the third factor as was done in the investigation of the two-
way interaction. In Excerpt 5.20, we see that the A × B × C interaction was
significant at the .005 level. After setting up the table of means, the researchers tested for simple effects by comparing the two instructional
methods (inductive vs. deductive) at each separate level of the other two
factors. The first test involved comparing 57.40 and 45.45, and a significant
difference was obtained. The remaining three comparisons (49.73 vs. 46.93;
47.00 vs. 51.56; and 50.68 vs. 49.70) did not result in significant differences.

EXCERPT 5.20 USE OF A SIMPLE EFFECTS ANALYSIS TO INVESTIGATE A SIGNIFICANT THREE-WAY INTERACTION IN A THREE-WAY ANOVA

TABLE 2 UNWEIGHTED MEANS ANALYSIS OF VARIANCE

SOURCE	df	MS	F
Instructional methods (A)	1	439.35	4.85*
Subject matters (B)	1	67.33	
Science interest levels (C)	1	1.13	
A × B	1	1116.94	12.34**
A × C	1	111.83	
B × C	1	225.92	
A × B × C	1	760.03	8.39***
Error	222	90.55	
Total	229		

* $p < .05$.
** $p < .001$.
*** $p < .005$.

TABLE 3 MEAN CRITERION TEST SCORES FOR EACH CELL IN THE EXPERIMENTAL DESIGN

GROUP	TRANSPORTATION TECHNIQUE		AIRCRAFT RECOGNITION	
	INDUCTIVE	DEDUCTIVE	INDUCTIVE	DEDUCTIVE
High scientific interest	57.40	45.45	47.00	51.56
Low scientific interest	49.73	46.93	50.68	49.70

Interpretation of the data was further complicated by the significant Instructional Methods × Subject Matters × Interest Levels interaction. To assist in interpreting this interaction, the mean criterion-test scores were computed for each "cell" in the experimental design. These data are presented in Table 3. For the Transportation Technique courses, these data showed that the inductive instructional method was superior to the deductive method. The between-method differences, however, were large and statistically significant ($p < .01$) only for the high scientific interest group. They were small and not statistically significant ($p > .20$) for the low scientific interest group.

Data for the Aircraft Recognition courses showed that the inductive method was more effective for the low scientific interest group but that the deductive method was more effective for the high scientific interest group, although neither of these differences was statistically significant.

SOURCE: G. K. Tallmadge and J. W. Shearer. Relationships among learning styles, instructional methods, and the nature of learning experience. *Journal of Educational Psychology*, 1969, 60, 222–230. Copyright 1969 by the American Psychological Association. Reprinted by permission.

Excerpt 5.20 shows very clearly why an interpretation of main effects, in the presence of significant interactions, may prove to be misleading. As the summary table indicates, the main effect for instructional methods was significant. If the researchers had interpreted this main effect (without conducting an investigation of simple effects), they would have concluded that the inductive instructional method was more effective than the deductive instructional method. As the simple effects analysis showed, however, this conclusion holds true only for the high scientific interest group of Ss taking the transportation technique courses. With respect to the aircraft recognition courses and the low scientific interest Ss in the transportation technique courses, the inductive method was not found to be more effective than the deductive method.

A FINAL COMMENT ON INTERACTIONS

The final paragraph in the preceding section should not be taken to mean that interactions, when found to be significant, are "bad" or that a factorial ANOVA without any significant interactions is better than one which has some. As Kerlinger points out, one of the main advantages of a factorial ANOVA is its ability to reveal interactions.

> One of the most significant and revolutionary developments in modern research design and statistics is the planning and analysis of the simultaneous operation and interaction of two or more variables. Scientists have long known that variables do not act independently. Rather, they often act in concert. The virtue of one method of teaching contrasted with another method of teaching depends on the teachers using the methods. The educational effect of a certain kind of teacher depends, to a large extent, on the kind of pupil being taught. An anxious teacher may be quite effective with anxious pupils but less effective with non-anxious pupils. Different methods of teaching in colleges and universities may depend on the intelligence and personality of both professors and students.[3]

In many journal articles, you will find that the main research question of the investigator is related to whether or not an interaction exists between two or more variables. And if such an interaction turns out to be significant, we must evaluate the study as having been successful.

REVIEW TERMS

cell	ordinal
disordinal	power
factor	second-order interaction
factorial ANOVA	simple main effects
first-order interaction	2×3 ANOVA
level	$3 \times 4 \times 5$ ANOVA
main effect	

[3] F. N. Kerlinger. *Foundations of Behavioral Research.* New York: Holt, Rinehart and Winston, 1964, p. 213.

REVIEW QUESTIONS

1. In a factorial ANOVA, will the number of main effect F ratios always be equal to the number of factors in the ANOVA? *no*

2. What is the minimum number of levels that can be included in a two-way ANOVA? *5 2 for main effects / for interaction*

3. Suppose a researcher uses a two-way factorial ANOVA in which there are two levels of one factor and four levels of the other factor. Further suppose that there are 10 Ss within each cell. How many degrees of freedom will there be in the ANOVA summary table for (a) the interaction and (b) within-groups (error)?

4. How many MS values will usually be included in a two-way ANOVA summary table? How are these MS values used to obtain F ratios? *4* *MS for each one + the MS for error*

5. If a researcher uses a 2×2 factorial ANOVA with 20 Ss in each cell, each of the main effect means will be based upon _____ Ss.

6. Suppose a researcher uses a 2×2 ANOVA. The F ratios for the main effects turn out to be significant, but the interaction does not. Should the researcher apply a multiple comparison post hoc test (e.g., Tukey) to find out what caused the main effects to be significant. *no*

7. Suppose a researcher uses a two-way factorial ANOVA to analyze his data. Near the end of the article, he states: "The results indicate that the differential effectiveness of the counseling methods varies according to the sex of the client." Which of the F ratios in the ANOVA summary table would provide the statistical support for this statement? *the interaction*

8. Suppose a researcher obtains a significant interaction from his 3×3 factorial ANOVA. If he graphs the interaction, how many lines will there be in the graph? How will the ordinate be labeled? How will the abscissa be labeled?

9. Usually, only significant interactions are graphed. Suppose, however, that a researcher were to graph the interaction from a two-way ANOVA in which there is absolutely *no* interaction. How would the lines in the graph be arranged? *parallel =*

10. When will a researcher usually conduct an investigation of simple main effects?

11. Suppose in a three-way ANOVA summary table the degrees of freedom for all main effects and interactions are equal to 1, while the within-groups *df* is equal to 72. Given this information, how many Ss were there in the study?

12. How many second-order interactions are there in a $2 \times 3 \times 4$ factorial ANOVA?

13. Suppose a researcher obtains a significant A $\times$ B interaction from a $2 \times 2 \times 2$ factorial ANOVA. Nothing else turns out to be significant. If the researcher graphs the significant interaction, how many lines will there be in the graph?

14. How many lines would it take to graph a significant A × B × C inter-
    action from a 2 × 2 × 2 factorial ANOVA?
15. How many first-order interactions will there be in a 2 × 3 × 4 × 5 fac-
    torial ANOVA?

CHAPTER 6

ANALYSIS OF VARIANCE WITH REPEATED MEASURES

In many research studies each of the various subjects is measured just once, with the resulting data being analyzed with a t test, a one-way ANOVA, or some other statistical procedure. However, in some studies the Ss are measured two, three, or more times. Studies in which each subject provides more than one piece of data can be subdivided into two broad categories: (1) studies in which there are several dependent variables, with the data from each of these variables being subjected to a separate analysis and (2) studies in which there is only one dependent variable, with Ss measured across all levels of one (or more) of the independent variables (factors). The following two hypothetical studies will help to make clear the distinction between these two categories.

As an example of the first type of study, suppose a researcher is interested in finding out whether secretarial speed and accuracy in clerical tasks is related to the time interval since the last meal. His hunch is that a secretary who is too hungry or too full will not be as efficient as one who is moderately full or hungry. To measure clerical efficiency, a simple test is selected that includes very elementary addition and subtraction problems (e.g., $2 + 3 = 5$; $6 - 3 = 3$; etc.) administered under a restrictive time limit such that no one can possibly finish the test. In conducting the study, our hypothetical researcher might randomly assign one-third of his pool of available secretaries to each of the three treatment groups: group I would take the test immediately before lunch, group II would take the test immediately after lunch, and group III would take the test in the middle of the afternoon. The clerical test might be scored in terms of the secretary's speed (how far she got through the test) and accuracy (percent of the attempted problems answered correctly). In this hypothetical study there is one independent variable (time-since-last-meal) and two dependent variables (speed and accuracy), and the data corresponding to each of the dependent variables might be analyzed by a separate one-way ANOVA. Note that each of the Ss in this study is measured at only one level of the independent variable.

As an example of the second type of study, suppose that a different researcher is interested in the same question concerning secretarial efficiency in clerical tasks as related to hunger, but he has only a small number of secretaries available for his study. Instead of dividing the Ss into three different groups, he decides to test each secretary under all three conditions. On Monday all Ss take the clerical test immediately before going to lunch. On Tuesday they all take the test immediately after returning from lunch. And on Wednesday the test is administered to all Ss in the middle of the afternoon. In scoring the test each day, our second researcher does something different—he subtracts the number of errors made by each secretary from the total number of problems attempted. Thus, he uses only one measure of efficiency rather than two, as was the case in the first study. Our second researcher has a study in which there is still only one independent variable, but the Ss are measured successively at each level of this variable. Because of the nature of his study, a regular one-way ANOVA would not be appropriate. Instead, the researcher would have to use a special form of analysis of variance which would take into consideration the fact that the same Ss were measured at different levels of the independent variable[1]

Although many different types of ANOVA's fall into the second of the above two categories, attention in this chapter will be focused on the two most frequently used: the Lindquist Type I ANOVA and the Lindquist Type III ANOVA. These specific ANOVA's, and others from the same category that are more complex, can be referred to as repeated measures ANOVA's since in each case the subjects are measured across the levels of one (or more) of the factors. In addition, the terms *split-plot* ANOVA and

---

[1] If this study were actually to be conducted, differences between levels of the independent variable would be confounded with possible differences between days. To circumvent this undesirable confounding, the researcher could randomly arrange the order in which the Ss are exposed to the three levels of the independent variable, making sure that one-third of the Ss take the test at each of the three time periods on each day of the experiment.

mixed design are sometimes used by authors to describe this group of ANOVA's.

LINDQUIST TYPE I ANOVA'S

Suppose a researcher is interested in comparing three methods of teaching statistics in terms of retention rates. His study might involve (1) randomly assigning 30 Ss to three groups, (2) exposing each group to a different teaching method, and (3) administering a comprehensive examination four times—at the end of the academic term and then again after periods of 6, 12, and 18 weeks. In this hypothetical study there are two factors (teaching method and time of exam) with repeated measures across all levels of the second factor, that is to say, all of the Ss are measured at each level of the time-of-exam factor. A diagram for this study is presented in Figure 6.1.

TIME OF EXAM

TEACHING METHOD	Ss	END OF TERM	AFTER 6 WEEKS	AFTER 12 WEEKS	AFTER 18 WEEKS
Teaching Method I	S_1 · · · S_{10}				
Teaching Method II	S_{11} · · · S_{20}				
Teaching Method III	S_{21} · · · S_{30}				

FIGURE 6.1. *Diagram of the hypothetical study involving a Lindquist Type I ANOVA.*

Alternative Terms for a Lindquist Type I ANOVA

A Lindquist Type I ANOVA always involves two factors and repeated measures across the levels of one of the factors. For this reason, some authors refer to this procedure as a two-factor ANOVA with repeated measures on one factor. Other authors might refer to our hypothetical study as a split-plot factorial 3.4 ANOVA. In this notation the number following the decimal point indicates the number of levels in the repeated measures factor, while the number preceding the decimal point indicates the number of levels in the factor that does not have repeated measures. Thus, if there had been only two teaching methods and three administrations of the exam, split-plot factorial 2.3 ANOVA would have been an appropriate descriptive term. Some authors might have described the hypothetical study as a 3 × 4 mixed design.

Finally, a few researchers might refer to our study as having one between-subjects factor and one within-subjects factor. The *within-subjects factor* is synonymous with the repeated measures factor, and in the hypothetical study this would correspond to the time of exam factor. A *between-subjects factor*, on the other hand, does not have repeated measures, and each subject in the study is exposed to only one level of this factor.

In summary, four different terms are used by researchers to mean the same thing as the term Lindquist Type I ANOVA. The reader of journal articles must become familiar with all four terms if he desires to understand the published literature within his field. Excerpts 6.1–6.4 demonstrate the alternative labels.

EXCERPTS 6.1–6.4 ALTERNATIVE TERMS TO DESIGNATE A LINDQUIST TYPE I ANOVA

The statistical design used to test for treatment differences was a two-factor mixed design with repeated measures on one factor.

source: F. J. Hedquist and B. K. Weinhold. Behavioral group counseling with socially anxious and unassertive college students. *Journal of Counseling Psychology*, 1970, *17*, 237–242. Copyright 1970 by the American Psychological Association. Reprinted by permission.

A 3 × 4 factorial analysis of variance with repeated measures on one factor was utilized.

source: R. F. Haase and D. J. DiMattia. Proxemic behavior: counselor, administrator, and client preference for seating arrangement in dyadic interaction. *Journal of Counseling Psychology*, 1970, *17*, 319–325. Copyright 1970 by the American Psychological Association. Reprinted by permission.

A split-plot factorial 3.4 ANOVA was performed on the AACL anxiety scores (obtained after the first, second, sixth, and tenth showings).

source: P. O. Davidson and S. F. Hiebert. Relaxation training, relaxation instruction, and repeated exposure to a stressor film. *Journal of Abnormal Psychology*, 1971, *78*, 154–159. Copyright 1971 by the American Psychological Association. Reprinted by permission.

The statistical model was for one between-subject and one within-subject factor. The between-subjects variable had two levels corresponding to the two groups, reinforced and nonreinforced. The within-subjects factor represented the pre-post measure of reading skill.

source: R. L. Bednar, P. F. Zelhart, and S. Weinberg. Operant conditioning principles in the treatment of learning and behavior problems with delinquent boys. *Journal of Counseling Psychology*, 1970, *17*, 492–497. Copyright 1970 by the American Psychological Association. Reprinted by permission.

Three Research Questions

In Chapter 5 we saw that a two-way ANOVA allows the researcher to answer three questions: (1) Is there a significant main effect for the first factor? (2) Is there a significant main effect for the second factor? (3) Is there a significant interaction between the two factors? In a Lindquist Type I ANOVA there are also three research questions and these questions also deal with two main effects and one interaction. However, the interpretation of these research questions is somewhat different due to the existence of repeated measures.

The first research question deals with the main effect of the factor that does not have repeated measures across its levels (the between-subjects factor). In the hypothetical study the question would be, "Is there a significant difference among the three teaching methods?" To answer this question, an average score is obtained for each subject by adding together all of his scores (across the repeated measures variable) and then dividing by the number of times he was measured. (In Figure 6.1 each subject's average score could be placed in an additional column to the right of the last column in the diagram.) Now, if we averaged these scores for the Ss in the top row of the diagram (i.e., all Ss who were exposed to teaching method I), we would have a single score which would correspond to the main effect mean for teaching method I. A main effect mean could be found in a similar manner for each of the other two rows of the diagram. The first research question involves a comparison of these three means to see if there are any significant differences among them. If there were such a significant difference, it would mean that the teaching methods differentially affect student performance on the exam, when each subject's scores are averaged across the four testing periods.

The second research question deals with the main effect of the repeated measures (within-subjects) factor. In the hypothetical study this question would be, "Did the 30 Ss perform equally well on the comprehensive examination each time it was administered?" To answer this question, an average score for the first administration of the exam is found. (The resulting mean could be placed beneath the first column in Figure 6.1.) A mean score is computed in a similar manner for each of the other three levels of the repeated measures factor. Thus, there are four means associated with this main effect, each one representing the average score for all Ss on a separate administration of the test. If the statistical analysis revealed a significant difference among these means, it would indicate that performance on the test, for all 30 Ss considered as a group, did not remain constant across the four testing periods. If this study were actually undertaken, we would probably find this main effect to be significant, with the mean performance on the test highest on the first administration (at the end of the academic term) and lowest on the last administration (18 weeks later).

The third research question deals with a possible interaction between the two factors. In the hypothetical study the third question would be, "Is the trend of test performance across the four testing periods similar for all three groups?" To answer this question, a mean score for each cell of Figure 6.1 is computed. Then, the pattern of the means across the top row is compared with the patterns for the other two groups. If our hypothetical study were actually conducted, we would probably find that each successive mean score for any group would be lower than the preceding mean (due to forgetting). If the rate of forgetting were similar for each of the three groups (even though they may have had different means at the end of the academic term), there would not be an interaction between teaching method and time of testing. On the other hand, if one teaching method led to a slow rate of forgetting (with the four means for that group of Ss being relatively constant), while a different teaching method led to a fast rate of forgetting (with the four means for this group of Ss being quite different), the statistical analysis would indicate a significant interaction.

Reporting the Results of a Lindquist Type I ANOVA

The Lindquist Type I Summary Table. Although a Lindquist Type I ANOVA and a regular two-way ANOVA are similar in having three research questions, the summary tables for these two ANOVA's are not the same. Table 6.1 presents a summary table for our hypothetical study involving the three teaching methods and the four administrations of the comprehensive exam. With the exception of the various degrees of freedom, the numbers in this table are imaginary.

TABLE 6.1 SUMMARY TABLE FOR
THE HYPOTHETICAL LINDQUIST TYPE I ANOVA

SOURCE	df	SS	MS	F
Between-subjects	29			
Teaching methods	2	48	24	6
Error-between	27	108	4	
Within-subjects	90			
Time of exam	3	24	8	4
Interaction	6	24	4	2
Error-within	81	162	2	
Total	119	366		

As Table 6.1 indicates, the various sources in a Lindquist Type I summary table are divided into two sections, between-subjects and within-subjects. Note that these two sources are used solely for the purpose of labe the two sections of the table, and neither of them has a corresponding F ratio. The between-subjects part of the summary table contains two sources. The first is the main effect of the factor that does not have repeated measures (teaching method). The second is the error-between. The MS for teaching methods is divided by the MS for error-between in order to get the first of the three F ratios of the table.

The within-subjects part of the summary table contains three sources: the main effect of the repeated measures factor (time of exam), the interaction between the two factors involved in the study, and the error-within. The MSs for the repeated measures main effect and the interaction are each divided by the error-within MS to obtain the other F ratios in the table.

With one-way, two-way, or three-way ANOVA's the total degrees of freedom was always equal to one less than the number of subjects involved in the study. In a repeated measures ANOVA this rule-of-thumb no longer applies. To find the number of subjects, we must add 1 to the df associated with the between-subjects source. If we add 1 to the total df, we now get the total number of data observations (scores) involved in the study. Notice that we can check the accuracy of the within-subjects df by subtracting the between-subjects df from the total df ($119 - 29 = 90$). Since the between-subjects and within-subjects sources are used solely to label the top and bottom parts of the table, the df values for these two sources are sometimes indented (as in Table 6.1) or put in parentheses.

The df for the main effect of teaching method is equal to the number of levels in that factor minus 1, and the df for error-between can be found by subtracting the df for teaching method from the between-subjects df ($29 - 2 = 27$). The df for the main effect of the repeated measures factor is equal to the number of levels in this factor minus 1. The interaction df is found by multiplying the df for teaching method and the df for time of exam ($2 \times 3 = 6$). Finally, the df for error-within can be found by subtracting the df for time of exam and for the interaction from the within-subjects df ($90 - 3 - 6 = 81$).

Although we can use the above knowledge to check the accuracy of the degrees of freedom that appear in published Lindquist Type I summary tables, there is no way that we can determine whether the numerical values for the sums of squares (SS) are accurate. The researcher uses a complicated formula to get these numbers, and we should consider them to be accurate. As was the case with all previously discussed ANOVA's, the numerical value for any mean square (MS) is found by dividing the SS on that row of the table by the df on the same row. The only difference in getting the F ratios in Table 6.1 as compared with one-way or factorial ANOVA's is that two error terms are used in the Lindquist Type I summary table, not just one.

Different Formats for the Lindquist Type I Summary Table. All authors do not use the same format for their Lindquist Type I summary tables. For example, although the summary table in Excerpt 6.5 is very similar to

Table 6.1, notice that (1) in the source column the word subjects has been abbreviated to Ss; (2) there is no row for total at the bottom of the table; (3) the interaction source is stated as A × B; and (4) SS values are not given in the table.

EXCERPT 6.5 TYPICAL LINDQUIST TYPE I SUMMARY TABLE

TABLE 2 ANALYSIS OF VARIANCE OF
NUMBER OF CORRECT RESPONSES IN
NINE CONCEPT PROBLEMS FOR TWO
RULE CONDITIONS

SOURCE	df	MS	F
Between Ss	135		
Rule condition			
(A)	1	82.54	1.85
Error between	134	44.64	
Within Ss	1088		
Problems (B)	8	392.75	64.24*
A × B	8	8.33	1.36
Error within	1072	6.11	

* $p < .001$.

SOURCE: J. L. Dunham and C. V. Bunderson. Effect of decision-rule instruction upon the relationship of cognitive abilities to performance in multiple-category concept problems. *Journal of Educational Psychology*, 1969, *60*, 121–125. Copyright 1969 by the American Psychological Association. Reprinted by permission.

In spite of these slight differences, the table still tells us that there were 136 subjects involved in the study, that these subjects were divided into two groups which differed in terms of rule conditions, and that there were nine levels of the repeated measures variable (problems). In addition, the results of the three *F* tests indicate that (1) the subjects in the first rule condition group did not perform significantly better or worse than the subjects in the second rule condition, when performance was averaged across all levels of the repeated measures variable, (2) the average score for all subjects was

---

EXCERPT 6.6 ALTERNATIVE NAMES FOR THE TWO
ERROR SOURCES IN A LINDQUIST TYPE I SUMMARY TABLE

TABLE 1 ANALYSIS OF VARIANCE FOR THE
INTERVIEW GROUP ACQUISITION DATA

SOURCE	df	MS	F
Between Ss			
Conditioning (C)	1	6.63	4.67*
Error	12	1.42	
Within Ss			
Experimenters (E)	1	0.39	1.95
C × E	1	2.62	13.10**
Error	12	0.20	

* $p < .055$.
** $p < .005$.

SOURCE: R. L. Vitalo. Effects of facilitative interpersonal functioning in a conditioning paradigm. *Journal of Counseling Psychology*, 1970, *17*, 141–144. Copyright 1970 by the American Psychological Association. Reprinted by permission.

significantly higher on some problems than on others, and (3) the problems that were easy (or difficult) for the first rule condition group were also the ones, in general, that were easy (or difficult) for the second rule condition group.

In Excerpts 6.6 and 6.7 we see two more Lindquist Type I summary

---

EXCERPT 6.7 ALTERNATIVE NAMES FOR THE TWO ERROR SOURCES IN A LINDQUIST TYPE I SUMMARY TABLE

TABLE 1 ANALYSIS OF VARIANCE OF DIFFERENCES BETWEEN COUNSELORS, CLIENTS, AND ADMINISTRATORS ACROSS FOUR PROXEMIC SEATING ARRANGEMENTS

SOURCE	df	SS	MS	F
Between subjects	29	2090.97		
Groups (A)	2	169.12	84.56	1.19
Subjects within groups	27	1921.85	71.18	
Within subjects	90	6223.00		
Arrangements (B)	3	1022.70	340.90	6.39**
A × B	6	875.75	145.96	2.73*
B × Subjects within groups	81	4324.55	53.39	

* $p < .025$.
** $p < .001$.

SOURCE: R. F. Haase and D. J. DiMatti. Proxemic behavior: counselor, administrator, and client preference for seating arrangement in dyadic interaction. *Journal of Counseling Psychology*, 1970, *17*, 319–325. Copyright 1970 by the American Psychological Association. Reprinted by permission.

---

tables that differ from the summary table we just looked at in the names given to two sources, error-between and error-within. In Excerpt 6.6 the word *error* is used. Of course, the MS for the error-between (1.42) was used to get the F ratio for the main effect of conditioning, while the MS for the

---

EXCERPT 6.8 ADDITIONAL WAYS OF LABELING THE TWO ERROR SOURCES

TABLE 2 ANALYSIS OF VARIANCE OF CORRECT RESPONSES IN 12 LEARNING TRIALS (in Blocks of Two Trials)

SOURCE	df	MS	F	p
Between-Subjects	36			
Groups (B)	2	25.88	3.74	< .05
Error (b)	34	6.91		
Within-Subjects	185			
Trials (A)	5	28.66	24.07	< .01
A × B	10	1.41	1.18	ns
Error (w)	170	1.19		
Total	221			

SOURCE: S. Muehl. The effects of visual discrimination pretraining on learning to read a vocabulary list in kindergarten children. *Journal of Educational Psychology*, 1960, *51*, 217–221. Copyright 1960 by the American Psychological Association. Reprinted by permission.

error-within was used for the main effect of experimenters and the C × E interaction. In Excerpt 6.7 the term *subjects within groups* is used in place of error-between and B × subjects within groups is used instead of error-within.

Excerpt 6.8 is a summary table that differs from the previous ones in two respects. First, the items listed under between-subjects and within-subjects are not indented. Thus, the reader who is not aware of the fact that these two sources are used to label two parts of the table might think that the author made a mistake with respect to the total *df*. Second, the terms error (b) and error (w) are used instead of error-between and error-within.

In Excerpt 6.9 the author omits the between-subjects and within-subjects labels and separates the two parts of the table by a space. He uses the terms error (a) and error (b) rather than error-between and error-within. Comparing Excerpts 6.8 and 6.9, we see that some authors use the term error (b) to designate the between-subjects error while other authors use the same term to designate the within-subjects error. Although this inconsistency is at first quite confusing, the position of the term error (b) within the table allows us to know its meaning.

EXCERPT 6.9 ADDITIONAL WAYS OF LABELING THE TWO ERROR SOURCES

TABLE 3 SUMMARY OF ANALYSIS OF VARIANCE

SOURCE	SS	df	MS	F
Treatments	222.13	2	111.06	< 1
Error (a)	12,401.37	69	179.73	
Trials	84.94	1	84.94	4.40*
Treatments × Trials	34.38	2	17.19	< 1
Error (b)	1,331.31	69	19.29	
Total	14,074.12	143		

* $p < .05$.

SOURCE: T. J. D'Zurilla. Persuasion and praise as techniques for modifying verbal behavior in a "real life" group setting. *Journal of Abnormal Psychology*, 1966, 71, 369–376. Copyright 1966 by the American Psychological Association. Reprinted by permission.

Results of a Lindquist Type I ANOVA Presented Without a Summary Table. In Chapters 4 and 5 we saw that authors sometimes present the results of their one-way or factorial analysis of variance within the text of the article, not in an ANOVA summary table. In a similar manner many researchers who use a Lindquist Type I ANOVA delete the traditional summary table from their articles. In Excerpt 6.10 we see how this is usually done.

The first *F* value (21.16) in Excerpt 6.10 is for the main effect of treatments. The *df* value preceding the slash is equal to 4, since there were five

---

EXCERPT 6.10 RESULTS OF A LINDQUIST TYPE I
ANOVA PRESENTED WITHOUT A TABLE

A mixed analysis of variance with five levels of the treatment variable (between) and four levels of trials (within) was performed on these data. The effect due to experimental treatments was significant ($F = 21.16$, $df = 4/70$, $p < .01$). In addition, the analysis yielded significant effects due to blocks of trials ($F = 80.39$, $df = 3/210$, $p < .01$), and to the interaction between treatments and blocks of trials ($F = 3.05$, $df = 12/210$, $p < .01$).

SOURCE: F. J. DiVesta and R. T. Walls. Transfer of solution rules in problem solving. *Journal of Educational Psychology*, 1967, *58*, 319–326. Copyright 1967 by the American Psychological Association. Reprinted by permission.

---

levels in this factor, and the *df* value following the slash (70) is for the error-between. If we add these two degrees of freedom together, we get 74, which is equal to the between-subjects *df*. Knowing this, we can determine that 75 subjects were involved in this study. The next two *F* values (80.39 and 3.05) are for the main effect of the repeated measures variable and the interaction. The second *df* associated with each of these *F* values is the same

---

EXCERPT 6.11 GRAPH OF AN INTERACTION
FROM A LINDQUIST TYPE I ANOVA

The statistical design used to test for treatment differences was a two-factor mixed design with repeated measures on one factor (Bruning & Kintz, 1968).

The results indicated graphically in Figure 1 show the average frequency of verbal assertive responses emitted by the experimental and control subjects outside of treatment for six consecutive 1-week periods.

Following the fourth week of the experiment the experimental subjects dramatically increased the frequency of their verbal assertive responses over those of the control subjects. By the fifth week, the experimental subjects were emitting almost four more assertive responses on the average than the control subjects. During the last week of the experiment, the experimentals decreased their rate with the behavior rehearsal subjects responding less than they did during the fourth week, but continuing to respond more frequently than the first, second, or third weeks.

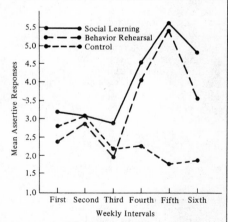

FIGURE 1. *Mean frequency of verbal assertive responses for six consecutive weekly intervals (treatment period).*

SOURCE: F. J. Hedquist and B. K. Weinhold. Behavioral group counseling with socially anxious and unassertive college students. *Journal of Counseling Psychology*, 1970, *17*, 237–242. Copyright 1970 by the American Psychological Association. Reprinted by permission.

(210), since both of these F's were obtained by using the MS for error-within.

Graphing the Interaction in a Lindquist Type I ANOVA. To help convey the meaning of the interaction F ratio, be it significant or nonsignificant, some authors include a graph to show how each separate group performed across the levels of the repeated measures variable. In such a graph the dependent variable is listed along the ordinate (vertical axis), the levels of the repeated measures variable are listed along the abscissa (horizontal axis), and the different groups are designated by lines within the graph.

As an example of such a graph, consider Excerpt 6.11. This graph corresponds to a 3×6 Lindquist Type I ANOVA in which the interaction F ratio was found to be significant at the .001 level. A nonsignificant F ratio would produce a graph in which the lines designating the groups would run almost parallel to one another. As the actual graph indicates, however, the trend of the six weekly means for the groups led to divergent lines.

EXCERPT 6.12 USE OF A MULTIPLE COMPARISON PROCEDURE TO INVESTIGATE A SIGNIFICANT MAIN EFFECT

An ANOVA for repeated measures (4 groups of subjects × 3 categories of affect for names) was performed. Since each of the 4 groups was treated identically during the initial learning phase, no differences were expected among them, and only affect was predicted to be significant. Tables 1 and 2 summarize the results of the analysis, which clearly support the prediction, and present relevant descriptive statistics. A Newman-Keuls test applied to the means indicated that reliably more errors were made in learning to associate neutral names with trigrams than in learning to associate either liked or disliked names ($p < .05$, in each case), but that the difference between the liked and disliked conditions, although in the right direction, that is, fewer errors for the liked names, was short of statistical significance.

TABLE 1 NUMBER OF ERRORS MADE IN LEARNING TO ASSOCIATE NAMES OF DIFFERENTIALLY-LIKED PUBLIC FIGURES WITH TRIGRAMS

TRIGRAMS PAIRED WITH	M ERRORS	s
Liked persons	2.75	2.9
Disliked persons	3.13	2.7
Neutral persons	3.92	3.5

TABLE 2 SUMMARY OF ANALYSIS OF VARIANCE

SOURCE	SS	df	MS	F
Between Ss				
Mediation groups (A)	30.04	3	10.01	< 1
Ss within groups (B)	1293.83	56	23.10	
Within Ss				
Affect for names (C)	42.43	2	21.22	7.58°
A × C	22.86	6	3.81	1.36
C × B	314.04	116	2.80	

° $p < .001$.

SOURCE: A. J. Lott, B. E. Lott, and M. L. Walsh. Learning of paired associates relevant to differentially liked persons. *Journal of Personality and Social Psychology*, 1970, 16, 274–283. Copyright 1970 by the American Psychological Association. Reprinted by permission.

Follow-up Analyses for a Lindquist Type I ANOVA

If one (or both) of the main effects is significant and if there are three or more levels associated with that factor, the researcher will probably apply a multiple comparison test to find out where the significant differences lie among the main effect means. Any of the tests discussed in Chapter 4 (e.g., Tukey, Newman-Keuls, Scheffé) might be used for this follow-up analysis. In Excerpt 6.12 we see how one of these multiple comparison tests was applied to a significant three-level main effect for the repeated measures factor. If there had been only two levels associated with the significant main effect, a multiple comparison test would not have been needed, for the researcher could simply have looked at the two main effect means to determine which was significantly larger than the other.[2]

If the interaction is significant, the researcher will probably apply tests of simple main effects to find out the cause of the observed interaction. Each

---

EXCERPT 6.13 USE OF A SIMPLE MAIN EFFECT ANALYSIS
TO INVESTIGATE A SIGNIFICANT INTERACTION

The performance curves for the control, stress, and interference groups are presented in Figure 1. An analysis of variance for repeated measures on the recall scores of the three groups on the three tests indicated a significant group effect ($F = 3.97$, $df = 2/42$, $p < .025$), a significant trials effect ($F = 105.45$, $df = 2/84$, $p < .001$), and, of most importance, a significant Groups $\times$ Trials interaction ($F = 9.87$, $df = 4/48$, $p < .001$), which indicates that the groups were differentially affected by the various treatments. Further analyses indicated that there were no significant differences between the groups prior to the different treatments (Test 1— $F = 1.59$, $df = 2/42$) or after the debriefing (Test 3—$F = 1.61$, $df = 2/42$), but that there were significant differences between the groups after the experimental manipulation (Test 2—$F = 11.05$, $df = 2/42$, $p < .001$). A Duncan multiple-range test on data from the second test indicated that there were no significant differences between the interference and ego-threat groups but that both of these groups differed significantly ($p < .001$) from the control group.

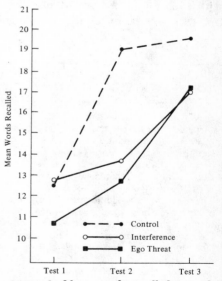

FIGURE 1. *Mean words recalled in each condition on each test.*

SOURCE: D. S. Holmes and J. R. Schallow. Reduced recall after ego threat: repression or response competition. *Journal of Personality and Social Psychology,* 1969, *13*, 145–152. Copyright 1969 by the American Psychological Association. Reprinted by permission.

---

[2] In Table 2 of Excerpt 6.12, the *df* for C $\times$ B should be 112, not 116.

of these tests may be likened to a one-way ANOVA which compares the various levels of one factor at a particular level of the second factor. In Excerpt 6.13 tests of simple main effects were used to compare the three groups at each level of the repeated measures variable. Three of these tests were conducted (since there were three levels of the trials factor), with a significant F being obtained for Test 2. Because there were three groups involved in the study, a multiple comparison test (Duncan) was needed following the significant simple main effect.[3]

In conducting tests of simple main effects, some authors compare the various groups in the study at each separate level of the repeated measures variable (as in Excerpt 6.13), while other researchers apply the test for each group, comparing that group's performance across the repeated measures variable. A few researchers conduct tests of simple main effects in *both* directions. This approach was used in Excerpt 6.14. In this study, tests of simple main effects indicated that significant differences existed among the three trials means for the first group (.61, 4.54, 6.31), among the three group means on Trial 2 (4.54, .50, .79), and among the three group means on Trial 3 (6.31, .70, 1.07). Since there were more than two means associated with each of these simple main effect comparisons, a multiple comparison test was needed to identify which means were significantly different from

---

EXCERPT 6.14 USE OF SIMPLE MAIN EFFECTS ANALYSES TO INVESTIGATE A SIGNIFICANT INTERACTION

A 3 × 3 repeated-measures analysis of variance indicated that both main effects as well as the interaction were statistically significant. The Groups × Trials interaction . . . indicated that the only significant performance changes were associated with the models-reinforced-aware group. There were no significant differences among the three groups on Trial 1, indicating that the groups were essentially the same in operant rate of "-ing" words. The means for the model-reinforced-aware groups increased significantly after each observation session, and on Trials 2 and 3 were significantly larger than the means for the models-reinforced-not-aware and models-not-reinforced condi-

tions. These latter two groups at no time differed significantly from each other, and neither showed any significant increase across trials.

TABLE 2 MEAN NUMBER OF "-ING" WORDS FOR MODELS-REINFORCED-AWARE, MODELS-REINFORCED-NOT-AWARE, AND MODELS-NOT-REINFORCED GROUPS

	TRIALS			
GROUP	1	2	3	M
Models-reinforced–aware	.61	4.54	6.31	3.82
Models-reinforced–not-aware	.30	.50	.70	.50
Models-not-reinforced	.57	.79	1.07	.81
M	.51	2.03	2.81	1.78

source: D. L. Hamilton, J. J. Thompson, and A. M. White. Role of awareness and intentions in observational learning. *Journal of Personality and Social Psychology*, 1970, *16*, 689–694. Copyright 1970 by the American Psychological Association. Reprinted by permission.

---

[3] In the third set of parentheses in Excerpt 6.13, the *df* following the slash should be 84, not 48.

one another within each group of means. Although the authors do not tell us which specific multiple comparison test was used, the results of these comparisons are discussed in the last three sentences of the excerpt.

TEACHING METHOD	ABILITY LEVEL	Ss	TIME OF EXAM			
			END OF TERM	AFTER 6 WEEKS	AFTER 12 WEEKS	AFTER 18 WEEKS
Teaching Method 1	High Ability	S_1 . . . S_{10}				
	Low Ability	S_{11} . . . S_{20}				
Teaching Method 2	High Ability	S_{21} . . . S_{30}				
	Low Ability	S_{31} . . . S_{40}				
Teaching Method 3	High Ability	S_{41} . . . S_{50}				
	Low Ability	S_{51} . . . S_{60}				

FIGURE 6.2. *Diagram of the hypothetical study involving a Lindquist Type III ANOVA.*

LINDQUIST TYPE III ANOVA'S

We have seen that a Lindquist Type I ANOVA is similar to a two-way factorial ANOVA in terms of (1) the number of F ratios that appear in the summary table, (2) the procedure for graphing the interaction, and (3) the procedures for follow-up analyses for both multiple comparisons and simple main effects. Now we will direct our attention to the Lindquist Type III ANOVA. As we discuss this second repeated measures ANOVA, the reader should look for a similarity between a Lindquist Type III and a three-way factorial ANOVA. Although there are distinct differences between these two ANOVA's, they are identical in terms of the number of F ratios, procedures for graphing interactions, and application of follow-up analyses.

As an example of a Lindquist Type III ANOVA suppose a researcher is interested in comparing three methods of teaching statistics in terms of their effects on (1) high and low ability students and (2) retention after the academic term. His study might involve randomly assigning 10 high ability Ss and 10 low ability Ss to each of the three treatment groups, exposing each group of 20 Ss to a different teaching method, and administering a comprehensive examination four times—at the end of the academic term, and then again after periods of 6, 12, and 18 weeks. In this hypothetical study there are three factors (teaching method, ability level, and time of exam) with repeated measures across all levels of the third factor. A diagram for this study is presented in Figure 6.2.

Alternative Terms for a Lindquist Type III ANOVA

Whenever a Lindquist Type III ANOVA is used, three factors are involved in the study and repeated measures are made across the levels of one of the factors. For this reason, some authors describe this form of ANOVA as a three-factor ANOVA with repeated measures on one factor. Other authors might refer to our hypothetical study as a split-plot factorial 32.4 ANOVA. The numbers 3 and 2 indicate the number of levels in the two factors which do not have repeated measures. The number which follows the decimal point (4) indicates the number of levels in the repeated measures. Another phrase which could be used to describe the hypothetical study is $3 \times 2 \times 4$ mixed design. Finally, a few researchers might refer to the ANOVA of Figure 6.2 as having two between-subjects factors and one within-subjects factor. Excerpts 6.15–6.18 demonstrate three of the alternative labels used by some authors in place of the term Lindquist Type III ANOVA.

Seven Research Questions

In Chapter 5 we saw that a three-way ANOVA allows the researcher to answer seven research questions: three concerned with main effects, three with first-order (two-way) interactions, and one with second-order (three-

---

EXCERPTS 6.15–6.18 ALTERNATIVE TERMS TO
DESIGNATE A LINDQUIST TYPE III ANOVA

The results were analyzed . . . by a three-way analysis of variance (student grade × student sex × concept) with repeated measures on the last dimension.

SOURCE: K. Yamamoto, E. C. Thomas, and E. A. Karns. School-related attitudes in middle-school age students. *American Educational Research Journal*, 1969, *6*, 191–206. Copyright 1969 by the American Educational Research Association, Washington, D.C.

To test the difference between the experimental conditions (auditory versus visual modes of presentation, and good versus poor types of input), an analysis of variance using a three-factor mixed design with repeated measures on one factor (Bruning & Kintz, 1968) was employed.

SOURCE: R. Oaken, M. Wiener, and W. Cromer. Identification, organization, and reading comprehension for good and poor readers. *Journal of Educational Psychology*, 1971, *62*, 71–78. Copyright 1971 by the American Psychological Association. Reprinted by permission.

The intentional learning data were analyzed in a three-factor design with two two-level between-subjects factors (learning task and type of punishment) and one four-level repeated measures factor (successive blocks of items).

SOURCE: A. G. Greenwald. Difficulty of associative performance following training with negative instances: a note on punishment effects. *Journal of Educational Psychology*, 1970, *61*, 255–259. Copyright 1970 by the American Psychological Association. Reprinted by permission.

Hypothesis 3 was tested by means of a repeated-measures design described by Lindquist (1953) as a Type III mixed design.

SOURCE: M. D. Merrill. Specific review versus repeated presentations in a programmed imaginary science. *Journal of Educational Psychology*, 1970, *61*, 392–399. Copyright 1970 by the American Psychological Association. Reprinted by permission.

---

way) interaction. The same questions are answered with a Lindquist Type III ANOVA, but the interpretation of the questions is somewhat different due to the existence of repeated measures.

Each of the three main effects in a Lindquist Type III ANOVA corresponds to a different factor in the study. In reference to the hypothetical example presented earlier (see Figure 6.2), the research question related to the main effect of the teaching method factor could be stated, "Do the three teaching methods lead to equal student performance, when all data associated with each teaching method are combined for high and low ability students and then averaged across the four administrations of the test?" The research question related to the main effect of the ability level factor asks, "Do high ability students perform the same as low ability students, when all data associated with the three subgroups of high ability (or low ability) students are combined and then averaged across the four administrations of the test?" The research question related to the main effect of the time-of-exam factor asks, "Do the 60 students in the study (disregarding their classification as to teaching method or ability level) perform equally well on each administration of the exam?"

The three first-order interactions in a Lindquist Type III ANOVA involve the three possible combinations of two factors taken together. With respect to our hypothetical study, the research question related to the method-by-time interaction (TM × TE) could be stated as, "Do the 20 students in each teaching method group (the high- and low-ability groups combined) have a similar trend or pattern of test performance over the four administrations of the examination?" A significant interaction would indicate that the increase or decrease in mean test performance from one testing period to the next varies according to the teaching method involved. The research question related to the ability-by-time interaction (AL × TE) asks, "Is the trend of test performance for the 30 high ability students across the four tests administrations similar to the trend for the 30 low ability students?" A difference in the increase or decrease from one test period to the next for high ability and low ability students would result in a significant interaction. The research question related to the method-by-ability interaction (TM × AL) asks, "Is the difference between the overall performance (averaged across testing periods) of high and low ability students the same magnitude within each of the three teaching method conditions?" A significant interaction here would indicate that the degree to which high and low ability students differ is dependent upon their teaching method.

As in a three-way ANOVA, the second-order interaction in a Lindquist Type III ANOVA involves all three factors of the study. With respect to the hypothetical study, the research question related to the method-by-ability-by-time interaction (TM × AL × TE) is, "Is the performance on the exam for any subgroup of 10 students influenced by the unique combination of their ability level, the teaching method they received, and the specific time at which they are tested?" If this triple interaction were found to be significant, it would indicate that the trend of test performance across the four testing periods is not the same for the six subgroups of students.

Reporting the Results of a Lindquist Type III ANOVA

The Lindquist Type III Summary Table. Although a Lindquist Type III ANOVA and a regular three-way ANOVA are similar in having seven research questions, the summary tables for these two ANOVA's are quite different. Table 6.2 presents a model summary table for our hypothetical study. (As in our other models, except for the various degrees of freedom, the numbers in the table were selected to make the math of the table easy to follow.)

As was the case for a Lindquist Type I ANOVA, the various sources in Table 6.2 are divided into two sections, between-subjects and within-subjects. In the between-subjects part of the summary table, we find two main effects and one first-order interaction. In the within-subjects part of the

TABLE 6.2 SUMMARY TABLE FOR
THE HYPOTHETICAL LINDQUIST TYPE III ANOVA

SOURCE	df	SS	MS	F
Between-subjects	59			
Teaching method (TM)	2	36	18	9
Ability level (AL)	1	24	24	12
TM × AL	2	20	10	5
Error-between	54	108	2	
Within-subjects	180			
Time of exam (TE)	3	18	6	12
TM × TE	6	12	2	4
AL × TE	3	9	3	6
TM × AL × TE	6	6	1	2
Error-within	162	81	0.5	
Total	239	214		

table, we find one main effect, two first-order interactions, and one second-order interaction. A simple rule will help the reader to remember which sources belong in the lower part: Any main effect or interaction which involves a repeated measure variable will be found in the within-subjects part of the summary table.

In a Lindquist Type III summary table the various degrees of freedom associated with the three main effects and the four interactions are found in the same manner as in a regular three-way ANOVA, and the procedure for finding the degrees of freedom for the remaining sources is the same as in a Lindquist Type I ANOVA. Again similar to a Lindquist Type I ANOVA, the *MS* for error-between is used to obtain the *F* ratios in the top part of the table, while the *MS* for error-within is used to calculate the *F* ratios in the lower part of the table. Because of these characteristics of a Lindquist Type III summary table, it might be appropriate to think of it as a cross between the summary tables from a regular three-way ANOVA and a Lindquist Type I ANOVA.

The information contained in a Lindquist Type III summary table can be very helpful in telling the reader about a study. If you were to see Table 6.2 in a journal article, you could determine, without reading a word of the author's text, (1) the names of the three factors, (2) which factor has repeated measures, (3) how many levels there are in each factor, (4) how many subjects there are in the study, and (5) the basic answer for each of the seven research questions. Therefore, the summary table should be used

EXCERPT 6.19 A FORMAT FOR THE
LINDQUIST TYPE III SUMMARY TABLE

TABLE 2 ANALYSIS OF VARIANCE OF
ACHIEVEMENT DATA

SOURCE	df	MEAN SQUARE	F
Between Ss	99		
Remote Associates Test (A)	1	9804.1	20.48°
Response mode (B)	1	1704.1	3.56
A × B	1	5.6	< 1
Error between	96	478.6	
Within Ss	200		
Subtests (C)	2	125.6	1.21
A × C	2	1.0	< 1
B × C	2	1467.2	14.16°
A × B × C	2	154.7	1.49
Error within	192	103.6	
Total	299		

° $p < .001$.

SOURCE: S. Tobias. Effect of creativity, response mode, and subject matter familiarity on achievement from programmed instruction. *Journal of Educational Psychology*, 1969, *60*, 453–460. Copyright 1969 by the American Psychological Association. Reprinted by permission.

to facilitate an understanding of the study whenever it is included in the journal article.

Different Formats for the Lindquist Type III Summary Table. Although many researchers use a Lindquist Type III ANOVA in their research studies,

EXCERPT 6.20 ANOTHER FORMAT FOR
THE LINDQUIST TYPE III SUMMARY TABLE

TABLE 1 ANALYSIS OF VARIANCE FOR
CHOICE BEHAVIOR ON TRIALS 1–8

SOURCE	df	MS	F
Between Ss			
Games (A)	2	.820	1.23
Kinds of input (B)	1	59.501	89.34°
A × B	2	.507	—
Error (b)	114	.666	
Within Ss			
Trials (C)	7	.656	6.64°°
A × C	14	.207	2.09°
B × C	7	1.782	18.04°°
A × B × C	14	.185	1.87°
Error (w)	798	.099	

° $p < .05$.
°° $p < .01$.

SOURCE: D. G. Pruitt. Motivational processes in the decomposed prisoner's dilemma game. *Journal of Personality and Social Psychology*, 1970, *14*, 227–238. Copyright 1970 by the American Psychological Association. Reprinted by permission.

they do not use the same format for presenting their results in a summary table. Consider the tables in Excerpts 6.19 and 6.20. Although they are similar to the model summary table, notice that (1) both tables delete the SS column, (2) the notation for the two errors is different, (3) Excerpt 6.19 provides *df* values for between Ss and within Ss, while Excerpt 6.20 does not, (4) Excerpt 6.20 does not have a total row, and (5) Excerpt 6.19 contains all seven *F* values while Excerpt 6.20 does not present *F* values if they are smaller than 1.

The summary table in Excerpt 6.21 uses the term subjects within groups instead of error-between and C × subjects within groups instead of error-within. This table does have a column of SS values, but again only some of the *F*s were computed.

EXCERPT 6.21 ALTERNATIVE NAMES FOR THE TWO ERROR SOURCES IN LINDQUIST TYPE III SUMMARY TABLE

TABLE 2 SUMMARY OF ANALYSIS OF VARIANCE

SOURCE OF VARIATION	SUM OF SQUARES	*df*	MEAN OF SQUARES	*F*-RATIO
Between	100.58	47		
Training (A)	35.88	1	35.88	29.90*
Test (B)	3.79	1	3.79	3.15
A × B	7.93	1	7.93	6.61**
Subjects within Groups	52.98	44	1.20	
Within	85.25	144		
Trials (C)	4.89	3	1.63	2.86
A × C	.84	3	.28	
B × C	1.86	3	.62	
A × B × C	1.39	3	.46	
C × Subjects within Groups	76.27	132	.57	

   * $p < .01$.
   ** $p < .05$.

SOURCE: B. J. Gaines and L. M. Raskin. Comparison of cross-modal and intra-modal form recognition in children with learning disabilities. *Journal of Learning Disabilities*, 1970, *3*, 243–246.

Additional variations in format for Lindquist Type III summary tables are shown in Excerpts 6.22 and 6.23. In Excerpt 6.22 the results of three separate Lindquist Type III ANOVA's (one for each experiment) are presented in a single summary table, the descriptive labels between-subjects and within-subjects have been omitted, and the terms error 1 and error 2 are used instead of error-between and error-within. In Excerpt 6.23 the results of two separate analyses are presented in one table and the two errors are placed together at the bottom of the table.

Presenting the Results of a Lindquist Type III ANOVA Without a Summary Table. As we have seen previously, researchers often present the results

EXCERPT 6.22 ANOTHER LINDQUIST TYPE III SUMMARY TABLE

TABLE 2 ANALYSES OF VARIANCE OF FOOD CONSUMPTION

		EXP. I		EXP. II		EXP. III	
SOURCE	df	MS	F	MS	F	MS	F
Box (B)	1	36.00	12.90*	4.25		12.28	3.00
Order (O)	1	143.00	51.25*	5.49	1.27	7.30	1.78
B × O	1	43.91	15.74*	4.84	1.12	1.33	
Error 1	16	2.79		4.31		4.09	
Drive (D)	1	8.53	5.05†	0.11		0.09	
D × B	1	24.57	14.54*	1.82	2.53	0.02	
D × O	1	5.16	3.05	18.37	25.51*	48.33	25.44*
D × B × O	1	21.71	12.85*	0.08		0.18	
Error 2	16	1.69		0.72		1.90	

* $p < .01$.
† $p < .05$.
SOURCE: E. L. Wike, C. Cour, and R. L. Mellgren. Establishment of a learned drive with hunger. *Psychological Reports*, 1967, *20*, 143–145. Reprinted by permission of author and publisher.

of their statistical analyses within the text of their articles, not in tables. In Excerpt 6.24 the authors discuss the outcome of their Lindquist Type III ANOVA without including a summary table in the article.

Graphing Interactions in a Lindquist Type III ANOVA. Authors sometimes include a graph of a significant interaction in a Lindquist Type III ANOVA in order to help the reader understand the nature of the interaction.

EXCERPT 6.23 ANOTHER VERSION OF
A LINDQUIST TYPE III SUMMARY TABLE

TABLE 1 ANALYSIS OF VARIANCE FOR SOUND-DISCOMFORT RATINGS AND SKIN-CONDUCTANCE MEASURES

		SOUND-DISCOMFORT		SKIN-CONDUCTANCE	
SOURCE	df	MS	F	MS	F
Sex (A)	1	3.57	5.45*	.00740	.09
Choice (B)	1	.70	1.07	.20259	2.55
Escape (C)	1	.60	4.81*	.04100	16.60**
A × B	1	.02	.03	.03040	.38
A × C	1	.09	.74	.00640	2.60
B × C	1	5.67	45.80**	.01993	8.07**
A × B × C	1	.19	1.54	.00509	2.06
Error between	36	.65		.07959	
Error within	36	.12		.00247	

* $p < .05$.
** $p < .01$.
SOURCE: N. L. Corah and J. Boffa. Perceived control, self-observation, and response to aversive stimulation. *Journal of Personality and Social Psychology*, 1970, *16*, 1–4. Copyright 1970 by the American Psychological Association. Reprinted by permission.

EXCERPT 6.24 RESULTS OF A LINDQUIST TYPE III
ANOVA PRESENTED WITHOUT A SUMMARY TABLE

The measure of performance in this experiment was the percent occurrence of left-turn responses over training and extinction trials. The mean percent response for each of the four subgroups is presented in Table 1 in blocks of three trials for the ten blocks of training and four blocks of extinction trials. Figure 1 summarizes these results by comparing the experimental group (N-18) and the control group (N-18), deprivation disregarded, over training and extinction trials. The training and extinction data, presented in Table 1, were treated separately by Type III analyses of variance (Lindquist, 1953). Three significant effects were obtained from the analysis of the training data: Treatments (F = 45.25; df = 1/32; p < 0.001), Trials (F = 2.70; df = 9/288; p < 0.01), and Treatments × Trials (F = 5.86; df = 9/288; p < 0.001).

Although an examination of the group means presented in Table 1 suggests a separation of the two experimental groups over the last four blocks of training trials, with more responding in the high deprivation group than for the low deprivation group, this effect was not reliable. Neither the main effect of deprivation nor its interaction with other variables was significant.

The analysis performed on the extinction data revealed two significant effects: Treatments (F = 45.26; df = 1/32; p < 0.001), and Treatment × Deprivation (F = 4.36; df = 3/96; p < 0.010). The extinction analysis suggested a trials effect (F = 2.69; df = 3/96; p < 0.10), but apparently not enough trials were available for this analysis to produce a striking effect.

TABLE 1 MEAN PERCENT RESPONSE FOR EACH OF THE FOUR EXPERIMENTAL GROUPS OVER THE TEN BLOCKS OF TRAINING TRIALS AND FOUR BLOCKS OF EXTINCTION TRIALS (Three Trial Blocks)

CONDITIONS	TRAINING TRIALS										EXTINCT. TRIALS			
	1	2	3	4	5	6	7	8	9	10	11	12	13	14
High Depriv.														
Exp.	30	44	58	63	67	55	63	89	85	89	85	70	52	67
Con	30	26	41	22	18	41	30	15	33	22	26	22	07	15
Low Depriv.														
Exp.	30	44	41	55	63	59	52	70	72	78	70	55	70	44
Con	37	15	30	26	41	22	15	15	30	18	15	18	30	07

SOURCE: E. Siqueland and L. P. Lipsitt. Conditioned headturning in human new-borns. *Journal of Experimental Child Psychology*, 1966, 3, 356–376.

The procedure for graphing a first-order interaction from a three-factor ANOVA was discussed in Chapter 5, and this procedure is the same in a Lindquist Type III ANOVA. In essence, the data of the study are averaged across the variable that is not involved in the interaction, with the resulting graph appearing as though it came from an ANOVA involving two factors instead of three. In Excerpt 6.25 we see a graph which corresponds to the treatment × trials interaction from Excerpt 6.24 for the training data and also for the extinction data. The first of these was significant (note that the two lines which represent training are not parallel), while the second was not significant (note that the two extinction lines are parallel).

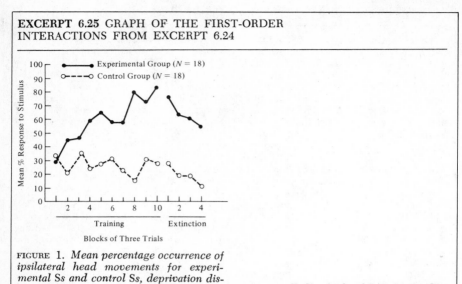

EXCERPT 6.25 GRAPH OF THE FIRST-ORDER
INTERACTIONS FROM EXCERPT 6.24

FIGURE 1. *Mean percentage occurrence of ipsilateral head movements for experimental Ss and control Ss, deprivation disregarded, over 10 blocks of 3 training trials and 4 blocks of 3 extinction trials (Experiment 1).*

SOURCE: E. Siqueland and L. P. Lipsitt. Conditioned headturning in human new-borns. *Journal of Experimental Child Psychology*, 1966, *3*, 356–376.

The procedure for graphing the second-order (three-way) interaction from a Lindquist Type III ANOVA is similar to the procedure for graphing the second-order interaction from a regular three-way ANOVA (see Chapter 5). Excerpt 6.26 contains a graph of a significant three-way interaction from a Lindquist Type III ANOVA in which the three factors were (1) sex of the student, (2) grade level of the student, and (3) curricular area. In this study each student indicated his preference for four curricular areas on a semantic differential rating scale that produced several scores, one of which was termed certainty. The graph of the three-way interaction clearly indicates that the preferences of individuals in different grades for the four curricular areas vary according to whether the individuals are boys or girls.

Follow-up Analyses for a Lindquist Type III ANOVA

If the results of a Lindquist Type III ANOVA indicate a significant main effect and if there are three or more levels associated with that particular factor, then the researcher will probably use a multiple comparison test (e.g., Duncan, Tukey, etc.) to find out where the significant differences lie among the main effect means. In Excerpt 6.27 the authors obtained a significant difference for the main effect of the repeated measures factor, and since there were three levels associated with this factor, the Newman-

EXCERPT 6.26 GRAPH OF A SECOND-ORDER
INTERACTION FROM A LINDQUIST TYPE III ANOVA

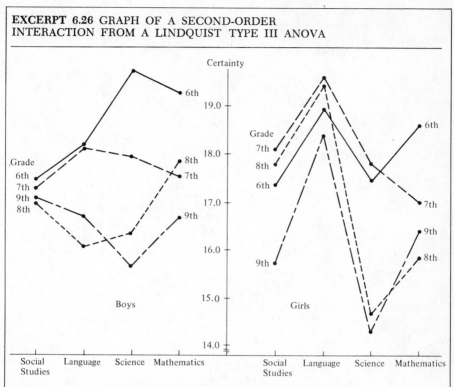

FIGURE 1. *Curriculum-by-sex-by-grade interaction on certainty factor.*

SOURCE: K. Yamamota, E. C. Thomas, and E. A. Karns. School-related attitudes in middle-school age students. *American Educational Research Journal*, 1969, *6*, 191–206. Copyright 1969 by the American Educational Research Association, Washington, D.C.

Keuls test was used to identify which of the three means (in Table 3) were significantly different from one another.

If a first-order interaction is significant, researchers will often disregard the results related to the main effects of the two factors that are involved in the interaction. Instead, they will probably average the data of the study across the variable that is not involved in the interaction and then apply tests of simple main effects. In Excerpt 6.28 a $2 \times 2 \times 4$ Lindquist Type III ANOVA was used to analyze the acquisition data, with the three factors being diagnosis (schizophrenic Ss vs. neurotic Ss), method (reinforcement vs. no reinforcement), and trials (four blocks of 20 card presentations). Since the trials-by-diagnosis interaction was significant, the data of the study were averaged across the two levels of the methods factor, thus resulting in the graph in Figure 1. A test of simple main effects was used to compare the four acquisition means of the schizophrenic Ss (i.e., the four points on the dashed line), with results indicating no significant difference among the means. A second test of simple main effects could have been used (but was not) to compare the four acquisition means of the neurotic Ss (the four points on the solid line).

EXCERPT 6.27 USE OF A MULTIPLE COMPARISON
TEST TO INVESTIGATE A SIGNIFICANT MAIN EFFECT

RESULTS

Learning Trigram-Name Pairs

As in Study I, the number of errors made by each subject, before reaching the criterion, was tallied separately for the liked, neutral, and disliked trigram-name pairs, and a $2 \times 2 \times 3$ (Sex $\times$ Mediation Group $\times$ Affect for Names) ANOVA for repeated measures was performed. The errors made on the trigram-name and name-trigram trials were summed. A summary of the analysis and relevant descriptive statistics are given in Tables 3 and 4.

Liking (affect for persons) was highly significant. As predicted, subjects made fewest errors in learning to associate liked names with trigrams, most errors for neutral names, and an intermediate number of errors for disliked names. A Newman-Keuls test applied to the means indicates that liked-disliked and liked-neutral differences were reliable ($p < .01$), but that the disliked-neutral difference was short of statistical significance.

TABLE 3 NUMBER OF ERRORS MADE IN LEARNING TRIGRAM ASSOCIATES FOR NAMES OF LIKED, DISLIKED, AND NEUTRALLY REGARDED ACQUAINTANCES

TRIGRAMS PAIRED WITH	M	s
Liked persons	1.23	1.3
Disliked persons	2.10	2.2
Neutral persons	2.31	2.4

TABLE 4 SUMMARY OF ANALYSIS OF VARIANCE

SOURCE	SS	df	MS	F
Between Ss				
Sex (A)	22.32	1	22.32	2.67
Mediation group (B)	6.99	1	6.99	<1
A × B	2.82	1	2.82	<1
Ss within groups (C)	401.90	48	8.37	
Within Ss				
Affect for names (D)	33.86	2	16.93	9.20°
A × D	8.91	2	4.46	2.42
B × D	4.24	2	2.12	1.15
A × B × D	.63	2	.32	<1
D × C	177.02	96	1.84	

° $p < .001$.

SOURCE: A. J. Lott, B. E. Lott, and M. L. Walsh. Learning of paired associates relevant to differentially liked persons. *Journal of Personality and Social Psychology*, 1970, *16*, 274–283. Copyright 1970 by the American Psychological Association. Reprinted by permission.

In Excerpt 6.29 a significant second-order interaction was obtained from the Lindquist Type III ANOVA. To further investigate this interaction, the researcher applied tests of simple interactions, with the possible interaction between two of the three factors being analyzed at each separate level of the third factor. In addition to the tests of simple interactions, the author conducted tests of simple main effects and, as stated in the last sentence of the excerpt, these tests substantiated the results of the initial tests of simple interactions.

EXCERPT 6.28 USE OF A SIMPLE MAIN EFFECTS TEST TO INVESTIGATE A SIGNIFICANT FIRST-ORDER INTERACTION

The 16 Ss in each diagnostic category were randomly divided into two equal groups. One group in each diagnostic category was reinforced for any sentence begun with I or We, and the other group for similar use of He or They. During the acquisition phase (the first 80 cards), E said "Good" in a flat, unemotional tone following any sentence which S began with the pronouns chosen to be reinforced. In order to determine a base level without reinforcement, however, reinforcement was not administered until after the first 20 card presentations.

RESULTS

In the treatment of the data, the number of reinforced pronouns occurring within 8 successive blocks of 20 cards each were used as the criterion measures, taking the first four blocks as the acquisition and the last four as the extinction phases. A summary of the analysis of the acquisition trials is shown in Table 1. Neither Methods nor any of its interactions with other factors approaches reliable significance. It may therefore be safely assumed that the Ss conditioned equally to both classes of pronouns, I-We and He-They. The F values for Trials and Trials-by-Diagnosis are each significant at the 2% level. It seems, therefore, that conditioning did occur, but it occurred differentially for the two diagnostic groups. Since it appeared from inspection of Figure 1 that the curve for the schizophrenic Ss was not ascending, the data for this group alone were further analyzed. The two methods of conditioning were combined, since the initial analysis had shown no differences attributable to methods. The F value for Trials was clearly not significant, allowing the inference that no reliable conditioning occurred for the schizophrenic group. It may then be concluded, because the over-all Trials effect

for all groups is significant, that conditioning occurred in the neurotic group.

TABLE 1 SUMMARY OF ANALYSIS OF ACQUISITION DATA

SOURCE OF VARIATION	df	ms	F
Between Subjects	31		
Diagnosis	1	82.89	
Methods	1	2.82	
Diagnosis × Methods	1	.38	
Error (b)	28	34.89	
Within Subjects	96		
Trials	3	32.15	3.96*
Trials × Diagnosis	3	28.57	3.52*
Trials × Method	3	5.38	
Trials × Diagnosis × Method	3	7.70	
Error (w)	84	8.12	
Total	127		

* Significant 2% level.

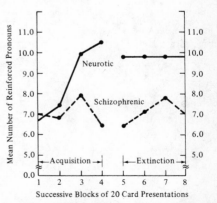

FIGURE 1. *Mean numbers of reinforced pronouns for successive blocks of 20 card presentations.*

SOURCE: E. Cohen and B. D. Cohen. Verbal reinforcement in schizophrenia. *Journal of Abnormal and Social Psychology*, 1960, **60**, 443–446. Copyright 1960 by the American Psychological Association. Reprinted by permission.

EXCERPT 6.29 FOLLOW-UP INVESTIGATION OF
A SIGNIFICANT SECOND-ORDER INTERACTION

In view of the significant three-way interaction, simple interactions were studied further. Analyses of simple interactions revealed three significant F ratios (cf. Winer, 1962, p. 340): (a) The Preference $\times$ Assignment interaction was significant for the gold dust problem ($F = 9.894$, $df = 1/62$, $p < .005$) but not the divisible problem; group superiority over individuals among those initially favoring individual work was significantly greater than among those initially favoring group work; (b) the Preference $\times$ Problem interaction was significant for those assigned to groups ($F = 5.681$, $df = 1/62$, $p < .025$), but not for individuals; although groups of persons preferring to work alone

were faster on the average at the gold dust problem, groups composed of those actually inclined to group work were faster at the divisible problem; (c) the Assignment $\times$ Problem interaction was significant for those preferring individual work ($F = 6.910$, $df = 1/62$, $p < .025$), but not for those preferring groups; considering only subjects preferring to work alone, the mean group performance superiority over individuals was significantly greater (i.e., faster) for the gold dust than for the divisible problem. Further analyses of simple main effects only confirmed the pattern of internally consistent results just described.

TABLE 2 ANALYSIS OF VARIANCE OF MEAN SOLUTION TIMES

SOURCE	df	MS	F
Individual-group preferences (A)	1	81938.671	1.580
Group-individual assignments (B)	1	966838.4391	18.639**
A $\times$ B	1	155523.4702	2.998
Error	62	51871.9725	
Problem (C)	1	9789.2776	
A $\times$ C	1	22054.2514	
B $\times$ C	1	51816.1132	
A $\times$ B $\times$ C	1	384839.5976	7.374*
Error	62	52191.6734	

* $p < .01$.
** $p < .001$.

SOURCE: J. H. Davis. Individual-group problem solving, subject performance, and problem type. *Journal of Personality and Social Psychology*, 1969, 13, 362–374. Copyright 1969 by the American Psychological Association. Reprinted by permission.

REVIEW TERMS

between-subjects
error (b)
error (w)
Lindquist Type I ANOVA
Lindquist Type III ANOVA

mixed design
split-plot 3.4 ANOVA
subjects within groups
trials
within-subjects

REVIEW QUESTIONS

1. How many factors are there in a Lindquist Type I ANOVA? Of these, how many are between-subjects factors and how many are within-subjects factors?

2. In a repeated measures ANOVA, is the trials factor considered to be a between-subjects factor or a within-subjects factor?

3. How many research questions can be answered by a 3×5 factorial ANOVA with repeated measures on one factor?

4. Suppose a researcher uses a split-plot factorial 2.3 ANOVA. A total of 20 Ss are used in the study. How many degrees of freedom will there be in the summary table for (a) error-between, (b) the interaction, and (c) error-within?

5. How many MS values are usually found in a Lindquist Type I ANOVA?

6. In a repeated measures ANOVA, can we figure out how many Ss were in the study by adding 1 to the total number of degrees of freedom in the summary table?

7. When will a researcher use the term "error (b)" to stand for "error-within," rather than "error-between"?

8. Which of the following Fs, if significant, would indicate a different trend of performance over trials for two treatment groups: the main effect F for trials, the main effect F for treatment groups, or the trials-by-treatment groups interaction F?

9. In graphing the interaction from a two-factor mixed design, will the levels of the repeated measures factor (a) be marked off along the vertical axis, (b) be marked off along the horizontal axis, or (c) correspond to the various lines in the graph?

10. Does the term subjects within groups mean the same thing as error-within or error-between?

11. How many factors are there in a Lindquist Type III ANOVA? Of these, how many are between-subjects factors and how many are within-subjects factors?

12. In a three-way ANOVA with repeated measures on one factor, how many of the interaction terms will be found in the top part of the summary table?

13. Suppose a researcher uses a $2 \times 2 \times 2$ Lindquist Type III ANOVA. All of the main effects turn out to be significant, while all of the interactions turn out to be nonsignificant. How many times should the researcher apply a multiple comparison test to further investigate the significant main effects?

14. In Chapter 5, we saw that tests of simple main effects are sometimes used to investigate a significant interaction in a factorial ANOVA. Is it appropriate for a researcher to use tests of simple main effects after getting a significant interaction in a repeated measures ANOVA?

15. Suppose a researcher uses a three-way analysis of variance with repeated measures on two factors. How many F ratios would be in the between-subjects portion of the ANOVA summary table?

CHAPTER 7

THE
ANALYSIS
OF
COVARIANCE

In the previous three chapters we discussed several different types of analysis of variance. Now our attention is directed to one of the most useful and sophisticated parametric statistical techniques—the analysis of covariance.

INTRODUCTION

Suppose a researcher had an interest in comparing two methods of teaching a course in basic statistics. In conducting his study, our hypothetical researcher might give a 40-item pretest covering the course material on the first day of the statistics class. Next, he might assign the students in the class to two treatment groups, a lecture group and a discussion group. Students in the first of these groups would hear formal lectures covering the course material. Students in the second group would find that the instructor assumed the role of a moderator for a discussion in which all students were encouraged to participate. Both of these treatment groups would have the same instructor for their class meetings and both groups would meet the same number of times during the academic term. At the end of the course a common final examination (identical to the pretest) would be administered to both groups of students.

In Table 7.1 we see some imaginary data to correspond with our hypothetical study. As these data indicate, the lecture group had mean scores of 14.5 on the pretest and 34.8 on the final, while the discussion group had mean scores of 9.5 on the pretest and 32.1 on the final. These data indicate clearly that each of the two groups performed better on the final examination than they did on the pretest. However, the study was conducted to find out whether one of the two methods of teaching facilitated learning more than the other. Therefore, it is necessary to make a direct comparison of the lecture group and the discussion group in terms of the amount of information learned during the academic term.

TABLE 7.1 DATA FOR THE HYPOTHETICAL STUDY COMPARING TWO TEACHING METHODS

LECTURE GROUP			DISCUSSION GROUP		
STUDENT	PRETEST SCORE	FINAL SCORE	STUDENT	PRETEST SCORE	FINAL SCORE
Bill	10	32	Stan	6	29
Jane	16	38	Lucy	14	34
Walt	14	35	Dave	10	32
.	.	.	.	.	.
.	.	.	.	.	.
.	.	.	.	.	.
Mary	13	33	Ruth	8	28
	$\bar{X}_{pre} = 14.5$	$\bar{X}_{fin} = 34.8$		$\bar{X}_{pre} = 9.5$	$\bar{X}_{fin} = 32.1$

To compare the lecture group against the discussion group, our hypothetical researcher might initially decide to use an independent samples t test to see if the mean score on the final exam for the Ss in the lecture group ($\bar{X} = 34.8$) is significantly different from the mean score on the final exam for Ss in the discussion group ($\bar{X} = 32.1$). The results of such a t test might indicate that the lecture group Ss performed significantly better on the final exam than did the discussion group Ss.

After thinking about his research question, our researcher would probably realize that the t test mentioned above is not a good way to analyze his data. The researcher is interested in finding out whether the two teaching methods differ in the extent to which they facilitate student learning, and since the two groups did not start the academic term with the same amount of knowledge about the course material (as shown by the discrepancy between the two pretest means, 14.5 and 9.5), a comparison of the two final exam means does not represent a legitimate way to answer the question.

Students in the lecture group started out the academic term with more knowledge about the course material than the students in the discussion group. To be fair in our comparison of the two teaching methods, the final exam means need to be adjusted to take into account the initial differences between the pretest means. In a nonscientific manner, our researcher could make this adjustment by first averaging the two pretest means to find out the mean score for all Ss, disregarding group membership, on the pretest. This would result in an overall pretest mean of 12.0. Since the lecture group had a pretest mean that was 2½ points higher than the overall average, this group's final exam mean must be reduced by 2½ points to account for the fact that the students in this group began the course with a head start. Thus, the adjusted final exam mean for the lecture group becomes equal to 34.8 minus 2.5, or 32.3. On the other hand, the discussion group had a pretest mean that was 2½ points below the overall average; therefore, this group's final exam mean must be increased by 2½ points to account for the fact that the students in this group began the course with a disadvantage. Thus, the adjusted final exam mean for the discussion group becomes equal to 32.1 plus 2.5, or 34.6.

It is very important for our researcher to have these adjusted final exam means, because a t test comparison of the unadjusted final exam means would have caused the researcher to draw the wrong conclusion from his data analysis. On the basis of a t test comparison the researcher would have decided that the lecture method of teaching was better, while a comparison of the adjusted final exam means shows that, after initial differences were taken into consideration, it was the discussion method that really caused the most student learning.

As we mentioned earlier, the procedure for calculating the adjusted final exam means for the two treatment groups in the hypothetical study was somewhat nonscientific. Our researcher would not actually follow this procedure. Instead, he would use a statistical procedure called the analysis of covariance, which accomplishes the same objective, but in a more scientific, precise manner.

Two Purposes of the Analysis of Covariance

The analysis of covariance is used most often by researchers to compare group means on a dependent variable, after these group means have been adjusted for differences between the groups on some relevant *covariate* (*concomitant*) variable. The hypothetical study described above falls into this category. The dependent variable was the score on the final exam, the covariate was the score on the pretest, and the analysis of covariance would (1) adjust the final exam means on the basis of the covariate (pretest) means and then (2) compare these adjusted final exam means to see if they are significantly different from one another. It is important to note that the adjustment is on the dependent variable means. The covariate means are never adjusted. Although many authors use the analysis of covariance for the purpose of adjusting for initial group differences without stating this in their article, Excerpt 7.1 shows how some authors indicate clearly that they were using the analysis of covariance because of its adjusting properties.

EXCERPT 7.1 USE OF COVARIANCE TO
ADJUST FOR INITIAL DIFFERENCES

The use of the analysis-of-covariance design was to control statistically any initial differences in the students which might have been present and which might confound differences between the two groups of students.

SOURCE: G. Sax and M. Reade. Achievement as a function of test difficulty. *American Educational Research Journal*, 1964, *1*, 22–25. Copyright 1964 by the American Educational Research Association, Washington, D.C.

The second purpose of the analysis of covariance involves the issue of power. In Chapter 5, we said that a statistical test is powerful if it is sensitive to differences among the groups that are being compared. In other words, a statistical test is powerful if it is likely to pick up significant differences. Other things held constant, researchers prefer to use the more powerful of two available statistical tests.

Suppose we return to the hypothetical study involving the lecture and the discussion methods of teaching the statistics course. Imagine for a moment that the pretest means for both groups turned out to be identical to one another. Let us say that both groups had a pretest mean equal to 12.0. With identical means on the pretest, no adjustment would be needed in terms of the final exam means. Thus, there might be a tendency to think that the covariate scores are useless, and some researchers might even disregard the pretest scores and use only the posttest scores in the data analysis. This would be a mistake, because use of the covariate data within an analysis of covariance would provide a more powerful (sensitive) statistical analysis than would the analysis of just the final exam data with the covariate data omitted.[1]

The Nature of the Covariate

In both uses of covariance, scores on the covariate variable and the dependent variable are often measured by means of the same measuring instrument. When this is done, the covariate can be referred to as a *pretest*, while the dependent variable can be referred to as a *posttest*. In the hypothetical study both variables were measured by the administration of the same exam, the scores before the course being the covariate or pretest and

---

[1] There is one circumstance in which it would be better for a researcher to disregard the covariate data. This is when the correlation between the covariate and the dependent variable is very low.

the scores on the same exam after the course being the dependent variable or posttest. Excerpt 7.2 is another example of a pretest used as the covariate.

EXCERPT 7.2 USE OF A PRETEST AS A COVARIATE

The six posttest variables were separately analyzed using an . . . analysis of covariance design with the appropriate pretest as a covariate in each analysis.

SOURCE: T. V. Busse et al. Environmentally enriched classrooms and the cognitive and perceptual development of Negro preschool children. *Journal of Educational Psychology*, 1972, *63*, 15–21. Copyright 1972 by the American Psychological Association. Reprinted by permission.

It is possible, however, for the covariate and dependent variables to be obtained by different measuring instruments, that is, the covariate and the dependent variable are not always the same variable. For example, a researcher might use the analysis of covariance to compare groups in terms of undergraduate grade point average (the dependent variable) after the GPA means are adjusted for group differences in terms of scores on the ACT entrance exam (the covariate). In many studies the dependent variable will be a measure of academic achievement, while the covariate will be a measure of general intelligence.

DIFFERENT TYPES OF ANALYSIS OF COVARIANCE

In the previous three chapters we discussed several types of analysis of variance which differ from one another in terms of the number of F ratios that are provided, the format of the summary table, the use of repeated measures, and in other ways. For each of these ANOVA's, there is a comparable analysis of covariance. Thus, a one-way analysis of covariance is similar to a one-way ANOVA, the main difference being that the former includes a covariate variable while the latter does not. Hence, it may be helpful to think of each analysis of covariance that will be discussed as an extension of its ANOVA counterpart to incorporate a covariate.

One-Way Analysis of Covariance

In a one-way analysis of covariance, there are (1) two or more comparison groups which differ from one another along a single dimension, (2) scores on the dependent variable, and (3) scores on the covariate variable. The null hypothesis associated with this analysis is that there is no difference among the adjusted means on the dependent variable. A significant F will be found if these adjusted means are far enough apart from one another.

Excerpt 7.3 contains a summary table and a table of means from a report of a one-way analysis of covariance. The format of the summary table is similar to the one-way ANOVA summary tables discussed in Chapter 4. The asterisk next to the F value of 5.10 and the explanation of the asterisk beneath the summary table indicate that the four adjusted criterion VP means (10.60, 10.26, 11.76, and 5.80) contain at least two means which are significantly different from one another. A multiple comparison follow-up test (e.g., Newman-Keuls, Tukey, Scheffé) would be applied to the adjusted means to find out exactly where the significant differences are.

EXCERPT 7.3 TWO TABLES CORRESPONDING TO A ONE-WAY ANALYSIS OF COVARIANCE

SUMMARY OF ANALYSIS OF COVARIANCE

SOURCE	SS	df	MS	F
Conditions	491.79	3	163.93	5.10*
Error	2,924.60	91	32.14	
Total	3,416.39	94		

* $p < .01$.

MEAN PRE-VP AND MEAN AND ADJUSTED MEAN CRITERION-VP SCORES

		PRE-VP		Criterion VP OBTAINED		ADJUSTED
CONDITION	N	M	SD	M	SD	M
Pn	24	11.38	11.50	12.25	11.69	10.60
Pe	24	8.23	7.60	9.50	7.43	10.26
Nl	24	8.31	9.35	11.06	10.14	11.76
CI	24	8.96	9.32	5.60	6.44	5.80

SOURCE: T. J. D'Zurilla. Persuasion and praise as techniques for modifying verbal behavior in a "real life" group setting. *Journal of Abnormal Psychology*, 1966, *71*, 369–376. Copyright 1966 by the American Psychological Association. Reprinted by permission.

With respect to the table of means in Excerpt 7.3, it is possible to see why the adjusted criterion means are different from the obtained criterion means. The first group (Pn) started out with a pre-VP mean that was substantially higher than the overall pre-VP mean based upon all four groups of Ss. Thus, the criterion VP mean for the Pn group was adjusted downwards. On the other hand, the three remaining groups all had below average pre-VP means, and their criterion VP means are all adjusted upwards. From one point of view, the adjusted means for the four groups represent the best possible estimates as to what the groups would have actually obtained for criterion VP means if the four groups had started out with identical pre-VP means.

Finally, note that the total *df* in the summary table is equal to 2 less than the number of subjects in the study. Although the *df* for the top row of the summary table is still equal to the number of groups minus 1, the use of the covariate within the analysis has "used up" one of the error degrees of freedom. Therefore, to figure out, from the summary table, how many Ss were in a one-way analysis of covariance study, you must add 2 to the total degrees of freedom (rather than 1 as is the case with a one-way ANOVA summary table).

Before leaving the topic of a one-way analysis of covariance, let us look at one more summary table. The specific summary table in Excerpt 7.4 contains a tremendous amount of information about the study, even though the SS column and the row for total have been omitted. The title of the table indicates what the dependent variable was; the first footnote beneath the table explains what the covariate variable was; the degrees of freedom allows us to figure out that there were 120 subjects in the study distributed among three groups; and the second footnote tells us that a significant difference (at the .05 level) was found among the three group means. The only thing that the table does not indicate is which specific group means are significantly different from one another, and a multiple comparison follow-up test would be needed to find out the answer to this question.

EXCERPT 7.4 SUMMARY TABLE FOR A ONE-WAY ANALYSIS OF COVARIANCE

TABLE 5 ANALYSIS OF COVARIANCE: COMPARISON OF GROUPS ON HOMONYM/ STANDARD ENGLISH ITEMS.[a]

SOURCE OF VARIATION	df	MEAN SQUARES	F
Between groups	2	.58	4.13[b]
Within groups	116	.14	

[a] Peabody Picture Vocabulary test scores are covariate factors.
[b] $p < .05$.

SOURCE: R. L. Gottesman. Auditory discrimination ability in Negro dialect-speaking children. *Journal of Learning Disabilities*, 1972, 5, 94–101.

Factorial Analyses of Covariance

A two-way analysis of covariance is similar to a two-way ANOVA in that both involve two factors, each of which contains at least two levels. In each case, the summary tables will contain two main effects (one for each factor) and one interaction.

A two-way analysis of covariance is similar to a one-way analysis of covariance in three respects. First, a covariate variable is always associated with both a one-way and a two-way analysis of covariance. Second, the three F ratios in the summary table pertain to the adjusted means on the dependent variable. Third, one df is used up by the covariate, thus making the within-groups (error) and total degrees of freedom one less than would be the case had the covariate not been used. (Note that from a different perspective, these points represent differences between a two-way analysis of covariance and a two-way analysis of variance.)

In Excerpt 7.5 we see a summary table that presents the results of two separate two-way analyses of covariance, each of which was conducted using a different dependent variable. The first dependent variable was relaxation rate and the second was accuracy of self-estimation. In each analysis there were two levels of the first factor (feedback method) and three levels of the second factor (achievement discrepancy). There were 56 subjects associated with the first analysis and 48 associated with the second.

EXCERPT 7.5 SUMMARY TABLE CONTAINING THE RESULTS OF TWO SEPARATE TWO-WAY ANALYSES OF COVARIANCE

TABLE 2 ANALYSIS OF COVARIANCE OF RELAXATION RATE AND ACCURACY OF SELF-ESTIMATION

SOURCE	RELAXATION RATE			ACCURACY OF SELF-ESTIMATION		
	ADJUSTED df	ADJUSTED MS	F	ADJUSTED df	ADJUSTED MS	F
Feedback method (A)	1	.00227	6.56*	1	22.62	4.57*
Achievement discrepancy (B)	2	.00038	1.08	2	4.22	.85
A × B	2	.00050	1.46	2	6.10	1.23
Within	49	.00035		41	4.95	
Total	54			46[a]		

[a] Scores in the analysis of self-estimation measures were based on 16 overachievers, 16 predictable achievers, and 16 underachievers. Eight predictable achievers with test scores not similar in level to the scores of over- or underachievers were not included in this part of the analysis because dissimilar levels of scores could have easily resulted in differences caused by artifacts.
* $p < .05$.
SOURCE: J. R. Forster. Comparing feedback methods after testing. *Journal of Counseling Psychology,* 1969, *16*, 222–226. Copyright 1969 by the American Psychological Association. Reprinted by permission.

When the analysis of covariance is used, some authors use the terms adjusted *df*, adjusted *SS* or *SS'*, and adjusted *MS* or *MS'* rather than the usual *df*, *SS*, and *MS* that we have seen in all summary tables in the previous three chapters. This is the case in Excerpt 7.5. The use of the word adjusted in a column label in a summary table is an indication that a covariate was used within the context of the statistical analysis. However, it should be noted that many authors do not use the additional terminology to label the columns and the use of the word adjusted in column label is not restricted to a two-way analysis of covariance—it may appear with other forms of analysis of covariance (e.g., a one-way analysis of covariance).

Just as a two-way analysis of covariance is similar to a two-way analysis of variance, a three-way analysis of covariance is similar to a three-way analysis of variance. Both involve three factors, each of which will contain at least two levels, and the summary tables for both analyses provide information regarding three main effects, three first-order (two-way) interactions, and one second-order (three-way) interaction.

A three-way analysis of covariance differs from a three-way ANOVA in three ways: (1) the analysis of covariance involves a covariate variable, (2) it provides F ratios that pertain to adjusted means, and (3) it has one less df because of the inclusion of the covariate data. Furthermore, all follow-up analyses in a three-way analysis of covariance deal with the adjusted means on the dependent variables.

In Excerpt 7.6 we see a three-way analysis of covariance summary table. Notice that the authors indicate, in the bottom row of the table, that the total df (70) was found by subtracting 2 from the number of Ss in the study. In most articles the author does not provide this statement, and you will have to remember that the number of subjects is 2 greater than the total df.

EXCERPT 7.6 SUMMARY TABLE FOR A THREE-WAY ANALYSIS OF COVARIANCE

TABLE 1 ANALYSIS OF COVARIANCE: FINAL SCORES ADJUSTED BY INITIAL SCORES

SOURCE	df	MS	F
Self-esteem (A)	1	7.78	2.058
Certainty (B)	1	6.43	1.701
Responsibility (C)	1	24.75	6.55°
A × B	1	3.29	<1
A × C	1	45.65	12.08°°
B × C	1	37.98	9.25°°
A × B × C	1	18.45	4.88°
Within error	63	3.78	
Total $(N - 2)$	70	—	—

° $p < .05.$
°° $p < .01.$

SOURCE: J. Maracek and D. R. Mettee. Avoidance of continued success as a function of self-esteem, level of esteem certainty, and responsibility for success. *Journal of Personality and Social Psychology*, 1972, 22, 98–107. Copyright 1972 by the American Psychological Association. Reprinted by permission.

Repeated Measures Analysis of Covariance

In Chapter 6 we examined two frequently used repeated measures analyses of variance, the Lindquist Type I ANOVA (which involves two factors with repeated measures on one of the factors) and the Lindquist Type III ANOVA (which involves three factors with repeated measures on one of the factors). If a covariate variable is used in either of these designs, the researcher will have a Lindquist Type I or a Lindquist Type III analysis of covariance.

There are two possible ways in which scores on the covariate can be obtained in a repeated measures analysis of covariance. First, it is possible for the researcher to collect the covariate data prior to measuring Ss on the dependent variable along the repeated measures factor. In this way, just one covariate score per subject is used to make adjustments of the means that are associated with all main effects and interactions. Second, it is possible for the researcher to collect data on the covariate variable just prior to *each* measurement on the dependent variable. If this is done, there will be as many covariate scores per subject as there are levels of the repeated measures variable. The distinction between these two ways of using the covariate are related to a difference in the summary tables.

As you remember from the previous chapter, both the Lindquist Type I and the Lindquist Type III summary tables are subdivided into a between-subjects part at the top and a within-subjects part at the bottom, and each of these two portions of the summary table has a separate error source. The summary table for the analysis of covariance with either of these repeated measures designs will look the same as if covariance had not been used, except for the degrees of freedom. If only one covariate score is used per subject, then one of the degrees of freedom from the between-subjects error is used up. Thus, we must add 2 to the *df* for between-subjects in order to figure out how many subjects were involved in the study. If several covariate scores are collected for each subject (one for each level of the repeated measures factor), then 1 *df* is used up from both of the errors. Hence, we would again have to add 2 to the *df* for between-subjects to determine how many subjects were in the study, and we would have to add 3 to the total degrees of freedom in order to find out how many pieces of data on the dependent variable were involved in the data analysis.

To help explain these differences between the summary tables associated with a repeated measures ANOVA and a repeated measures analysis of covariance, let us look again at two summary tables that were presented in Chapter 6. Excerpt 7.7 (Excerpt 6.8) corresponds to a Lindquist Type I ANOVA while Excerpt 7.8 (Excerpt 6.19) corresponds to a Lindquist Type III ANOVA.

If a covariate had been used in Excerpt 7.7 and if there had been just one covariate score per subject (probably collected prior to the six trials), 1 *df* for error (b) would have been used up in the data analysis, making the *df* for between-subjects 35. We would have to add 2 to this number in order to figure out how many Ss were in the study. If several covariate scores had been collected from each subject (one paired with each of the trials), then both error (b) and error (w) would lose 1 *df*, resulting in values of 33 and 169, respectively.

The same adjustments would be necessary in Excerpt 7.8. If one covariate score had been used per subject, then the *df* for error-between would decrease from 96 to 95. As a consequence, the *df* for between Ss and total would become 98 and 298, respectively. If a separate covariate score had been collected from each subject for each of the three levels of the repeated measures factor (Subtests), then both the error-between and the error-within would lose 1 *df*.

Of the two possible ways of using the analysis of covariance in con-

EXCERPT 7.7 SUMMARY TABLE FOR LINDQUIST TYPE I ANOVA

TABLE 2 ANALYSIS OF VARIANCE OF
CORRECT RESPONSES IN 12 LEARNING
TRIALS (in Blocks of Two Trials)

SOURCE	df	MS	F	p
Between-Subjects	36			
Groups (B)	2	25.88	3.74	<.05
Error (b)	34	6.91		
Within-Subjects	185			
Trials (A)	5	28.66	24.07	<.01
A × B·	10	1.41	1.18	ns
Error (w)	170	1.19		
Total	221			

SOURCE: S. Muehl. The effects of visual discrimination pretraining on learning to read a vocabulary list in kindergarten chi'dren. *Journal of Educational Psychology*, 1960, *51*, 217–221. Copyright 1960 by the American Psychological Association. Reprinted by permission.

junction with a repeated measures design, most researchers use the first method, which involves only one covariate score per subject. Regardless of which method is used, the interpretation of the F values must be tied to the adjusted means, not the actual means obtained by Ss on the dependent variable, and procedures for follow-up analyses refer to these adjusted means. The results are then interpreted in the same way as repeated measures ANOVA's which do not involve any covariate scores.

EXCERPT 7.8 SUMMARY TABLE FOR LINDQUIST TYPE III ANOVA

TABLE 2 ANALYSIS OF VARIANCE OF
ACHIEVEMENT DATA

SOURCE	df	MEAN SQUARE	F
Between Ss	99		
Remote Associates Test (A)	1	9804.1	20.48°
Response mode (B)	1	1704.1	3.56
A × B	1	5.6	<1
Error between	96	478.6	
Within Ss	200		
Subtests (C)	2	125.6	1.21
A × C	2	1.0	<1
B × C	2	1467.2	14.16°
A × B × C	2	154.7	1.49
Error within	192	103.6	
Total	299		

° $p < .001$.

SOURCE: S. Tobias. Effects of creativity, response mode, and subject matter familiarity on achievement from programmed instruction. *Journal of Educational Psychology*, 1969, *60*, 453–460. Copyright 1969 by the American Psychological Association. Reprinted by permission.

ASSUMPTIONS FOR THE ANALYSIS OF COVARIANCE

Since each of the different types of analysis of covariance can be thought of as an extension of its ANOVA counterpart, the assumptions that were discussed in the previous three chapters must also be met in the analysis of covariance in order for the researcher (and his reading audience) to have confidence in the covariance results. For example, whenever the analysis of covariance is used to compare groups that differ in size, a test of the assumption of homogeneity of variance is appropriate. With groups having the same number of Ss, however, the analysis of covariance is robust to this assumption and it need not be subjected to empirical testing.

In addition to the more general assumptions that are appropriate to both the analysis of variance and the analysis of covariance, there are a few assumptions that are specifically associated with the analysis of covariance.[2] One of these special assumptions concerns the requirement that the various comparison groups have what is called a *common slope*. A detailed explanation of the rationale underlying this assumption is beyond the scope of this book. Nevertheless, the reader of journal articles should know, first, that a test of this assumption *is* important to the analysis of covariance (even if the groups are all the same size) and, second, that the researcher hopes that his data will satisfy the assumption, thus resulting in a nonsignificant difference. Excerpt 7.9 shows how two authors reported the results of their test of this assumption.

EXCERPT 7.9 TEST OF THE ASSUMPTION OF A COMMON SLOPE

RESULTS

Because empathy scores on the post measures were considered to be, in part, a function of initial empathy level as well as a function of treatment effects, the analysis of covariance seemed most appropriate as a statistical control for differences in initial level of counselor empathy. Since an initial test indicated that a common slope could be assumed ($F = 1.71$, *ns*) other steps in the analysis were conducted.

SOURCE: P. A. Payne and D. M. Gralinski. Effects of supervisor style and empathy on counselor learning. *Journal of Counseling Psychology,* 1968, *15,* 517–521. Copyright 1968 by the American Psychological Association. Reprinted by permission.

Unfortunately, all authors who use the analysis of covariance do not refer to this prerequisite as an assumption of a common slope. Although they mean the same thing, certain researchers indicate that they are testing the assumption of *homogeneous regression coefficients* or *homogeneity of regression*. In Excerpts 7.10 and 7.11 we see examples of how authors use these alternative names for the common slope assumption.

Another assumption is important to the analysis of covariance but not

---

[2] Only two of these assumptions will be discussed in this book. The others, although important, are not discussed by authors in their journal articles.

EXCERPTS 7.10–7.11 ALTERNATIVE TERMS
FOR THE ASSUMPTION OF COMMON SLOPE

An analysis of covariance was used to compare the performance of the two groups on the quiz, with scores on a previous examination being used as the concomitant variable. Prior to using the covariance analysis, a check was made on the assumption of homogeneous regression coefficients. This preliminary test indicated that the assumption of equal regression slopes was tenable, thus permitting the use of the conventional analysis of covariance.

SOURCE: S. W. Huck and J. D. Long. The effect of behavioral objectives on student achievement. *Journal of Experimental Education*, 1973, *42*, 40–41.

Analyses of covariance were performed to determine if there were any statistically significant differences between the means of the two groups on any of the dependent variables. Tests for homogeneity of regression indicated parallel regression slopes.

SOURCE: G. P. Cartwright. The relationship between sequences of instruction and mental abilities of retarded children. *American Educational Research Journal*, 1971, *8*, 143–150. Copyright 1971 by the American Educational Research Association, Washington, D.C.

---

the analysis of variance. The *assumption of linearity* requires that the relationship between the covariate variable and the dependent variable be linear (rather than curvilinear). This simply means that an increase of a specified number of points on the covariate is related to about the same increase on the dependent variable, regardless of where the increase is on the continuum of possible covariate scores. For example, the relationship between height and weight is linear because an increase in two inches in height is related to about the same increase in weight no matter whether we go from 60 to 62 inches, 70 to 72 inches, or 80 to 82 inches. On the other hand, age and strength are not linearly related because as people get older their strength increases to a point and then it begins to decrease.

In most instances the nature of the researcher's data allows him to have confidence that the assumption of linearity is not violated. Therefore, it is unusal to find a journal article in which this assumption is discussed. However, Excerpt 7.12 shows how authors may indicate that they tested the assumption of linearity. The symbols X and Y stand for the covariate and the dependent variables, respectively. The authors plotted the data in a scattergram and then decided the assumption was not violated on the basis of the

---

EXCERPT 7.12 TEST OF THE ASSUMPTION OF LINEARITY

The basic assumption of analysis of covariance, linearity of regression of Y on X, was satisfied by the use of a scatter diagram for both stages of the study.

SOURCE: V. R. Morris, B. B. Proger, and J. E. Morrell. Pupil achievement in a nongraded primary plan after three and five years of instruction. *Educational Leadership*, 1971, *4*, 621–625.

pattern the data formed. Instead of such a visual (or "eye-ball") test, an author may choose to use a statistical test of this assumption in which he computes a calculated value and then compares it against an appropriate critical value.

USING TWO OR MORE COVARIATES

We have been discussing the analysis of covariance as if there is always one covariate variable and one dependent variable. However, a researcher may use two or more covariates. When this is done, the mean scores of the comparison groups on the dependent variable are adjusted to account for differences between the groups on each of the covariate variables. Furthermore, there is an increase in power when multiple covariates are used as long as (1) there is a high correlation between each covariate and the dependent variable and (2) there is a low correlation between each pair of covariates.

In Excerpt 7.13 we see the results of a one-way analysis of covariance in which there were two covariates. The two adjusted posttest means represent a scientific guess as to how the two groups would have performed on the posttest assuming that they had identical pretest means and identical interest means.

EXCERPT 7.13 ONE-WAY ANALYSIS OF COVARIANCE WITH TWO COVARIATES

TABLE 3 ANALYSIS OF COVARIANCE OF ELECTRONICS CLASSES (TEACHERS VERSUS NONTEACHERS) POSTTEST PERFORMANCE, USING PRETEST SCORES AND PUPILS' EXPRESSED INTEREST IN ELECTRONICS AS COVARIATES

SOURCE	df	SS	MS	F
Between	1	5.3	5.3	.72
Within	28	205.9	7.3	
Total	29	211.2		

	CONTROL VARIABLES		CRITERION VARIABLE	
GROUP	PRETEST $\bar{X}$	INTEREST $\bar{X}$	UNADJUSTED POSTTEST $\bar{X}$	ADJUSTED POSTTEST $\bar{X}$
Teacher	6.9	3.6	24.3	23.9
Nonteachers	6.8	3.5	22.7	23.1

SOURCE: W. J. Popham. Performance tests on teaching proficiency: rationale, development, and validation. *American Educational Research Journal*, 1971, *8*, 105–107. Copyright 1971 by the American Educational Research Association, Washington, D.C.

In Excerpt 7.14 we see an example of a study in which three covariates were employed within the context of a factorial analysis of covariance.

REVIEW TERMS

adjusted means   homogeneity of regression
common slope   linearity
concomitant variable   *MS'*
covariate   *SS'*

REVIEW QUESTIONS

1. In an analysis of covariance, is it possible for the covariate and the
dependent variable to be different variables?
2. Which means are adjusted in the analysis of covariance: group means on
the concomitant variable, group means on the dependent variable, or
group means on the independent variable?
3. The analysis of covariance is a useful procedure because it statistically
equates the comparison groups in terms of mean scores on the concomitant variable. What other purpose does a covariance analysis serve?
4. Suppose you come across an analysis of covariance summary table in
which the between-groups *df* is equal to 2 and the within-groups *df* is
equal to 71. How many groups were compared by the researcher?
Assuming that only one covariate variable was involved, how many Ss
were used in the study?
5. Sometimes a researcher will use the term adjusted *MS* to label a column
in an analysis of covariance summary table. Is the procedure for finding
F values the same in this situation as it would be had the column been
labeled *MS*?
6. How many factors are there in a 3 × 4 analysis of covariance?
7. Suppose a researcher uses a Lindquist Type I analysis of covariance. A
single covariate score is obtained from each of the Ss prior to measuring the Ss on the levels of the repeated measures factor. Would a degree
of freedom be used up from (a) just the error-between, (b) just the
error-within, or (c) both of the error sources?
8. Is the analysis of covariance robust to the assumption of homogeneous
regression slopes?

9. In a one-way analysis of covariance, there are three variables: the concomitant variable, the independent variable, and the dependent variable. Which two of these variables are assumed to have a linear relationship?

10. Suppose a researcher uses an analysis of covariance in which there are two covariate variables. To maximize the power of the statistical analysis, should the correlation between the two covariate variables be (a) close to +1.00, (b) close to 0.00, or (c) close to −1.00?

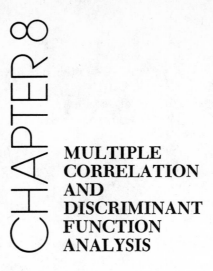

CHAPTER 8

MULTIPLE
CORRELATION
AND
DISCRIMINANT
FUNCTION
ANALYSIS

If you were asked to predict what is going to happen in the future, you could go out and try to find a crystal ball that would provide some specific answers. However, you might be on a futile search, since there just aren't too many crystal balls left in the world that really do work. Fortunately, two statistical techniques have been developed which can be used for predictive purposes, and these techniques are being used more and more by applied researchers. In this chapter we shall examine both of these prediction devices.

Before we begin our discussion of these two techniques—multiple correlation and discriminant function analysis—it should be noted that their capacity to predict is somewhat limited as compared with a good crystal ball. Whereas a well-shined crystal ball could be used to predict anything and everything, these two statistical procedures are normally used to predict how an individual will perform with respect to some measured variable such as grade point average or whether a person will pass (or fail) a certain course. Neither of these statistical procedures can be used to predict events such as an earthquake, a stock market crash, the assassination of a president, the arrival of a handsome stranger, and so forth.

To predict these things, we will have to continue our search for a good crystal ball.

This chapter is divided into three main sections. The first section is an introduction to concepts of prediction. It provides the foundation for two following sections on multiple regression and two different types of discriminant function analysis.

AN INTRODUCTION TO PREDICTION

Suppose you have just been hired as the Director of Admissions at a relatively small liberal arts college. Every year, about twice as many individuals apply to this school as can be admitted and, therefore, the admissions office is forced to reject about 50 percent of the people who apply. The previous admissions director had a simple method for deciding whom to admit and whom to reject—he simply drew names out of a hat. However, this procedure had two undesirable results: some well-qualified individuals were denied admission whereas others who were admitted flunked out after their first academic term. The Board of Trustees, after firing the old admissions director, hires you and indicates that your job is to admit people who will succeed.

Instead of admitting people at random, you would probably try to discover a predictor variable that is highly correlated with college grades. After doing some research on the freshman class, you might discover that the students who earned high grades during their first semester in college were typically the same individuals who earned A's and B's in high school, while the people who obtained low grades in their first set of college courses tended to be the ones who had a C or D average in high school. In other words, there seems to be a high positive correlation between high school grades and first-semester college grades. Suppose the Pearson correlation coefficient between these two variables turns out to equal +.80.

A Simple Prediction Equation

Given the data on these two variables for your group of college freshmen, it would be possible to develop a *prediction equation* which has the following form:

$$Y' = .15 + .70X$$

In this prediction equation the symbol Y' stands for the predicted first-semester grade point average (GPA) at your school, X stands for a student's high school GPA, and .15 is the constant term of the equation. (In general, the variable that is being predicted, Y', is called the *criterion* or *dependent variable*, while the variable that is used to make the prediction, X, is called the *predictor* or *independent variable*.) The ultimate purpose of the prediction equation would be to get an estimate of how well each applicant would do academically in his first semester at college. For example, if an applicant had a high school GPA of 3.0 (a straight B average), by substituting in the equation, you would find that his predicted college GPA is 2.25 (.15 + .70 × 3.0). In other words, this applicant would be predicted to earn a C+ average.

For each member of the original group of college freshmen used to develop the prediction equation, you would have information on his high school GPA (X), his actual first-semester college GPA (Y), and, by using the prediction equation, his predicted first-semester GPA (Y'). You could then compare each person's predicted college GPA with his actual college GPA to see if the prediction equation works. If there is a close similarity between the predicted and actual GPA's, then you would have confidence in using the prediction equation in the future.[1]

The Correlation Between Y and Y' Is Equal to the Correlation Between X and Y

To measure scientifically the success of the prediction equation, it would be possible to correlate (using the Pearson correlation method) the predicted and actual GPA scores for the group of individuals used to develop the prediction equation. If you did this, however, the resulting correlation coefficient would turn out to be .80, that is, identical to the correlation that you already knew existed between high school GPA and first semester college GPA. In other words, the degree to which the prediction equation is successful is contingent upon the magnitude of the correlation between the predictor and criterion variables. If the correlation between X and Y is .65, then the correlation between Y and Y' would also be .65. Thus, it is possible for a researcher to judge the success of a prediction equation without having to (a) develop the equation, (b) substitute each person's X score into the equation to obtain a Y score, and (c) correlate the Y and Y' scores.

As Director of Admissions, you would want to use the best possible predictor variable, that is, you would want to minimize the discrepancy between the predicted first-semester college GPA and the actual GPA that freshmen earn at your school. Knowing that the correlation between any possible X variable and the Y variable is identical to the correlation between Y and Y' (after we have used that particular X in a prediction equation to get Y' scores), the task of selecting the best predictor variable becomes easy. All you would need to do is identify that particular predictor variable that has the highest correlation with the criterion variable. Thus, if the correlation between IQ and college GPA was to equal .90, IQ would be a better predictor variable than high school GPA. On the other hand, if the correla-

---

[1] This assumes, of course, that future applicants to the college are similar to the ones in the group used to develop the prediction equation.

tion between height and college GPA was equal to .20, you would know that height is not a good predictor variable.[2]

Interpretation of r^2 as the Percentage of Criterion Score Variance That Is Predictable

No matter what variable is used as X in the prediction equation, the square of the correlation between that predictor variable and the criterion variable is referred to as the *coefficient of determination*. Thus, if the correlation between high school GPA and college GPA is equal to .80, then the coefficient of determination is equal to .64. In short, the coefficient of determination indicates the proportion (or percentage, if we move the decimal point two places to the right) of the variance among the criterion scores that can be explained by differences in the predictor variable. Therefore, if the correlation between high school GPA and college GPA is equal to .80, then 64 percent of the differences among people in terms of their college GPA is predictable on the basis of differences in their high school GPA's.

MULTIPLE CORRELATION

We have been talking about predicting college GPA on the basis of just one predictor variable. It should be obvious that you would be able to predict more successfully if you had additional information about the individuals who enter your college. In other words, by using two or more predictor variables, you would be able to decrease errors of prediction.

Let us suppose that in examining the records of the current freshman class at your school, you discover that there is a high positive correlation ($r = .72$) between the scores earned on the college entrance examination and first-quarter college grades. In addition, suppose you find out that there is a moderate negative correlation ($r = -.47$) between absence rate in high school and first-quarter college grades. Therefore, you conclude that by using these two predictor variables along with high school GPA, it may be possible to predict more accurately how a student will perform during his first term in college.

After submitting the data to a computer, the results of the statistical analysis would provide (on the computer printout) several pieces of information about the study. To find out whether it is possible to predict college GPA on the basis of the three predictor variables, you would be particularly interested in three types of information included in the computer printout—the multiple regression prediction equation, the coefficient of multiple correlation, and the beta weights. Since these are also the three

---

[2] It should be noted that it is the strength, not the nature, of the correlation which is important. A strong negative correlation, such as $-.80$, would be a better predictor variable than a moderate positive correlation such as $+.60$.

main pieces of information that authors include in their journal articles, let us now examine what each of these things means.

Three Important Elements of a Multiple Correlation Study

The Multiple Regression Prediction Equation. In the prediction equation we used above there was one predictor variable (X). A *multiple regression prediction equation* has the same basic form, but several predictor variables are involved, each represented by a different X symbol (X_1, X_2, X_3, etc.). We have said that there might be three predictor variables in our hypothetical college admissions study. Therefore, the multiple regression prediction equation provided by the computer might be

$$Y' = .20 + .50X_1 + .002X_2 - 5X_3$$

where Y' once again stands for the predicted college GPA, X_1 is the high school GPA, X_2 is the score on the entrance exam, X_3 is the absence rate in high school, and .20 is the constant. To obtain the predicted college GPA, an applicant's scores on the three predictor variables would be substituted for X_1, X_2, and X_3. For example, if an applicant to our college had a high school GPA of 3.00 (a straight B average), a score on the entrance exam of 600, and an absence rate of .02 (indicating he missed 2 days out of every 100), his Y' score would be equal to $.20 + .50(3.00) + .002(600) - 5(.02) = .20 + 1.50 + 1.20 - .10 = 2.80$. In other words, he would be predicted to obtain a B− average in his first quarter of college work.

In Excerpts 8.1 and 8.2 we see two different multiple regression equations that actually appeared in journal articles. In the first the criterion variable was the score on the Cooperative French Test and there were seven predictor variables. In the second the criterion variable was college grade point average and there were just two predictor variables (scores on the

---

EXCERPTS 8.1–8.2 TWO MULTIPLE
REGRESSION PREDICTION EQUATIONS

The regression equation associated with the seven-test battery is as follows:

$$Y' = 14.18 + .43X_1 + 2.72X_2 + .18X_3 + .96X_4 - .17X_5 + .53X_6 + .06X_7$$

SOURCE: P. Pinsleur. Predicting success in high school foreign language courses. *Educational and Psychological Measurement*, 1963, *23*, 349–357.

The multiple regression equation utilizing both GRE scores is the best equation possible for this grouping. The predicted grade point average is obtained by substituting the GRE scores of a student into the following formula:

Predicted GPA = .002242 (GRE–V) + .001725 (GRE–Q) + 4.255760

SOURCE: G. E. Madaus. Departmental differentials in the predictive validity of the Graduate Record Examination aptitude tests. *Educational and Psychological Measurement*, 1965, *25*, 1105–1110.

verbal and quantitative portions of the Graduate Record Examination). Notice that these two multiple regression equations have the same basic form even though (1) they involve different numbers of predictor variables, (2) one formula involves symbols while the other involves words, and (3) the constant is the first term to the right of the equal sign in the first equation and the last term in the second equation.

The Coefficient of Multiple Correlation. The second important piece of information concerning multiple correlation is R, the *coefficient of multiple correlation*. In essence, R provides an index of the accuracy of the prediction equation. Think back to our hypothetical study. For each person in your freshman class, you would have scores on the three predictor variables plus an actual first-quarter college GPA. By substituting predictor scores for X_1, X_2, and X_3, it would be possible to obtain a predicted first-quarter college GPA for each of your freshmen. If the prediction equation is working successfully, the predicted GPA score for the members of your original group of students should be similar to their actual GPA scores. Of course, we would not expect the predicted GPA and actual GPA for any particular student to be identical (since perfect prediction is not realistic), but we would have to find a high degree of correspondence between these two variables to say that the prediction equation is successful.

The coefficient of multiple correlation, R, can be thought of as a simple Pearson correlation between the actual Y scores and the predicted Y scores if the prediction equation were to be used to obtain a Y' score for each of the individuals in the original group from which the prediction equation was developed. Thus, if R turned out to equal $+1.00$, it would indicate that the predicted scores correlate perfectly with actual criterion scores. As we just

---

EXCERPTS 8.3–8.4 COEFFICIENTS OF MULTIPLE CORRELATION

A multiple R between GPA and the two predictor variables was computed to be .22.

SOURCE: G. E. Madaus. Departmental differentials in the predictive validity of the Graduate Record Examination aptitude tests. *Educational and Psychological Measurement*, 1965, *25*, 1105–1110.

A previous study by Watley and Martin (2) investigated the effectiveness of a large number of intellectual and nonintellectual variables for predicting academic achievement for male freshmen in a college of business administration. Five predictors were identified as the best combination of variables for this purpose. These were the mathematics and verbal scores of the Scholastic Aptitude Test (SAT) of the College Entrance Examination Board, the Restraint and Thoughtfulness trait scores of the Guilford-Zimmerman Temperament Survey (GZTS), and high school rank (HSR). The multiple correlation coefficient between these five predictor variables and the criterion (first-year grades) was .82.

SOURCE: D. J. Watley and J. C. Merwin. The effectiveness of variables for predicting academic achievement for business students. *Journal of Experimental Education*, 1964, *33*, 189–191.

mentioned, however, it is highly unlikely that R will ever reach $+1.00$, no matter how many predictor variables are included in the multiple regression equation. In practice, Rs of .80 or .90 are considered to be quite high. In Excerpts 8.3 and 8.4 we see two different multiple correlation coefficients, one of which is relatively low ($R = .22$) and the other quite high ($R = .82$).

In our discussion of prediction based upon a single predictor variable, we indicated that the correlation between the predictor and criterion variables turns out to be exactly equal to the correlation between actual criterion scores and predicted criterion scores. Furthermore, we noted that the square of the correlation coefficient tells us the percent of the variance of Y that is predictable on the basis of information on X. The interpretation of R and R^2 is analogous. If the coefficient of multiple correlation is equal to .80, then the correlation between actual criterion scores and predicted criterion scores will also be equal to .80. Under these circumstances the coefficient of determination is equal to .64 and the researcher is likely to state that 64 percent of the Y variance is predictable on the basis of the set of predictor variables. For example, consider Excerpt 8.5 in which the criterion variable was graduate quality point average (GQPA), with the predictor variables being undergraduate quality point average (UQPA), scores on the Purdue English Test (PET), and scores on four different subscales of the National Teachers Examination.

EXCERPT 8.5 INTERPRETATION OF R^2 AS PERCENT OF CRITERION SCORE VARIANCE ACCOUNTED FOR BY PREDICTOR VARIABLES

The equation for the prediction of GQPA was as follows:

$$\text{GQPA} = 1.770 + .285 \text{ UQPA} + .002 \text{ PET} + .001 \text{ TAE} \\ + .004 \text{ PES} - .005 \text{ WEE} - .009 \text{ SLF}$$

The multiple R for this equation was .554 which accounted for about 30.7 percent of the variance in the criterion variable.

SOURCE: J. B. Ayers. Predicting quality point averages in Master Degree programs in education. *Educational and Psychological Measurement*, 1971, *31*, 491–495.

Regression Coefficients and Beta Weights. In a multiple regression prediction equation the numbers that precede each of the predictor variables are called *regression coefficients*. It is important to realize that these regression coefficients cannot be compared against one another in an attempt to determine which of the various predictor variables is the best predictor, because the predictor variables in the actual prediction equation are not on the same scale of measurement. Trying to make a direct comparison of regression coefficients for the different predictor variables would be like trying to compare 3 years with 60 inches or, more succinctly, like trying to compare apples with bananas.

It is possible to convert the regression coefficients into comparable units called *beta weights*. There will be one beta weight for each predictor variable, and the beta weights can be thought of as the regression coefficients that would have been obtained if the various predictor variables were equal to one another in terms of means and standard deviation. The predictor variable that has the largest beta weight, disregarding whether the beta weight is positive or negative, is the best predictor; conversely, a small beta weight indicates that the corresponding predictor variable is not contributing to successful prediction as much as the other predictor variable(s).

In Excerpt 8.6 we see a table that contains two sets of beta weights. In this study an attempt was made to determine whether school grades in reading and language for students in fourth grade could be successfully predicted on the basis of four subscores on the California Achievement Test (RV, RC, AR, and AF) and two IQ scores obtained with the California Test of Mental Maturity (CTMM). The two IQ scores were Language (LM) and Non-Language (NLM). A separate multiple regression prediction equation was developed for each of the two criterion variables, grades in reading and grades in language.

EXCERPT 8.6 USE OF BETA WEIGHTS
INSTEAD OF REGRESSION COEFFICIENTS

Multiple correlations and beta weights are shown in Table 2. The relative predictive value of the CTMM variables is greater with Language than with Reading, but RV and RC remain the best relative predictors.

TABLE 2 SETS OF BETA WEIGHTS AND MULTIPLE CORRELATION COEFFICIENTS FOR EACH CRITERION VARIABLE

CRITERION VARIABLE	*R* MULTIPLE CORRELATION	*BETA WEIGHTS FOR PREDICTORS*					
		LM	NLM	RV	RC	AR	AF
Reading	.598	.044	.094	.292	.472	−.142	−.016
Language	.574	.158	.158	.174	.453	−.166	−.009

SOURCE: H. E. Anderson. The prediction of reading and language from the California tests. *Educational and Psychological Measurement*, 1961, *21*, 1035–1036.

It should be noted that beta weights are not better than regression coefficients, or vice versa. They convey different types of information. In some situations beta weights are more helpful than regression coefficients, while in other situations the reverse is true. In general, if an author wants to provide a regression equation that can be used in similar studies to predict criterion scores (assuming that the new subjects are similar to the ones used by the author), then the regression coefficients will probably be included in the journal article. If the author wants to show which predictor variables contribute the most to successful prediction, then he will probably include the beta weights.

Tests of Significance in Multiple Correlation

Although many different types of significance tests can be conducted in relation to a multiple correlation study, three tests in particular are frequently seen in the published literature. To understand these tests, it is important to realize that the actual prediction equation that a researcher arrives at is unique to the specific sample of Ss used to develop the equation. If a second prediction equation (involving the same X and Y variables) were to be developed on the basis of a new group of Ss, neither the regression coefficients nor the coefficient of multiple correlation, R, would correspond exactly to those obtained from the first prediction equation. In fact, if there were 100 different prediction equations based upon different samples (that are considered to be drawn from the same population), it is highly unlikely that any two equations would be identical. We could assume that they would be similar to one another, but there would inevitably be variations in the regression coefficients and R.

Is the Researcher's Obtained R Significantly Different from Zero? In the first of the three popular tests of significance, the null hypothesis is that the coefficient of multiple correlation in the population is equal to zero. If a researcher tests this null hypothesis and rejects it, say at the .01 level, it indicates that the researcher's R is so far away from zero that the chances are less than 1 out of 100 that the population R is zero. From a different perspective, a rejection of this null hypothesis indicates that if the researcher were to repeat his study 100 times, with a different sample of Ss each time, and if the population R were really equal to zero, then only 1 of the calculated Rs would be likely to be as far away from zero as the R which the researcher actually obtained. But since he only had 1 sample rather than 100 different samples, he can conclude that the population R is probably not equal to zero. Thus, the null hypothesis is rejected.

EXCERPT 8.7 TESTING R TO SEE IF IT IS
SIGNIFICANTLY GREATER THAN ZERO

The criterion of creativeness and freedom in handling materials, a composite, was selected as the most representative single criterion. Therefore, this single criterion became the criterion variable for the computation of a multiple correlation for both sets of test measures. The multiple R for the semantic test measures and the criterion variable was .585 (significant at the .01 level). In the figural set of measures, the multiple R was .536 (significant at the .01 level). The two sets of test measures appeared to be predictors of the criterion variable, creativeness and freedom in handling materials.

SOURCE: C. A. Jones. Relationships between creative writing and creative drawing of sixth grade children. *Studies in Art Education*, 1962, 3, 34–43.

   A researcher who has obtained several different coefficients of multiple correlation in his study is likely to put the various Rs into a table. If each R is tested to see if it is significantly greater than zero and if the null hypothesis is rejected for any particular R, he might give us this information by putting an asterisk next to the numerical value for that R along with an explanatory footnote (e.g., $p < .05$). An author who obtains only one or two Rs in his study is likely to report whether or not they are significantly greater than zero within the text of the article rather than in a table. This procedure is exemplified in Excerpt 8.7.

   Is One R Significantly Different from a Second R? The second popular test of significance is used to determine whether the addition of one or more predictor variables will lead to a significant increase in the coefficient of multiple correlation. Although there will usually be an increase in R when additional predictor variables are used within the regression equation, the magnitude of the increase in R becomes smaller and smaller as more and more predictor variables are used. Since it is usually costly and time consuming to measure individuals with respect to each predictor variable, a researcher does not want to spend that time and money unless there is a payoff in terms of a significant increase in R. In Excerpt 8.8 we see how this significance test can be used. This study was conducted to determine whether a set of affective predictor variables (e.g., attitude toward teachers, academic interest, etc.) would increase the accuracy of prediction beyond that which was possible on the basis of traditional intellectual variables (e.g., scores on

---

EXCERPT 8.8 TESTING TWO Rs TO SEE IF THEY ARE SIGNIFICANTLY DIFFERENT FROM EACH OTHER

TABLE 3 F RATIOS OF DIFFERENCES BETWEEN R^2's FOR MALES

CRITERION VARIABLES	R^2		F
	APTITUDE	APTITUDE + AFFECTIVE	
Reading	.7921	.8100	6.0936*
Language	.7225	.7396	3.2000*
Arithmetic computation	.5700	.6084	5.1111*
Problem solving	.5776	.6241	6.1820*
Social studies	.5700	.6241	7.5018*
Science	.5050	.5929	3.5717*

   To assess the contribution of affective variables to the academic prediction, two types of multiple correlations were obtained; one with only the aptitude variables and the other with both affective and aptitude variables. The difference between the correlations for each criterion was tested for significance using the F ratio statistic (Guilford, 1956).

* $p < .01$, $df = 8/417$.

SOURCE: S. B. Khan. Affective correlates of academic achievement. *Journal of Educational Psychology*, 1969, 60, 216–221. Copyright 1969 by the American Psychological Association. Reprinted by permission.

verbal and mathematics aptitude tests). There were six criterion variables, each of which corresponded to a subtest of the Metropolitan Achievement Test. As the data analyses indicate, use of both types of predictor variables resulted in more accurate prediction than was possible when using just the aptitude test scores.

Is a Particular Beta Weight Significantly Different from Zero? As we mentioned earlier, the regression coefficients in the prediction equation are often converted into beta weights so that the researcher can compare the predictor variables with respect to their relative effectiveness as predictors. If a particular predictor variable is not helping very much to decrease the difference between predicted and actual criterion scores, then the beta weight for this predictor variable will be close to zero. On the other hand, a large beta weight (regardless of whether it has a plus or minus sign) indicates that the corresponding predictor variable is contributing a great deal to successful prediction.

After obtaining the multiple regression prediction equation and after converting the regression coefficients to beta weights, a researcher will sometimes apply a statistical test to each beta weight to see if it is significantly different from zero. In a sense, this test asks whether or not the beta weight for a predictor variable is large enough to justify including that variable as a predictor. If the researcher cannot reject the null hypothesis, then that particular predictor variable will probably be dropped out, and a new prediction equation will be developed on the basis of the remaining predictor variables.

In Excerpt 8.9 we see an actual study in which this third test of significance was employed. This study was conducted to see whether the achievement of graduate students majoring in education could be predicted. The criterion variable was graduate grade point average, and the original

---

EXCERPT 8.9 TESTING BETA WEIGHTS FOR SIGNIFICANCE

For elementary majors, two variables, UGPA and AE, yielded significant regression weights for predicting GGPA. This optimum regression function yielded a multiple correlation coefficient of .30. The MAT, which did not yield a significant degree of relationship with graduate success for these subjects, was deleted after the first step of analysis. No decrease in the multiple correlation coefficient occurred.

TABLE 2 BETA WEIGHTS AND MULTIPLE CORRELATION COEFFICIENTS BEFORE AND AFTER DELETION OF NON-SIGNIFICANT VARIABLES FOR ELEMENTARY MAJORS ($N = 111$)

	VARIABLE			
STATISTIC	MAT	UGPA	AE	$R_{1.23...j}$
Beta	−.0007	.2142*	.1811**	.30
Beta		.2141*	.1807**	.30

* $p < .01$.
** $p < .05$.

SOURCE: C. M. Eckhoff. Predicting graduate success at Winona State College. *Educational and Psychological Measurement*, 1966, *26*, 483–485.

predictor variables were undergraduate grade point average, score on the Miller Analogies Test (MAT), and score on the Advanced Education (AE) section of the Graduate Record Examination.

Cross-Validation

As stated earlier, a multiple regression prediction equation is usually developed on the basis of an existing group of people (for whom there are both predictor and criterion scores) so that it can be used later on to predict how other individuals will perform on the criterion variable. Even if the coefficient of multiple correlation is found to be significantly different from zero and even if R^2 turns out to be a large number (indicating that a large percent of criterion score variance is explained by the predictor variables), a researcher cannot have complete confidence that his regression equation will work successfully when it is used to predict criterion scores for a new group of individuals. In fact, it is highly likely that the regression equation will be less accurate when used with new people. Simply stated, the reason for this drop-off (or *shrinkage*) in predictive accuracy is due to the fact that the new group of people is not identical to the group that was used to develop the prediction equation. This means that the regression coefficients in the prediction equation will be different from what they optimally should be when the regression equation is used with the new group. The obvious result is a greater discrepancy between the predicted criterion scores for people in the new group and the actual criterion scores that these individuals will eventually obtain.

To find out whether the prediction equation has a chance of being successful when it is used with a new group of individuals, researchers often use a technique called *cross-validation*. Fortunately, this technique, which allows the researcher to make a scientific guess as to whether the prediction equation will work with a new group, does not require that the researcher wait around for the new group. He makes his scientific guess solely on the basis of people in the first group.

The technique of cross-validation involves four simple steps. (1) The original group of people (for whom both predictor and criterion scores are available) is randomly divided into two subgroups. (2) Just one of the subgroups is used to develop the prediction equation. (3) This equation is used to predict a criterion score for each person in the second subgroup (i.e., the subgroup that was not used to develop the prediction equation). (4) The predicted criterion scores for people in the second subgroup are correlated with their actual criterion scores. A high correlation (that is, significantly different from zero) means that the prediction equation works for people other than those who were used to develop the equation. If the individuals in future studies are not too much different from those in the cross-validation procedure, the researcher is justified in using the prediction equation for groups other than the original.

An excellent example of cross-validation is contained in Excerpt 8.10. The criterion variable was related to success in occupational training. As demonstrated in this study, a prediction equation will sometimes be shown to be relatively useless through a cross-validation analysis.

EXCERPT 8.10 CROSS-VALIDATION IN A MULTIPLE CORRELATION STUDY

Method. One group of trainees ($N = 224$) was used to determine which variables would make the best predictors and what optimum weight should be associated with each predictor. These trainees had ended training before April 1965 under MDTA in Muskegon, Michigan. The second group of trainees ($N = 96$) was made up of Muskegon MDTA trainees that ended training after April 1965. This second group of trainees was used to cross-validate results obtained with the first group of trainees.

The multiple regression equation for the initial group was:

$$X_1 = -.153 + .013x_2 + .124x_3 + .001x_4 - .001x_5 + .003x_6$$

X_1 is the new prediction score; x_2 is the age score; x_3 is the Unemployment Compensation score; x_4 is the Intelligence score; x_5 is the verbal score; and x_6 is the spatial score.

The resulting multiple correlation for the initial group was .325 with a standard error of estimate of .439. An F-test of this multiple correlation revealed $F = 4.22$ which was significant at the .01 level.

The same criterion and the same predictor equation were also used with the cross-validation group. The resulting multiple correlation for the cross-validation group was $-.06$. This correlation was not significant at the .05 level.

None of the predictor variables found to be statistically significant with the initial group continued to be significant when used with the cross-validation group.

Discussion. A cross-validation group was used to check the usefulness of the prediction equation developed with the initial group. If the equation had yielded a significant correlation, the moderate correlation coefficient obtained initially between the five prediction variables and actual success or failure in MDTA training would have been reaffirmed. Since the multiple correlation coefficient obtained with the cross-validation group was not significant, the equation cannot be used with future MDTA programs.

SOURCE: D. Sommerfeld and F. A. Fatzinger. The prediction of trainee success in a manpower development and training program. *Educational and Psychological Measurement*, 1967, **27**, 1155–1161.

DISCRIMINANT FUNCTION ANALYSIS

In many prediction studies the criterion (dependent) variable is continuous. This simply means that a predicted or obtained criterion score can logically fall anywhere along a continuum of possible scores. In the hypothetical example presented at the beginning of the chapter involving the prediction of college success, the criterion variable was continuous because a person can actually obtain a grade point average anywhere between 0.0 and 4.0. Height, yearly income, running speed—these are further examples of continuous variables because an individual's score can fall anywhere on the continuum. The statistical technique of multiple correlation is appropriate for the situation in which the criterion variable is continuous.

A criterion variable may be nominal rather than continuous, that is, it may involve group membership rather than a score along a continuum. For example, a college admission director might be interested in predicting whether or not an applicant is likely to be among that group of students who graduate after four years of college. In this case, the criterion variable is dichotomous, for a person must fall into one of two possible groups at the end of four years (those who graduate or those who don't). A nominal criterion variable can, of course, have more than two categories. For example, if we wanted to predict what a student will select as his major, there will be as many criterion categories as there are possible academic areas of specialization.

When a researcher conducts a prediction study with a nominal criterion variable, the statistical technique known as discriminant function analysis must be used instead of the multiple correlation technique used with continuous criterion variables. There are two main types of discriminant function analysis. One is appropriate for nominal criterion variables with two categories and the other for variables with three or more categories.

Two-Group Discriminant Function Analysis

When a researcher is interested in predicting to a dichotomous criterion variable, a *two-group* (or *simple*) *discriminant function analysis* is conducted. To describe the procedure of this statistical technique, let us return to the hypothetical example of the admissions director who has been charged with the responsibility of admitting students who will succeed. In this instance, however, let us assume that success is defined as graduating four years after being admitted. Thus, every student who is admitted can objectively be classified as successful or not successful.

Predictor variables are used in a two-group discriminant function analysis, as was the case in the multiple correlation. In fact, the same predictor variables can be used in both types of prediction studies. Let us assume, therefore, that the three predictor variables that were used to predict first-quarter college GPA are now used to predict whether or not a student will complete his undergraduate program in four years. As you recall, these predictor variables were high school GPA, score on an entrance exam, and rate of absence in high school.

Taking as his subjects the group of students who were admitted four years ago, our researcher could classify each subject in one of the two possible criterion groups, for the student either graduated or didn't graduate. For each person in these two groups, the researcher would also have information concerning the three predictor variables. After putting all of this information into the computer, the researcher would receive a computer printout containing three important pieces of information: (1) a discriminant function prediction equation, (2) a cut-off score, and (3) an *F* ratio which may or may not be significant.

The Discriminant Function Prediction Equation. The *discriminant function prediction equation* has about the same form as a multiple regression prediction equation. A symbol that stands for the predicted score is on the left side of the equation, and a symbol for each predictor variable (X_1, X_2, etc.) and a numerical coefficient for each X are on the right. When our hypothetical admissions director looks at his discriminant function prediction equation, it might look something like this:

$$Z = .75X_1 + .23X_2 + .53X_3$$

where X_1, X_2, and X_3 stand for high school GPA, score on the entrance exam, and rate of absence in high school. Although the coefficients for the predictor variables resemble the coefficients that are used in multiple correlation, the computer chooses these coefficients with a different purpose in mind. In multiple regression the regression weights are selected so as to minimize the difference between a person's predicted and actual criterion score. In discriminant function analysis the coefficients are selected so as to maximize correct classification when the researcher is predicting to a nominal criterion variable.

In Excerpt 8.11 we see a discriminant function prediction equation that actually appeared in a journal article. The study was conducted to determine if certain tests could be used to predict whether or not Pakistani graduate students would succeed in a postgraduate program in business administration. The criterion variable was simply whether a student succeeded or failed in the program, and the predictor variables were four subscales of an entrance examination. Note that the authors of Excerpt 8.11 chose to call the criterion variable D rather than Z. For some strange reason, in discriminant function analysis different authors use different symbols for the criterion variable, whereas in multiple regression analysis, almost all authors use Y to stand for the criterion variable.

EXCERPT 8.11 DISCRIMINANT FUNCTION PREDICTION EQUATION

The discriminant function is

$$D = .0683X_1 + .0097X_2 + .0359X_3 - .0444X_4$$

SOURCE: G. Crimsley and G. W. Summers. Selection techniques for Pakistani postgraduate students of business. *Educational and Psychological Measurement,* 1965, *25,* 1133–1142.

Let us return now to the hypothetical study involving college admissions. When an individual's scores on the three predictor variables are substituted for X_1, X_2, and X_3, Z will turn out to be a number. Instead of interpreting this number as a predicted score on a continuous criterion variable (as is done in multiple regression), the numerical value of Z is used for the purpose of predicting whether the individual in question will eventually become a member of the first criterion group or a member of the second criterion group. In our hypothetical example, the two criterion groups could be termed graduates and nongraduates.

The Critical or Cut-Off Score. The Z score obtained from the discriminant function prediction equation is compared against a *critical* or *cut-off score* which, like the equation, is contained in the computer printout. If the Z score for a new applicant to our hypothetical college is above the cut-off score, he would be predicted to be a member of the group which would graduate four years after being admitted. If the applicant's Z score were below the cut-off score, he would be predicted to be a member of the nongraduate group.

Usually the actual cut-off score is not included in the published journal article, but some authors do provide this information. For example, consider Excerpt 8.12. In this study the criterion variable (v) was passing or failing the introductory math course at Ottawa University. There was one predictor variable, score on the Cooperative Mathematics Test: Algebra III.

EXCERPT 8.12 CUT-OFF SCORE IN A DISCRIMINANT FUNCTION ANALYSIS

A discriminant analysis (Wert, Neidt, and Ahmann, 1954) was utilized to determine whether students who completed the M101 course successfully (grade of C or better) could be differentiated from those students who completed the M101 course unsuccessfully (grade of D or F).

The analysis resulted in the equation: $v = .001239x$. A critical value of $v = .011979$ was obtained; thus the critical Coop Algebra raw score was approximately 9.

SOURCE: C. B. Tatham and E. J. Tatham. A note on the predictive validity of the Cooperative Algebra III. *Educational and Psychological Measurement*, 1971, *31*, 517–518.

An F Test of the Discriminant Function. The third important piece of information provided by the computer analysis is the result of an F test. This test of significance is used to determine whether the discriminant function prediction equation is able to facilitate more accurate prediction than would be possible by chance alone. With a dichotomous criterion variable, we could predict group membership by flipping a coin, and we would be right approximately 50 percent of the time. When the discriminant function prediction equation is tested for significance, the result of the F test indicates whether there is a significant increase in accurate prediction above the 50 percent level that would be expected by chance.

Whereas authors frequently omit the actual discriminant function equation and the cut-off score from their journal articles, they usually report the results of the F test that assesses the degree to which the discriminant function technique provides accurate predictions. In Excerpt 8.13 we see how one author reported the results of this test.

EXCERPT 8.13 F TEST OF THE DISCRIMINANT FUNCTION

A STUDY IS PLANNED

This study investigated the possibility of predicting (with a meaningful degree of accuracy) which students entering agriculture would drop out during or immediately following their freshman year. Only measures (entrance scores and high-school rank) routinely available to college admission offices were used.

PROCEDURE AND RESULTS

Only 161 freshman agricultural students entering Kansas State University in September, 1961, had taken the American College Tests (ACT), covering English, Mathematics, Social Studies, and Natural Sciences (1). Approximately seventy-two percent of the 161 also had high-school ranks (in percentile form) on file at the university. After September 1962, dropouts from this group were determined. Dropouts were referred to as Group 1. Group 2 consisted of those students who enrolled again in September, 1962.

A two-group discriminate analysis (2), involving the four different test variables and the high-school rank measure, yielded an index of discrimination ($R = .472$). This index, tested for significance, yielded a highly significant F ratio ($F = 6.28$; $df = 5$ and 110; $p < .001$).

SOURCE: L. A. Stone. A discriminant analysis and prediction of dropouts for freshman year with agricultural students. *Journal of Educational Research*, 1965, 59, 36–38.

An Unusual Study Involving a Two-Group Discrimination Function Analysis. The two-group discriminant function analysis is probably used most frequently to predict whether or not individuals will succeed in academic or occupational situations. However, this statistical technique can be used in conjunction with any dichotomous criterion variable that is of interest to the researcher. Let us pause for a moment and consider one such study in which the criterion variable was somewhat unusual.

In a study recently conducted in the area of physical education, the criterion variable in the two-group discriminant function analysis was buoyancy. In this study, the investigators wanted to see if they could successfully predict whether or not a person would float in water. Their subjects were college women, and each woman was classified as a floater or a non-floater on the basis of what happened in the swimming pool when the subject was asked to stop moving her arms and legs. (In previous research, the subjects who didn't float were called sinkers.)

In an attempt to predict whether or not a subject would float or sink, the researchers measured all of the subjects on 17 anthropometric variables, such as standing height, sitting height, weight, hip width, and back fat. These 17 variables served as the predictor variables within the discriminant function analysis. The results of the investigation are presented in Excerpt 8.14.

EXCERPT 8.14 PREDICTING BUOYANCY WITH A
TWO-GROUP DISCRIMINANT FUNCTION ANALYSIS

A discriminant analysis was employed predicting floating or non-floating as a dichotomous variable from 17 different, continuous variables. . . . The following discriminant equation resulted:

$$V = 0.032X_1 - 0.441X_2 + 0.0035X_3 - 0.0282X_4 - 0.0179X_5 + 0.0053X_6$$
$$+ 0.0556X_7 - 0.0761X_8 - 0.0063X_9 - 0.0147X_{10} + 0.0301X_{11}$$
$$- 0.0070X_{12} - 0.0053X_{13} + 0.0147X_{14} - 0.0166X_{15}$$
$$+ 0.0006X_{16} + 0.0014X_{17}$$

The F-test was employed to determine the level of significance of the computed discriminant equation. The equation was not significant at the .05 level.

SOURCE: J. C. Mitchem and E. C. Lane. Buoyancy of college women as predicted by certain anthropometric measures. *American Association for Health, Physical Education, and Recreation Research Quarterly*, 1968, 39, 1032–1036.

Multiple Discriminant Function Analysis

If a researcher wants to determine whether a set of predictor variables can be used successfully to predict group membership and if the nominal-level criterion variable has three or more constituent categories, then the technique called *multiple discriminant function analysis* must be used. It should be noted that the term multiple is used in this instance to denote that the criterion variable has more than two categories, and it does not relate to the number of predictor variables involved. In both two-group and multiple discriminant function analysis, there can be two, three, or more predictor variables. (In multiple regression, however, the term multiple does refer to the inclusion of several variables as predictors.)

Although the statistical technique of multiple discriminant function is only slightly different from two-group discriminant function in terms of the original data that the researcher starts out with, there are three main differences between these two techniques with regard to the results provided by the statistical analyses. In multiple discriminant function analysis (1) a test called Wilks' lambda is likely to be applied as the first step in the data analysis process, and the results of this test will usually be presented at the beginning of the results section of the journal article, (2) there is a possibility that two or more discriminant function prediction equations will be associated with the same study, and (3) the author may use a picture to summarize the degree to which the discriminant equations facilitate successful predictions. Let us now look at each of these three parts of a multiple discriminant function study.

Wilks' Lambda Test of Significance. For each of the criterion groups, a mean score on each predictor variable could be computed by averaging the scores of the individuals in that particular criterion group. For example, if there were four predictor variables in the study and if there were three categories of the criterion variable, then four means could be computed on the basis of Ss in the first criterion group—their mean on the first predictor variable, their mean on the second predictor variable, etc. In a similar manner, four means could be computed on the basis of Ss in the second criterion group. If the predictor variables were GRE-verbal, GRE-quantitative, undergraduate grade point average, and height, and if the three criterion groups were defined so as to include, respectively, students who do not finish a graduate program, students who stop after obtaining a Master's degree, and students who complete a doctoral program, the various means could be arranged in a diagram as in Figure 8.1. (Of course, the means presented in this diagram are fictitious.)

| | | CRITERION GROUPS | | |
		NO DEGREE	MASTER'S DEGREE	PH.D. DEGREE
	GRE-V	480	520	650
PREDICTOR	GRE-Q	440	560	580
VARIABLES	U-GPA	2.84	3.07	3.35
	HEIGHT	69 in.	70 in.	68 in.

FIGURE 8.1. *Diagram of group means on four predictor variables.*

In Figure 8.1 there is a column of four means associated with each of the three criterion groups. Technically speaking, the four means associated with any one of the three criterion groups constitute that group's *mean vector.* Thus, there are three mean vectors in our hypothetical table of means. By definition, therefore, there will always be as many mean vectors as there are categories of the criterion variable, and within each mean vector there will be as many elements (i.e., means) as there are predictor variables. As should be obvious, the number of means in the first mean vector will be identical to the number of means in each of the other mean vectors.

The first step in a multiple discriminant function analysis involves determining whether the various predictor variables can, as a set, differentiate between the criterion groups. This is done by testing to see whether the mean vectors of the various criterion groups are different from one another. The null hypothesis is that the common elements in the various mean vectors are identical to one another. In other words, when a table of means is constructed in the same manner as we did for the hypothetical study involving three criterion groups and four predictor variables, the null hypothesis being tested is that the means that lie in any particular row of the table are identical. This does not mean, however, that the means in the first row (i.e., for the first predictor variable) have to be the same as the means in any other row (i.e., for a different predictor). Thus, the null hypothesis can be true without having identical (or even similar) means within the same mean vector.

The researcher, of course, hopes that the null hypothesis of identical mean vectors will be rejected, because if there is no difference between the criterion groups in terms of their mean scores on each of the predictor variables, then it should be clear that it would be a waste of time to develop prediction equations based upon these predictor variables. The prediction equation(s) would not facilitate accurate predictions of group membership, and the researcher could use his time more wisely by starting his study over again, this time using a new set of predictor variables.

The hypothesis of equal mean vectors in usually evaluated by means of the *Wilks' lambda* test. The application of this test will yield a calculated value, and most researchers convert the Wilks' lambda calculated value into a comparable F value. As in the analysis of variance and covariance, the calculated F value is then compared against a critical F value for the purpose of deciding whether the null hypothesis should be rejected or not rejected. In Excerpt 8.15 we see a section of an article in which the researchers were trying to determine if the College Interest Inventory (CII) could be used to predict whether a student was enrolled in a chemical, civil, electrical, or mechanical engineering program. There were 15 predictor variables (each being a different scale of the CII) and four criterion groups (each representing a different branch of engineering).

EXCERPT 8.15 APPLICATION OF THE WILKS' LAMBDA TEST

Before discriminant functions can be generated, it must be ascertained if, indeed, the four engineering curricular groups do differ significantly on the 15 curricular interest scales of the CII. This is a test of the equality of group centroids and is conventionally measured by the Wilks' lambda statistic (Wilks, 1932).

Wilks' λ was calculated to be .03168. This is equivalent to an F ratio of 24.69 with 45 and 506 degrees of freedom. The probability of obtaining an F this large by chance is less than .01. The 15 CII scales do discriminate among the engineering branches.

SOURCE: R. Neal and P. King. Comparison of a multivariate and a configural analysis for classifying engineering students. *Journal of Counseling Psychology*, 1969, *16*, 563–568. Copyright 1969 by the American Psychological Association. Reprinted by permission.

More Than One Discriminant Function Equation. Assuming that the Wilks' lambda test has indicated that the mean vectors associated with the criterion groups are not identical, the researcher can turn his attention to the discriminant function prediction equations that will appear on the computer printout. As we mentioned earlier, a multiple discriminant function analysis will result in two or more prediction equations. The maximum number of possible prediction equations is always equal to one less than the number of criterion groups or to the number of predictor variables, whichever of these is smaller.[3] If there are three criterion groups in the study and four predictor variables, two prediction equations will be provided by the computer analysis.

Within the context of the same study the discriminant equations will involve the same predictor variables. However, the numerical coefficients associated with the various predictor variables will not be the same in each of the discriminant equations. Thus, with three criterion groups and four predictors there will be two discriminant equations, and the first coefficient (corresponding to the first predictor variable) may be large in one equation and small in the other. Obviously, if the coefficients in the two equations were equal to one another for each predictor variable, it would be redundant to have two equations.

Normalized and Scaled Vectors. An author of a journal article could present the first discriminant function prediction equation in a form similar to this: $Z_1 = .71X_1 + 1.84X_2 + .13X_3$. Each successive discriminant equation would be similar except for the subscript attached to Z (which would be 2 for the second equation, 3 for the third equation, etc.) and the size of the coefficients associated with the various Xs, the predictor variables. Instead of doing this, however, most authors simply take the numerical coefficients out of each equation and put them into a table. The coefficients that are taken from each discriminant equation constitute what is technically called a *normalized vector*.[4] Therefore, if there are four discriminant equations in a study involving ten predictor variables, we can expect to see a table containing four normalized vectors, with each of the vectors containing ten numerical coefficients. Usually, the normalized vectors will be arranged vertically in a column, with the rows of the table labeled with the names of the predictor variables.

Most researchers will also present a *scaled vector* corresponding to each of the normalized vectors. The values within a particular scaled vector can be compared with one another so as to determine which of the predictor variables are most effective as predictors within the context of the corresponding discriminant equation. The values in the normalized vector cannot be compared in this manner. Thus, the values in each normalized vector are highly analogous to the regression coefficients in a multiple regression prediction equation, while the values in the corresponding scaled vector can be thought of as similar to beta weights.

---

[3] This rule of thumb should make it clear why there is only one prediction equation associated with a two-group discriminant function equation.

[4] In reality, the numerical coefficients in the normalized vector do not come directly from the discriminant equation. In the process of going from the discriminant equation to the normalized vector, the coefficients are adjusted, so that the sum of the squared coefficients is equal to 1.00. This is the technical meaning of the word *normalized* when used in this context.

In some articles, only the scaled vectors are presented. In Excerpt 8.16 both the normalized and the scaled vectors related to two discriminant function equations are presented.

EXCERPT 8.16 TABLE OF NORMALIZED AND SCALED VECTORS FOR TWO DISCRIMINANT EQUATIONS

Table 1 contains the coefficients of the 15 variables in the discriminant equation. The normalized vectors are the coefficients of the variables in the discriminant equations. The scaled vectors show the relative contribution of each variable to group discrimination. They indicate the relative importance of each of the scales in construction of the discriminant function.

TABLE 1 NORMALIZED AND SEALED VECTORS FOR DISCRIMINANT FUNCTIONS I AND II ON 15 VARIABLES (COLLEGE INTEREST INVENTORY SCALE SCORES)

| | Vectors | | | |
| | NORMALIZED | | SCALED | |
INTEREST SCALES	I	II	I	II
Agriculture	.0178	.0310	1.51	2.62
Home Economics	.1402	.0445	5.47	1.74
Literature and Journalism	.0758	.0460	5.16	3.13
Fine Arts	−.1151	−.0514	−10.87	−4.85
Social Science	−.1424	−.1276	−8.55	−7.65
Physical Science (Chemical Engineering)	.1781	−.1342	21.57	−16.25
Biological Science	−.0043	−.0283	−0.36	−2.36
Foreign Language	.2149	.1512	13.14	9.25
Business Administration	.0788	−.1376	7.45	−13.01
Accounting	.1307	.1874	7.18	10.30
Teaching	.1111	−.2038	7.69	−14.12
Civil Engineering	.8797	.1670	−78.63	14.92
Electrical Engineering	.1657	.6301	17.72	67.38
Mechanical Engineering	.1584	−.6433	16.20	−65.80
Law	.0283	−.0115	2.66	−1.08

SOURCE: R. Neal and P. King. Comparison of a multivariate and a configural analysis for classifying engineering students. *Journal of Counseling Psychology*, 1969, *16*, 563–568. Copyright 1969 by the American Psychological Association. Reprinted by permission.

Using the Various Discriminant Function Equations. To predict group membership in a multiple discriminant function study, we would substitute an individual's scores on the various predictor variables for the Xs in each of the discriminant equations, thereby leading to a predicted discriminant score (Z). The Z score from each equation would not be used by itself to predict group membership, but rather the whole set of Z scores would be used in combination to serve as the basis for one prediction. To illustrate how this is done, let us look at an imaginary study.

Suppose a researcher wanted to predict whether individuals are overweight, underweight, or about right in weight. These would be the three categories of the criterion variable. As predictor variables, suppose our researcher is allowed to measure people only with respect to (1) arm length, (2) waist circumference, (3) shoe size, and (4) heart rate. Since there are three criterion categories, there would be two discriminant equations provided by the computer. Each equation, however, would have four Xs, one for each of the predictor variables. The coefficients for the Xs would, of course, be different in the two equations. Suppose the equations turned out as follows.

$$Z_1 = 2.5X_1 + .27X_2 + 3.0X_3 + .03X_4$$
$$Z_2 = .16X_1 + 2.9X_2 + .11X_3 + 4.1X_4$$

In the first equation the predictor variables of arm length and shoe size have high coefficients.[5] Therefore, our hypothetical researcher might conclude that this first equation is mainly concerned with height. This seems logical since a person with long arms and big feet is likely to be a tall person. In the second equation the predictor variables of waist size and heart rate have large coefficients. Thus, the researcher might conclude that the second equation is mainly concerned with weight, since a person with a large waist and a fast heart rate is more likely to be a heavy individual.

To predict whether or not an individual is overweight, underweight, or normal, our researcher would need to use the joint information provided by both Z scores. Knowing something about a person's height is not sufficient information to permit an accurate prediction of group membership, since there would be tall and short people in each of the three criterion groups. Similarly, information about a person's weight is not enough, since there could be light and heavy people in each criterion category. Together, however, information provided by Z_1 and Z_2 should permit accurate classifications. If a person's score on Z_1 is much larger than his score on Z_2, a prediction would be made that this person is tall and light, and he would be classified as being underweight. If the score on Z_2 were much larger than the score on Z_1, the individual would be predicted to be in the overweight group. If the two Z scores were relatively equivalent, then a prediction would be made that the individual is neither overweight nor underweight.

Disregarding Discriminant Equations that Don't Help. Although a multiple discriminant function analysis will always yield at least two prediction equations, the various equations will not contribute equally to successful prediction of group membership. In other words, one of the prediction equations will be the most effective, a second will help but it won't be as good as the first equation, the third equation will be less helpful than the second, etc. In the jargon of statistics each of the discriminant equations will explain a certain percentage of the between-group variability. After the first

[5] Since our hypothetical researcher is comparing the coefficients within each discriminant function equation in an attempt to determine which predictor variables are most important, we must assume that the coefficients have come from the appropriate scaled vectors.

equation explains all that it can, the second equation tries to explain, or account for, the remaining between-group variability. The third equation attempts to explain that portion of the between-group variability that is not accounted for by the first two equations. In general, each successive discriminant equation tries to explain the between-group variability that has been left unexplained by the preceding equations.[6]

It should be obvious that the last prediction equation (or the last few equations) will not be able to explain very much of the between-group variance as compared with what has already been explained. How could it (they), if the first equations explain a large proportion (say 80 percent) of the variance? After determining how much variance is explained by each discriminant function equation, the researcher may decide that the last few equations do not help out enough to warrant their use. For example, consider Excerpt 8.17, in which the researcher decided to use only two equations since they accounted for almost 87 percent of the variance.

EXCERPT 8.17 ELIMINATION OF DISCRIMINANT
FUNCTION EQUATIONS THAT DO NOT HELP

Seven discriminant equations were obtained for the vocational groups; however, the first two were found to account for 86.7 per cent of the total discriminative power of the thirteen variables studied. The contribution of the remaining five equations toward group classification totaled less than 14 per cent of the discriminating power of the variables. It was decided to use only the first two functions for the validation study due to the minor discriminating contribution to be expected from functions III through VII. It may be noted that none of the remaining discriminant functions explained as much as 5 per cent of the among-groups variance.

TABLE 1 DISCRIMINATING POWER OF THE DISCRIMINANT FUNCTIONS

DISCRIMINANT FUNCTION		PER CENT OF TOTAL
I	1.509	69.9
II	0.363	16.8
III	0.104	4.8
IV	0.096	4.4
V	0.055	2.5
VI	0.021	1.0
VII	0.012	0.6

SOURCE: J. J. Doerr and J. L. Ferguson. The selection of vocational-technical students. *Vocational Guidance Quarterly*, 1968, *17*, 27–32. Copyright 1968 by the American Personnel and Guidance Association.

Although some researchers decide arbitrarily to discard one or more of the prediction equations when they don't appear to be contributing to successful prediction, other researchers apply a statistical test to each discriminant function equation prior to making the decision to keep or discard any particular equation. In the study connected with Excerpt 8.18 there were

[6] In this context the term *trace* means the same thing as between-group variability. Hence, some authors will state what percent of the trace has been explained by each discriminant equation.

three categories of the criterion group variable and thus a total of two discriminant equations. The results of the statistical test indicated that only the first equation was making a significant contribution to successful prediction.

EXCERPT 8.18 TESTING DISCRIMINANT EQUATIONS FOR SIGNIFICANCE

The discriminating power of the predictor test battery was determined by computation of Wilks' lambda. . . . The effectiveness of this discrimination was significant ($F = 2.25$, $df = 60/225$, $p < .001$). Chi square tests (Bartlett, 1941) were computed for each of the two derived discrimination functions . . . to determine the significance of discrimination along each dimension (Jones, 1964). The first discriminant function was found to be significant ($x^2 = 120$, $p < .001$), but the significance of the second vector failed to reach the necessary level ($p = .11$). The first vector accounted for 72 percent of the predictable group variation.

SOURCE: R. B. Vacchiano and R. J. Adrian. Multiple discriminant prediction of college career choice. *Educational and Psychological Measurement*, 1966, *26*, 985–995.

Graphing the Group Centroids. If there are only two discriminant function equations in a study, a researcher may present a graph of the *group centroids* in an attempt to show that the discriminant equations do work in differentiating between the various criterion groups. Furthermore, the graph of the centroids will help to explain how the two discriminant equations work in a joint way to facilitate accurate predictions.

The centroid for each group is found by means of a simple two-step process. First, the means for a particular criterion group on each predictor variable are substituted for the appropriate Xs in the first discriminant equation. After multiplying the group means by the coefficients, the result would be a mean predicted discriminant score which we could label $\bar{Z}_1$. In a similar manner, the same means for this group are substituted for the Xs in the second discriminant equation, and the result is $\bar{Z}_2$.

Second, the mean predicted discriminant scores from each equation for the first criterion group ($\bar{Z}_1$ and $\bar{Z}_2$) are used to define a point in a graph. The first mean ($\bar{Z}_1$) is used to indicate how far out the point is on the horizontal axis, and the second mean ($\bar{Z}_2$) is used to indicate how high the point is above the horizontal axis. Thus, there is only one point in the graph that corresponds to the two mean discriminant scores of the first criterion group. The term centroid refers to this single point.

The process described above is followed for each of the various criterion groups, and therefore there are always as many centroids as there are criterion groups (with each group centroid being defined by the pair of means $\bar{Z}_1$ and $\bar{Z}_2$). If the discriminant function equations are successful in facilitating accurate predictions of group membership, then the group centroids will be spread out in the graph. Conversely, if the discriminant equations don't work, the group centroids will be relatively close together.

In Excerpt 8.19 we see a graph of six group centroids. Table 4 contains the means that were used to plot the centroid points. It should be noted that since there were six criterion groups in this study, a total of five discriminant function equations were originally computed. However, since the first two equations accounted for 96 percent of the group differences, the remaining three equations were disregarded. As the authors explain in the text of the article and as the graph of the centroids shows, the first discriminant function equation discriminated between males and females, while the second equation discriminated between the three different types of college groups. This is a beautiful example to demonstrate how joint information between two (or more) discriminant function equations is sometimes necessary to adequately differentiate between the various criterion groups.

EXCERPT 8.19 AN EXAMPLE SHOWING
HOW CENTROIDS CAN BE GRAPHED

The surprising finding here is that the first and largest discriminant function separates the sexes, not the three college criterion groups. This can be better visualized in Figure 1. The horizontal axis is Discriminant Function 1, and the males are on the left and the females are on the right. The vertical axis separates the three college groups with the corresponding male and female groups being at approximately the same level on Discriminant Function 2. That is, Figure 1 is simply a plot of information of Table 4 showing locations of the different groups.

TABLE 4 CENTROIDS OF GROUPS IN DISCRIMINANT SPACE

	1	2
Non-College Males	11.89	29.10
Jr. College Males	11.78	33.43
College Males	11.27	39.87
Non-College Females	18.29	31.05
Jr. College Females	18.26	34.30
College Females	16.84	39.96

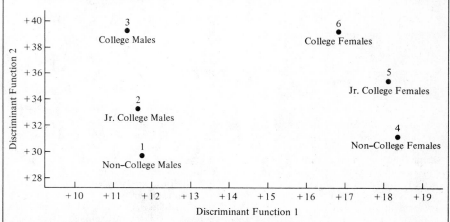

FIGURE 1. *Centroids of groups in ability discriminant space.*

SOURCE: W. W. Cooley and S. J. Becker. The junior college student. *Personnel and Guidance Journal*, 1966, *44*, 464–469. Copyright 1966 by the American Personnel and Guidance Association.

As you can readily understand, it is hard to graph a three- or four-dimensional figure on two-dimensional paper. Therefore, if there are three or more useful discriminant function equations in a particular study, we should not expect to see a graph of the group centroids. In this situation the author will probably present a table of the group means (i.e., the $\bar{Z}$s) that correspond to each discriminant equation. Within such a table, the various means for a particular criterion group (with each mean coming from a different discriminant equation) define that group's centroid.

Cross-Validation in Discriminant Function Studies

Earlier in this chapter, we explained that the statistical procedure called cross-validation is used to determine whether a prediction equation will work when it is used on a new group of individuals, that is, individuals who were not used to obtain the coefficients of the equation. Although our discussion at that time was limited to the topic of multiple correlation, the technique of cross-validation can also be used in discriminant function analysis. Not only is the purpose still the same, but so is the simple four-step procedure that we discussed. In short, discriminant function equations are developed on the basis of one group of people and then applied to a second group of people. If there is a high degree of correspondence between the actual group membership of the individuals in the second group and the predictions made for these people on the basis of the discriminant equations, then the cross-validation has been successful.

In Excerpt 8.20 we see a study in which a separate discriminant analysis was conducted in relation to three male groups and two female groups. For both the male part of the study and the female part, the results were cross-validated. In the excerpt presented here the authors use the term criterion group to refer to the initial group of Ss upon whom the prediction equation was developed.

EXCERPT 8.20 CROSS-VALIDATION IN
DISCRIMINANT FUNCTION ANALYSIS

Three male groups representing three academic areas of study, business, chemistry, and mathematics, and two female groups, education and nursing, served as criterion groups. These groups were comprised of a total of 245 students, 50 in each group, with the exception of chemistry which contained 45 students. Two additional male groups, 32 business and 36 mathematics students, and two additional female groups, 32 education and 38 nursing students, were utilized as cross-validation samples to test the classification efficiency of the discriminant functions derived from the criterion groups. Although the groups were pre-selected according to their area of study, students were assigned randomly to the criterion and cross-validation samples.

SOURCE: R. B. Vacchiano and R. J. Adrian. Multiple discriminant prediction of college career choice. *Educational and Psychological Measurement*, 1966, *26*, 985–995.

REVIEW TERMS

beta weights
centroid
coefficient of determination
coefficient of multiple correlation
criterion variable
critical (cut-off) score
cross-validation
dependent variable
independent variable
mean vector

multiple discriminant function
    analysis
normalized vector
prediction equation
predictor variable
regression coefficients
scaled vector
shrinkage
two-group discriminant function
    analysis
Wilks' lambda

REVIEW QUESTIONS

1. If the correlation between a predictor variable and a criterion variable is equal to .80, then the coefficient of determination will be equal to _____.

2. What is the minimum number of predictor variables that can be included in a multiple correlation study?

3. What symbol is used to designate the coefficient of multiple correlation? What is the range of numerical values this coefficient can assume?

4. Can beta weights be compared to determine which of the predictor variables are contributing the most to successful prediction? Can regression coefficients be used in this way?

5. There are three popular tests of significance associated with multiple regression. One of these answers the question: "Does the addition of one or more predictor variables result in significant increase in the coefficient of multiple correlation?" What questions are answered by the other two tests?

6. If a researcher subjects his obtained R to cross-validation, will the new R be larger or smaller than the original R? What is the technical term for this predicted increase or decrease in R when it is cross-validated?

7. In multiple correlation the criterion variable is continuous. In a two-group discriminant function analysis, the criterion variable is _____

8. In a multiple regression prediction equation there is a different X on the right side of the equation for each separate predictor variable, and each X is weighted by a numerical coefficient. Does this statement hold true for a two-group discriminant function analysis as well?

9. Suppose a researcher is using a two-group discriminant function analysis in an attempt to determine whether a set of predictor variables can be used to predict whether a medical student will go into (a) general

practice or (b) a specialized field. Data is provided by 50 Ss from each of the two criterion groups. Would the study be successful if 100 percent of the Ss were to be above the critical cut-off score?

10. In a multiple discriminant function analysis involving four criterion groups and ten predictor variables, how many mean vectors are there? How many elements are there within each mean vector?

11. Does the null hypothesis underlying Wilks' lambda test (as used in a multiple discriminant function analysis) require that all of the elements of a mean vector be identical to one another?

12. What is the maximum number of discriminant function equations that can be associated with a study involving five criterion groups and ten dependent variables?

13. Are the values in a normalized vector of a multiple discriminant function analysis analogous to the regression coefficients or the beta weights of a multiple correlation study?

14. In a multiple discriminant function analysis, will all of the discriminant equations contribute equally to successful prediction?

15. Suppose a researcher uses a multiple discriminant function analysis. If he graphs the various centroids, there will be as many points in the graph as there are (a) predictor variables, (b) discriminant function equations, or (c) criterion groups.

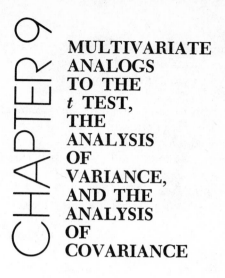

CHAPTER 9

MULTIVARIATE ANALOGS TO THE *t* TEST, THE ANALYSIS OF VARIANCE, AND THE ANALYSIS OF COVARIANCE

In Chapters 4 through 7 we discussed three main topics: (1) independent and correlated *t* tests, (2) one-way, factorial, and repeated measures analyses of variance, and (3) the analysis of covariance. Although these statistical procedures were developed to accommodate two or more independent variables and/or one or more covariate variables, they were originally designed for the situation where there is one and only one dependent (criterion) variable. Procedures involving only one dependent variable represent a *univariate analysis*.

There are several situations in which a researcher may have a simultaneous interest in two or more dependent variables. For example, it might be interesting to compare three different teaching methods with respect to (1) achievement on a final exam and (2) attitude toward the course material at the end of the academic term. Or, a researcher might like to use a 2 × 2 factorial design to investigate the effects of two different ways of arranging the items on a test (easy-to-hard vs. hard-to-easy) and two different time limits (long vs. short) on three dependent variables: (1) number of questions answered correctly, (2) test anxiety after taking the test, and (3) number of answers that are changed from right to wrong. Although there might be a temptation to use a separate univariate analysis for each dependent variable in these (or similar) studies, such an approach to data analysis would be incorrect.

It is inappropriate to use a separate univariate analysis for each of several dependent variables in one study for two reasons. First, correlations between the dependent variables are usually something other than zero, that is, people who score high on one dependent variable generally score high on the other dependent variables. Under this condition of correlated dependent variables, application of univariate tests—one for each dependent variable—will cause the probability of a Type I error to be higher than the level of significance that is used. This means that the set of univariate tests is *positively biased* in the sense that the null hypothesis will be rejected too often.

The second reason for avoiding a series of univariate tests is related to the fact that as the number of dependent variables increases, the probability of finding a significant difference by chance alone also increases, even if all correlations among the dependent variables are equal to zero. The rationale here is the same as that encountered in Chapter 4 when we tried to explain why multiple *t* tests are inappropriate as a post hoc multiple comparison procedure following a significant *F* in a one-way ANOVA (see p. 68).

Instead of using a separate univariate analysis for each dependent variable, a researcher should use a *multivariate analysis*. For each of the statistical tests discussed in Chapters 4–7, there is a multivariate analog. In each case, the multivariate analog answers the same number of research questions as the univariate analysis. Hence, a two-way multivariate analysis of variance will yield three *F* ratios, one for each main effect and one for the interaction. In a multivariate analysis, however, the initial results pertain to the entire set of dependent variables and have been adjusted for correlations among the dependent variables and the fact that having more than one dependent variable causes univariate tests to be positively biased. If a multivariate analysis indicates a significant difference between treatment groups, the researcher can be confident that the treatment groups differ with respect to at least one dependent variable and that the decision to reject the overall null hypothesis is made with the probability of a Type I error equal to the level of significance.

MULTIVARIATE *t* TESTS

Suppose a researcher decides to compare two different diet plans for losing weight. Further suppose that our hypothetical researcher has 50 subjects available for his study, each subject being an overweight individual who has volunteered to go on a diet. The 50 Ss are randomly assigned to two treatment groups. The 25 Ss in Group 1 are put on a diet plan which allows them to eat as much as they want at each meal, but the number of meals is limited to two a day for the first week and then one a day during the second week. The 25 Ss in Group 2 are put on a diet plan which allows them to have three meals and two snacks during each day, but the amount of food available at each eating period is limited so as to gradually decrease caloric intake over a two-week period.

The most obvious dependent variable in this hypothetical investigation is the amount of weight lost during the two-week duration of the study. However, suppose the researcher is also interested in comparing the two diets in terms of (1) the degree to which the Ss like their diet plans, (2) the moodiness among the dieters as evaluated by fellow workers, and (3) the ability to perform well on a series of clerical tasks involving simple arithmetic, checking for spelling errors, etc. Thus, there are really four dependent variables of interest to the researcher.

T^2 and F

Although there are several alternative (and statistically equivalent) multivariate tests that could be used to analyze the data from the hypothetical diet study, the most popular procedure is called *Hotelling's* T^2. This statistical test would compare the two groups of dieters and ask whether there is a difference between the groups on one or more of the four dependent variables, after adjustments have been made for possible correlations among the dependent variables and for the fact that more than just one dependent variable is involved. The null hypothesis being tested is that the two groups have the same population means for weight loss, that the two groups *also* have the same population means for the second dependent variable, etc. Stated in technical terms, the null hypothesis assumes that the two groups have identical population *mean vectors*. (If you have forgotten what a mean vector is, review the discussion of this term on page 166 of Chapter 8.)

The computational procedure that must be followed in getting a calculated value for T^2 is quite complex, and the assistance of a computer is required. Fortunately, the only thing that a reader of journal articles needs to know about getting a value for T^2 is that it is *not* found by squaring the value of a univariate t value. Beyond this simple point, it is much more important to know what it means to have a significant or nonsignificant T^2. Simply stated, if the calculated value for T^2 exceeds the critical value, say at the .05 level, the null hypothesis of identical mean vectors can be rejected. In the hypothetical study on diets a significant T^2 would indicate that the two groups differ significantly from each other on at least one of the dependent variables. A nonsignificant result would tell the researcher that the two diets are not related to group differences on any of the four dependent variables.

Since the most frequently used table of critical values in univariate analyses is the F table and since most statistics texts do not have a table of critical values for T^2, it is not hard to see why Hotelling went to the trouble to show that the calculated value of T^2 can be transformed into an F value. To transform T^2 to F, the following formula is used.

$$\frac{N - p - 1}{(N-2)p} \times T^2 = F$$

N is the total number of subjects and p is the number of dependent variables. The resulting calculated F value is compared against a critical F value having degrees of freedom equal to p and $(N - p - 1)$. The main point being made here is not the formula for getting from T^2 to F, but rather that the calculated value for a multivariate t test most often appears in the journal article as F, not T^2.

In Excerpt 9.1, we see an example of Hotelling's T^2 test. Notice how the results of the T^2 test were transformed into an F value.

EXCERPT 9.1 USE OF HOTELLING'S T^2

Forty-nine elderly men participating in an interdisciplinary study of human aging were classified into the diagnostic categories "senile factor present" and "no senile factor" on the basis of an intensive psychiatric examination. The Wechsler Adult Intelligence Scale had been administered to all subjects by an independent investigator, and certain subtests showed large differences between the two groups. It was proposed that a test be made of the hypothesis that the groups arose from populations with a common mean vector. . . .

The sample means are shown in Table 4.2. The two-sample T^2 statistic had the value 22.05; the associated F was 5.16, with degrees of freedom 4 and 44. Under the hypothesis of equal mean vectors the probability of exceeding such an F value would be less than 0.005, and we should reject that null hypothesis at the conventional 5 or 1 percent level.

TABLE 4.2 WAIS SUBTEST MEANS

	GROUP	
	NO SENILE FACTOR	SENILE FACTOR
SUBTEST	$N_1 = 37$	$N_2 = 12$
Picture completion	7.97	4.75
Information	12.57	8.75
Similarities	9.57	5.33
Arithmetic	11.49	8.50

SOURCE: D. F. Morrison. *Multivariate Statistical Methods*. New York: McGraw-Hill, 1967, pp. 126–127.

Follow-up Tests

A significant result in a univariate t test indicates that the two comparison groups are different with regard to the single dependent variable being analyzed and a follow-up test is obviously not necessary. A significant result in a multivariate t test of two groups requires further analysis to find out exactly which dependent variable(s) contributed to the overall significant difference. Thus, Hotelling's T^2 is a preliminary step in the multivariate process.

Let us return to the hypothetical diet study for a moment, and suppose that the Hotelling's T^2 procedure turned out to be significant at the .01 level. The researcher must now continue the analysis, comparing the two diet groups with respect to each of the separate dependent variables. Although it might seem as though four univariate t tests could be applied (one for each dependent variable), this procedure is not legitimate for the same reasons that make it inappropriate to apply univariate t tests in lieu of the preliminary multivariate test. Each of the univariate t tests does not take into consideration either correlations among the dependent variables or the fact that multiple t tests are being run, and this is true whether t tests are conducted as the first step in the data analysis procedure or as the second step following a significant T^2. In either case, the univariate t tests would tend to yield too many significant differences.

Two acceptable follow-up procedures are recommended by authorities in the field: *simultaneous confidence intervals* and *two-group linear discriminant function*. The first of these procedures involves setting up an equation, called a simultaneous confidence interval, for each dependent variable. An example of an interval is

$$0.69 \leqq \mu_{11} - \mu_{21} \leqq 4.37$$

The first and last terms of this equation are, of course, fictitious numbers. If our diet study were actually conducted and if the results of Hotelling's T^2 dictated a follow-up investigation, the numerical ends of the interval would be obtained by comparing the two groups on the first dependent variable, weight loss. The numbers used would not be the actual group means, but rather values based upon these means plus data from the other dependent variables as well. (If applied, a univariate t test would be based only upon the data of a single dependent variable. This is where the multivariate procedure makes its adjustments.)

The middle part of each simultaneous confidence interval corresponds to the null hypothesis for the dependent variable in question. The symbol μ_{11} stands for the population mean for the first group on the first dependent variable, and the symbol μ_{21} stands for the population mean for the second group on the first dependent variable. For the third dependent variable of the hypothetical diet study, the middle part of the confidence interval would be $\mu_{13} - \mu_{23}$. With respect to any particular dependent variable, the null hypothesis being tested is that $\mu_1 = \mu_2$ or, stated differently, $\mu_1 - \mu_2 = 0$. If the numbers on the ends of the confidence interval do not have zero between them—that is, if both numbers are positive (as in the case in the fictitious equation in the preceding paragraph) or both negative—then the null hypothesis can be rejected for that particular dependent variable. On the other hand, if the left end of the simultaneous equation is a negative number while the right end is a positive number (or vice versa), then the equation would be telling the researcher that the two population means might be identical.

In this latter case, the null hypothesis under consideration would be accepted.

In Excerpt 9.1 we saw a study in which T^2 was used to compare two groups of elderly men. There were four dependent variables, each corresponding to a different subtest from the WAIS. The null hypothesis of equal population mean vectors was rejected, and in Excerpt 9.2 we see how the researchers used simultaneous confidence intervals as a follow-up analysis.

EXCERPT 9.2 USE OF SIMULTANEOUS CONFIDENCE
INTERVALS AS A FOLLOW-UP TO A SIGNIFICANT T^2

Although we have rejected the null hypothesis, we still do not know which of the four mean differences may have contributed to the significant T^2 or for which it might be reasonable to conclude that their population means are equal. As we have noted, it would not be proper to test the four individual mean differences by univariate statistics, for we must have protection against the effects of positive correlations among the subtests as well as the tendency for individual differences to be significant merely by chance as more responses are included in the variate vectors. Simultaneous confidence intervals will be used to test the individual differences.

The resulting confidence interval for the difference in the arithmetic means of the senile-factor and no-senile-factor populations is

$$-0.79 \leq \mu_{31} - \mu_{32} \leq 6.77$$

Since zero is included in the interval, we conclude at the 5 percent joint significance level that the arithmetic means are not different. The other 95 percent simultaneous confidence intervals indicate different population means, although the picture-completion subtest stands out as the most significantly different:

Information	$0.10 \leq \mu_{11} - \mu_{12} \leq 7.54$	
Similarities	$0.16 \leq \mu_{21} - \mu_{22} \leq 8.32$	
Picture completion	$0.54 \leq \mu_{41} - \mu_{42} \leq 5.90$	

SOURCE: D. F. Morrison. *Multivariate Statistical Methods.* New York: McGraw-Hill, 1967, pp. 127–128.

The second acceptable procedure for making follow-up tests after obtaining a significant T^2 involves a two-group linear discriminant function. In Chapter 8 we discussed this statistical procedure in light of its ability to serve as a tool in predicting group membership. As a follow-up to a significant T^2, however, the role of the discriminant function procedure is completely different. In a sense, the two roles of discriminant function are exactly opposite. In both cases the weights associated with each of several X variables in the discriminant equation are chosen so as to maximize discrimination of the Z variable (group membership). When discriminant function analysis is used for prediction purposes, the X variables represent the various independent variables and the Z variable is the dependent variable. When it is used to probe a significant T^2, the Z variable becomes the independent variable while the X variables stand for the dependent variables.

If the value for T^2 in the hypothetical diet study turned out to be significant and if the researcher decided to use the discriminant function approach as a follow-up procedure, the resulting discriminant function equation[1] might look like this.

$$Z = 1.35X_1 + 0.03X_2 - 1.29X_3 + 0.17X_4$$

Of course, the four *X*s stand for the four dependent variables: weight loss, satisfaction, moodiness, and clerical skill. To determine where the two diet groups differ in terms of these variables, several authoritative texts indicate that the researcher should simply compare the relative size of the weights which precede each *X* (disregarding the plus or minus signs). Thus, the hypothetical discriminant equation given above indicates that the main cause of T^2 being significant is X_1 and X_3. In other words, the two diet groups differ far more in terms of weight loss and moodiness than they do in terms of the other dependent variables.

In most journal articles only the weights and not the actual discriminant equation will be presented. These weights are usually called *standard discriminant function coefficients*, and they should be thought of as analogous to the scaled vector that we discussed in the last chapter.

Hotelling's T^2 for Correlated Samples

In the hypothetical study involving two diet plans, the two groups of Ss were considered to be independent samples, for there was no a priori connection between a particular score in Group 1 and any of the 25 scores in Group 2. However, what if our hypothetical researcher had taken 25 pairs of overweight twins and then randomly assigned one member of each twin-pair to each diet group? In this situation the two groups of Ss would be correlated samples. As discussed in Chapter 4, there are two other situations involving correlated samples. In one, each member of the first group is matched with a member of the second group. In the other, the same group of Ss is measured under both treatment conditions (or before and after exposure to a single treatment).

Hotelling's T^2 can be applied to situations involving correlated samples. The computational procedure is different, but it is not essential for the reader of research reports to know the mathematics involved. What is important is that the multivariate null hypothesis and the meaning of a significant T^2 are identical for both the correlated samples and independent samples situations. In addition, a significant correlated samples T^2 must be probed with a follow-up test, just as is the case with an independent samples situation.

---

[1] Since there are only two comparison groups, there can be only one discriminant function equation. Remember our rule from Chapter 8—the maximum number of these equations is equal to the number of variables or one less than the number of groups, whichever is smaller.

Assumptions Underlying T^2

Hotelling's T^2 test is based upon the double assumption that the sample data have been drawn from multivariate normal populations and that these populations have equal *dispersion (covariance) matrices*. This complicated sounding assumption is analogous to the assumptions of normality and homogeneity of variance that are made in the univariate t test. In fact, the test for homogeneity of dispersion matrices is an extension of Bartlett's univariate chi-square test that we encountered in Chapter 4.

Unfortunately, the authorities who write books or articles on T^2 are not in agreement as to the need for testing this assumption. Even if the two samples sizes are equal, some authors claim the assumption should be tested,[2] while others indicate that T^2 is robust to violations of the assumption.[3] Thus, the published literature contains research articles in which this assumption is discussed and others in which it is not even mentioned. If the assumption is tested, it is important for you to know that a nonsignificant result (i.e., a failure to reject the null hypothesis) is what the researcher will be hoping for. Rejecting the null hypothesis obligates the researcher to use and to report a special version of T^2 that is designed for this situation.

MULTIVARIATE ANALYSIS OF VARIANCE

Although Hotelling's T^2 differs from a univariate t test in terms of the number of dependent variables that can be handled, both of these procedures are designed for the situation where there are just two treatment groups (i.e., just two levels of a single independent variable). If a researcher has two or more independent variables (factors) or more than two levels of a single independent variable, then an analysis of variance should be used. And if there are two or more criterion measures involved in the study, then a multivariate analysis of variance, abbreviated MANOVA, is required. To demonstrate the logic underlying MANOVA, let us now turn to a new hypothetical study.

Suppose a researcher is interested in finding out (1) if college students will get more out of a lecture if they are given a list of the important points to be made in the lecture and (2) if the distribution of such a list of main points will cause students to neglect the parts of the lecture that deal with minor points. On the day of the actual experiment, the teacher randomly distributes envelopes to the students in attendance that day, asking them to read the contents of the envelopes prior to the beginning of the lecture.

The subjects are randomly divided into three groups. Ss in the first group are given envelopes which contain a list of five questions and instructions which explain that these are the main questions that the student should be able to answer after the lecture. In reality, these five questions are related

[2] C. V. Kramer and D. R. Jensen. Fundamentals of multivariate analysis. Part II: inferences about two treatments. *Journal of Quality Control*, 1969, *1*, 189–204.

[3] D. F. Morrison. *Multivariate Statistical Methods*. New York: McGraw-Hill, 1967.

to the most important points to be made that day by the instructor. The second group of Ss receives envelopes containing a different set of five questions and the same instructions. Although the instructions claim that these are the main questions that the student should be able to answer on the basis of the day's lecture, these questions are *not* related to the most important points of the lecture. The third group of Ss receives envelopes which do not contain any questions, but only an announcement about a new course to be taught by the instructor for the first time during he next academic term. For the sake of argument let us assume that the students in the class do not catch on to the fact that there are three different types of envelopes.

The instructor concludes his lecture about five minutes before the end of the period, announces a "pop" quiz, and distributes a 10-item multiple-choice test to the entire group of students. The five odd-numbered test items are directly related to the questions on the list given to Group 1, while the five even-numbered items are related to the questions given to Group 2. The test papers are graded to provide one score for the odd-numbered items and one for the even-numbered items.

This hypothetical study has been set up to exemplify the most basic type of multivariate analysis of variance situation—a one-way MANOVA. There is one independent variable, or factor, which can be called simply treatment condition. There are three levels of this factor, each corresponding to a different group of Ss. And, finally, there are two dependent variables, the two scores based on the sets of five test items on the quiz.

The Null Hypothesis

The null hypothesis associated with a one-way MANOVA can be thought of as a direct extension of the null hypothesis that is tested by Hotelling's T^2; that is, instead of there being just two population means that are assumed to be equal for each of the dependent variables, there are now as many population means (for each dependent variable) as there are treatment groups. Thus, in our hypothetical study the null hypothesis would state that the three treatment groups do not differ in terms of mean scores on (1) the odd-numbered quiz items or (2) the even-numbered quiz items. From a more technical point of view, the null hypothesis in this study assumes that the population mean vector (i.e., a series of two means, one for each dependent variable) for each of the three treatment groups is the same. More generally, in a one-way MANOVA there will always be as many mean vectors as there are treatment groups and as many elements (means) within each vector as there are dependent variables.

Testing the Null Hypothesis

In our discussion of the multivariate *t* test we noted that there are alternative procedures that can be used instead of Hotelling's T^2. In a similar manner there are alternative methods that can be used to test the null hypothesis in an analysis of variance having multiple dependent variables. Of these alternative methods, the oldest and most frequently used procedure is Wilks' lambda. In Excerpt 9.3 we can see how an author might indicate that he used this particular test. In this study there were three dependent variables, each being a different type of error.

EXCERPT 9.3 USE OF WILKS' LAMBDA TO
TEST THE MULTIVARIATE NULL HYPOTHESIS

The multivariate analysis of variance . . . for redundant, equivalence, and noninformative errors was found to be significant ($F = 2.891$, $df = 3/82$, $p < .04$) by the Wilks' Lambda Criterion (Cooley & Lohnes, 1962).

SOURCE: J. D. McKinney. Developmental study of the acquisition and utilization of conceptual strategies. *Journal of Educational Psychology*, 1972, *63*, 22–31. Copyright 1972 by the American Psychological Association. Reprinted by permission.

The actual calculated value that is yielded by applying the Wilks' lambda test is labeled λ (or, sometimes, Λ). As opposed to the situation for t or F tests, a significant difference is obtained from Wilks' lambda by getting small calculated values. In other words, the greater the disparity between the mean vectors, the smaller the value for lambda. Since a complete table of critical values for lambda is not available, most researchers transform their calculated value for lambda into an F value.[4] Thus, the author of Excerpt 9.3 presents a value for F, not lambda.

It should be noted that there are three alternative procedures for testing the multivariate null hypothesis other than Wilks' lambda—Roy's largest root criterion, Hotelling's trace criterion, and the step down F procedure. Unlike the alternative approaches to T^2, which all yield identical results, Wilks' lambda and its alternatives are not equivalent. However, since these alternatives to Wilk's lambda are all reputable and since it is the researcher who must decide which procedure to use, there is no need for the reader of an article to try to understand the differences among these tests. The important thing for the reader to know is what it means to reject or accept the null hypothesis.

Follow-up Tests

If the null hypothesis in our hypothetical study were to be rejected, the researcher would know that the three treatment groups differ from one another in terms of mean scores on the odd-numbered test items and/or the mean scores on the even-numbered items. However, the researcher would not

---

[4] This transformation of lambda into F was developed by Rao, and the process is sometimes referred to as *Rao's approximation*. Another transformation procedure, considered to be inferior to Rao's, involves changing lambda into chi square, a nonparametric statistic that will be discussed in the next chapter.

know which specific levels of the treatment variable are different from one another nor would he know whether the treatment groups differ on just one dependent variable or both of them. Because these important questions still need to be answered, Wilks' lambda (or an alternative procedure) should be viewed as a preliminary test. If the results allow the researcher to reject the null hypothesis, a follow-up investigation is imperative.

Two follow-up procedures are recommended by authorities in the field: simultaneous confidence intervals and discriminant function analysis. In our discussion of T^2, we saw how simultaneous confidence intervals could be used to determine on which variable(s) the two groups differ. When used after a significant Wilks' lambda, the basic procedure is the same, but now an equation is set up to compare each possible pair of treatments on each dependent variable. In the hypothetical study there would be a total of six equations (comparing groups 1 vs. 2, 2 vs. 3, and 1 vs. 3, first on the odd-numbered items and then on the even-numbered items). As before, the inclusion of zero between the two numbers in this equation indicates that the two treatments being compared might have equal population means.

In Excerpt 9.4 we see an example of how simultaneous confidence intervals can be used to find out which groups differ from one another on

---

EXCERPT 9.4 USING SIMULTANEOUS
CONFIDENCE INTERVALS AS A FOLLOW-UP PROCEDURE

TABLE 4 SUMMARY OF THE MULTIPLE COMPARISONS FOR THE COLLEGE MAIN EFFECT

VARIATE	PROGRAM		
Predicted average	E	S	IM
Grade-point average	E	S	IM
Achievement	E	S	IM
Autonomy	E	IM	S
Change	E	S	IM
Cognitive structure	E	IM	S
Program commitment	E	S	IM
Program involvement	E	S	IM
Provide training	E	S	IM
Provide education	E	S	IM
Income	E	IM	S
Intellectual stimulation	E	IM	S

Note. E = Engineering; S = Science; IM = Industrial Management.

To determine which college differences on a given dependent variable contributed to the rejection of this hypothesis, the 95 per cent simultaneous confidence intervals on pairwise differences between the colleges for each component of the response vector were constructed. The results of these multiple comparisons are summarized in Table 4. In the table, the 12 dependent variables are listed along with the three colleges. A line under two or more colleges for a given variable indicates that the colleges do not differ at the .05 level with respect to that variable, that is, the confidence interval for that comparison contains 0.

SOURCE: E. Marks. Cognitive and incentive factors involved in within-university transfer. *Journal of Educational Psychology*, 1970, *61*, 1–9. Copyright 1970 by the American Psychological Association. Reprinted by permission.

what specific variable(s). In this study there were three groups: college students enrolled in engineering, science, and industrial management. There were 12 dependent variables (variates) related to the personality, ability, and program goals of the Ss. Notice how the author used the underlining method for presenting his results (and that he explained what the underlining means). Also, notice how the last sentence of the excerpt corresponds with our explanation of how the simultaneous confidence interval procedure works.

The second possible follow-up procedure involves the use of discriminant function analysis. If there are just two comparison groups in the study, the process is identical to the two-group discriminant function we discussed earlier in relation to the T^2 test. In Excerpt 9.5 we see how this form of discriminant function can be used as a follow-up procedure to find out where the significant differences lie after the multivariate null hypothesis has been rejected. In this study there were two groups (a married group and a single group) and five dependent variables. After getting a significant multivariate F, the author examined the coefficients from the discriminant function analysis (presented in the last column of the table) to determine which of the five variables were contributing to the significant F.

A multiple discriminant function analysis (as opposed to a two-group

---

EXCERPT 9.5 DISCRIMINANT FUNCTION ANALYSIS AS A FOLLOW-UP PROCEDURE IN MULTIVARIATE ANALYSIS OF VARIANCE

Marital status contrast. Data from this analysis are shown in Table 4. Significant differences were found when the effect of marital status, eliminating the effect of career, was tested. Multivariate F equals 3.45 ($df = 5/76$) and is significant ($p < .007$).

Standard discriminant function coefficients listed in this table show three large contributors: Knower, Competitor, and Homemaker.

TABLE 4 MULTIVARIATE ANALYSIS OF VARIANCE WITH MEASURES OF PERCEPTIONS OF THE FEMININE IDEAL AS THE DEPENDENT SET

SOURCE OF VARIATION	VARIABLE	MS BETWEEN GROUPS	UNI- VARIATE F[a]	$p <$	SDFC
Marital status, eliminating	Partner	212.125	9.716	.003	.279
the effect of career	Ingenue	86.111	3.700	.058	−.003
Married group—Single group	Homemaker	49.444	1.850	.178	−.661
	Competitor	292.609	9.974	.002	.642
	Knower	347.089	10.305	.002	.706

Note. Abbreviated: SDFC: Standardized discriminant function coefficients. Multivariate $F = 3.45$, $df = 5/76$, $p < .007$.
[a] $df = 1/80$.
SOURCE: P. Hawley. What women think men think: does it affect their career choice. *Journal of Counseling Psychology*, 1971, *18*, 193–199. Copyright 1971 by the American Psychological Association. Reprinted by permission.

discriminant analysis) should be used if there are three or more comparison groups involved in a study. In this situation, however, the computer will provide more than one set of discriminant coefficients. The precise number of possible discriminant function equations is equal to the number of dependent variables or one less than the number of comparison groups, whichever is smaller. In our hypothetical study concerning the lecture and the envelopes, there would be a possibility of two discriminant function equations (since there were three comparison groups and two dependent variables).

Each of the multiple discriminant function equations has the same form as a two-group discriminant equation. There are as many Xs on the right side of the equation as there are dependent variables, with a numerical weight attached to each X. The weights of the first equation are chosen (by the computer) so as to maximize the discrimination between the treatment groups. In other words, if we multiplied the scores on the dependent variables for each S by the appropriate weights and then if we summed the resulting values to get a Z score, there would be less overlap among the treatment groups with respect to the Z values than would be the case with any other possible set of weights for the Xs. The weights of the second (or any additional) equation are again chosen so as to maximize group discrimination, but there is now a requirement that Z scores from the new equation be uncorrelated with the Y scores from the previously formed equation(s).

Admittedly, it would be much simpler to have only one discriminant function equation. The rationale for having two or more such equations is as follows.

> . . . The very existence of more than one significant discriminant dimension puts the concept of group differences and resemblances in a different light from that based on a single [dependent variable]. The latter would, in comparing three groups (say), result in a conclusion of the form "Group 1 resembles Group 2 more than it does Group 3." But, with multiple dimensions of group differentiation, one would conclude, for instance, that "Group 1 resembles Group 2 more than it does Group 3 with respect to the first discriminant. However, Groups 1 and 3 are very similar with respect to the second discriminant, and both of these groups differ considerably from Group 2 in this respect."
> It might be thought that, carrying to the extreme the argument just presented, one would do best to speak of resemblances and differences with respect to each [dependent] variable separately. This is not so. For the original variables usually stand in such a complicated pattern of intercorrelations among one another that we cannot, without danger of redundancy and inconsistency, speak of group differences with respect to each of them separately. The several discriminant functions, on the

other hand, are uncorrelated with one another, and this fact enables us
to speak of differences and resemblances with respect to each of them.[5]

Assumption of Equal Dispersion Matrices

Earlier in this chapter we stated that Hotelling's T^2 test is based on the assumption of equal dispersion matrices. This assumption also underlies multivariate analysis of variance, the only difference being that in MANOVA's there will be as many dispersion matrices as there are levels of the independent variable. In our hypothetical study concerning the effect of stated objectives on the information conveyed in a lecture, there would be three dispersion matrices, one corresponding to each of the three treatment groups. As we stated earlier, this assumption can be thought of as being analogous to the homogeneity of variance assumption that underlies a univariate analysis of variance.

Some researchers do not go to the trouble to test this assumption because they feel that the MANOVA procedure is robust to minor violations of the assumption. Other authors bypass testing this assumption only when the sample sizes are equal and large. Still others believe that the assumption should always be tested, irrespective of the sample size. Thus, you can expect to see articles in which the assumption of equal dispersion matrices is tested, while other articles will not even mention dispersion matrices. Excerpt 9.6 reports a test of the assumption.

EXCERPT 9.6 TESTING THE ASSUMPTION
OF EQUAL DISPERSION MATRICES

A multivariate analysis of variance was computed to evaluate the hypothesis that these data might represent two independent random samples from a single multivariate normal population with reference to the two sets of nine means and the two variance-covariance matrices. The hypothesis of homogeneous dispersions was not tenable ($F = 1.99$, $df = 45/24537.436$, $p < .001$).

SOURCE: I. Sember and I. Iscoe. Structure of intelligence in Negro and white children. *Journal of Educational Psychology*, 1966, 57, 326–336. Copyright 1966 by the American Psychological Association. Reprinted by permission.

Requirements Concerning the Number of Variables and the Total Sample Size

There are two simple rules that must be followed by a researcher if he uses a multivariate analysis of variance. First, there should not be fewer dependent variables than there are treatment groups being compared. (We knowingly broke this rule in our hypothetical lecture experiment so that our example would be as simple as possible, but if we were actually to conduct

---

[5] M. M. Tatsuoka. *Discriminant Analysis: The Study of Group Differences*. Champaign, Ill.: Institute for Personality and Ability Testing, 1970, pp. 54–55.

this study, we would probably be interested in four or five dependent variables.) The second rule simply requires that the total number of subjects in the study be at least twice as large as the number of dependent variables.

FACTORIAL MANOVA'S

In Chapter 5 we discussed factorial analysis of variance for the univariate case, specifically, two-way and three-way ANOVA's. If a researcher has multiple dependent variables, a two- or three-way MANOVA is more appropriate than a univariate analysis for each dependent variable for two reasons. First, the univariate tests ignore intercorrelations between the dependent variables; second, they do not take into consideration the fact that the levels of a given factor are being compared more than once.

When a researcher uses a multivariate factorial analysis of variance, the same number of basic research questions will be answered as would have been the case had there been just one dependent variable. There will be two main effects and one interaction in a two-way MANOVA (see Excerpt 9.7); three main effects, three first-order interactions, and one second-order interaction in a three-way MANOVA; and so forth.

EXCERPT 9.7 PRELIMINARY RESULTS FROM A TWO-WAY MANOVA

A two-way MANOVA following the procedures outlined by Bock and Haggard (1968) was used to analyze the transformed data. The results indicate that there is a multivariate effect attributable to the Warning condition (F = 2.66, $p < .006$), while the multivariate difference for the test condition and the interaction between the Warning and Test conditions can be accounted for by chance alone (Fs = .62 and 1.03; $ps < .74$ and .42, respectively).

SOURCE: D. McQuarrie and A. Grotelueschen. Effects of verbal warning upon misapplication of a rule of limited applicability. *Journal of Educational Psychology*, 1971, 62, 432–438. Copyright 1971 by the American Psychological Association. Reprinted by permission.

Significant effects from a factorial MANOVA require further investigation to determine which of the variables are contributing to the rejection of the overall null hypothesis in question. For example, the authors in Excerpt 9.7 obtained a significant multivariate main effect for the factor called Warning condition. There were three levels of this factor, and there were also three dependent variables. To help portray the differences among the three warning groups, a discriminant function analysis was employed.

MULTIVARIATE ANALYSIS OF COVARIANCE

In Chapter 7 we talked about the analysis of covariance. We pointed out that a covariance analysis can be used to increase power and that it adjusts the comparison group means to account for differences between the groups on the concomitant variable (the covariate). We also indicated that for each type of analysis of variance (one-way, two-way, Lindquist Type I, etc.), there is a corresponding analysis of covariance. We also saw that there can be more than just one concomitant variable involved in any particular analysis of covariance.

A multivariate analysis of covariance is simply an analysis of covariance in which several dependent variables are analyzed at the same time. Thus, the ideas presented in the earlier parts of this chapter regarding multivariate analysis of variance and the ideas presented in Chapter 7 regarding the analysis of covariance should be synthesized when you think about a multivariate analysis of covariance. In Excerpt 9.8 the authors help us to make this synthesis by pointing out the advantage of a multivariate analysis in the first paragraph and by discussing one of the two advantages of covariance in the second.

EXCERPT 9.8 MULTIVARIATE ANALYSIS OF COVARIANCE

ANALYSIS

A single factor multivariate analysis of covariance (ANCOVA) with ten criteria and ten covariates was performed on the data (Bock, 1960). Univariate analyses for each of the separate criteria with its respective covariate were also performed following the multivariate analysis. This analysis procedure initially treated the ten criteria as a single collection (i.e., a generalized measure of self concept). The principal advantage of such a multivariate procedure over the traditional separate univariate F tests is that it permits a test of the possible interactions among multiple criteria that cannot be evaluated if each criterion variable is tested in isolation (Cooley & Lohnes, 1962; Jones, 1966).

This analysis procedure also statistically matched the Ss in the experimental and control groups on the pre-test scores. Consequently, post-test differences between the means of the experimental and control groups were analyzed after taking into account and making appropriate statistical adjustments for initial differences on the pre-test.

SOURCE: K. White and R. Allen. Art counseling in an educational setting: self-concept change among pre-adolescent boys. *Journal of School Psychology*, 1971, 9, 218–225.

It should be noted that the results of a multivariate analysis of covariance refer to the adjusted mean scores on the dependent variables, not the original unadjusted means. The authors of Excerpt 9.8 made this clear when they stated at another point in their article: "The multivariate analysis of covariance was significant ($p < .02$, $df = 10, 9$), indicating that the adjusted post-test scores on the ten scales of the Tennessee Self Concept Scale when considered together were significantly higher for the art counseling group than for the non-directed counseling group."

DEGREES OF FREEDOM IN MULTIVARIATE ANALYSES

When discussing univariate analyses of variance and covariance in previous chapters, we pointed out that the number of degrees of freedom associated with any main effect is equal to one less than the number of levels in that factor. We also explained that the *df* for any interaction is found by multiplying together the main effect degrees of freedom for the factors involved in the interaction. In multivariate analyses main effect and interaction degrees of freedom are found by, first, applying the two rules cited above and, second, multiplying the resulting degrees of freedom by the number of dependent variables involved in the study.

In Excerpt 9.9 we see a multivariate analysis of variance summary table. There were two levels of the first factor (race), two levels of the second factor (sex), and three levels of the third factor (test condition). This $2 \times 2 \times 3$ MANOVA was applied twice, once for the scores obtained from Form A of the criterion tests, and once for scores obtained from Form B of the same tests. In both analyses there were seven dependent (criterion) variables. In the *df* columns in the table the number preceding the slash mark corresponds to the *df* value for a main effect or interaction, while the number following the slash corresponds to the error (within-groups) *df*. Notice how the main effect and interaction *df* values in Excerpt 9.9 are equal to 7 (the number of dependent variables) times what would have been the univariate degrees of freedom.

EXCERPT 9.9 MAIN EFFECT AND INTERACTION *df* IN A 2 × 2 × 3 MANOVA

TABLE 3 MULTIVARIATE ANALYSIS OF VARIANCE OF CREATIVITY TESTS

SOURCE OF VARIATION	FORM A		FORM B	
	df	F	df	F
Race	7/150	1.31	7/150	.65
Sex	7/150	.72	7/150	.68
Test Condition	14/300	.70	14/300	.87
Race × Sex	7/150	.54	7/150	.52
Race × Test Condition	14/300	.84	14/300	.53
Sex × Test Condition	14/300	.47	14/300	2.47*
Race × Sex × Test Condition	14/300	1.07	14/300	1.09

* $p < .003$.

SOURCE: T. U. Busse, P. Blum, and M. Gutride. Testing condition and the measurement of creative abilities in lower-class preschool children. *Multivariate Behavioral Research*, 1972, 7, 287–298.

There is one additional difference between the *df* in a univariate analysis and the *df* in a multivariate analysis. When there is just one dependent variable, the *df* in an analysis of variance or analysis of covariance must be a whole number. In multivariate analyses, however, it is possible for the error (within-groups) *df* to be a decimal. The error *df* can be fractional only when there are three or more dependent variables *and* when there are four or more levels of the main effect (or one of the factors involved in an interaction). For example, consider Excerpt 9.10 in which the second column of *df* values contains the error *df* associated with each main or interaction effect.

EXCERPT 9.10 FRACTIONAL DEGREES
OF FREEDOM IN A 4 × 2 × 3 MANOVA

TABLE 2 MULTIVARIATE ANALYSIS OF VARIANCE (MANOVA) OF
SIX VARIABLES BY TRANSFER STATUS, SEX AND CLASSIFICATION

SOURCE	NUMERATOR df	DENOMINATOR df	F
Transfer status (A)	18.000	461.519	4.254**
Sex (B)	6.000	163.000	7.809**
Classification (C)	12.000	326.000	4.636**
AB	18.000	461.519	1.378 NS
AC	36.000	718.544	1.233 NS
BC	12.000	326.000	0.541 NS
ABC	36.000	718.544	1.123 NS

** $p < .01$.

SOURCE: B. M. Suddarth. A multivariate investigation of the academic achievement of transfer and native students. *Journal of College Student Personnel*, 1971, *12*, 133–137. Copyright 1971 by the American Personnel and Guidance Association.

REVIEW TERMS

discriminant function analysis
equal dispersion matrices
Hotelling's T^2
MANOVA
mean vector
multivariate analysis
positively biased

Rao's approximation
simultaneous confidence intervals
standardized discriminant function
  coefficients
univariate analysis
Wilks' lambda

REVIEW QUESTIONS

1. Suppose a researcher uses Hotelling's T^2 to compare two groups with respect to five relevant dependent variables. What is the main null hypothesis?
2. Can Hotelling's T^2 be used only with independent samples, only with correlated samples, or both?
3. The calculated value that a researcher obtains from Hotelling's T^2 is usually transformed into and reported as an _____ value.
4. After getting a significant T^2, is it appropriate for a researcher to use several univariate t tests in an attempt to find out which variables contributed to a rejection of the multivariate null hypothesis?
5. In using a multivariate analysis of variance, can the researcher have more dependent variables than treatment groups?

6. How many Ss are required in a study if a researcher intends to use a MANOVA with 10 dependent variables?
7. Do authorities agree on whether the assumption of equal dispersion matrices should be tested?
8. Suppose a researcher uses a one-way multivariate analysis of variance to compare three treatment groups with respect to four dependent variables. After obtaining a significant F, a discriminant function analysis is used to find out what caused the overall null hypothesis to be rejected. Will the resulting standardized discriminant function coefficients correspond to the independent variables of the study or the dependent variables?
9. Suppose the researcher in the previous question decided to use simultaneous confidence intervals rather than discriminant function analysis. How many confidence intervals would there be? If the numerical "ends" of the first confidence interval were 0.19 and 3.42, would this indicate a significant difference?
10. If there are two or more covariates involved in an analysis of covariance, is it appropriate to consider the statistical analysis to be multivariate?

CHAPTER 10

NONPARAMETRIC STATISTICAL TESTS

The statistical tests we have discussed in the last six chapters are all classified as parametric statistics and are the most popular and most frequently reported statistical tests in journal articles. In this chapter we will discuss another type of statistic—nonparametric or distribution-free procedures. Although there are many nonparametric procedures, the nine presented here are the ones most frequently reported in journal articles. Before examining these procedures, it might be helpful to explore some differences between the two types of statistical procedures and briefly explain why some researchers prefer to use one type of procedure rather than the other.

Parametric and nonparametric statistical procedures test different hypotheses involving different assumptions. *Parametric statistics* test hypotheses based on the assumption that the samples come from populations that are normally distributed. Also, parametric statistical tests assume that there is homogeneity of variance (that is, that variances within groups are the same). For example, a researcher collects sample data for two groups and applies a statistical test to estimate whether the means of the two populations from which the samples were drawn are equal. *Distribution-free* or *nonparametric statistical procedures* test hypotheses that do not specify normality or homogeneity of variance assumptions about the populations from which the samples were drawn. Some researchers prefer to use nonparametric statistics when they feel that these two assumptions are violated. Other researchers feel that most parametric statistics are robust against violations of normality and homogeneity and they prefer to use parametric tests in almost any situation. We cannot hope to resolve the issue of which test to use. Generally, it is agreed that unless there is sufficient evidence to suggest that the population is extremely non-normal and that the variances are heterogeneous, parametric tests should be used because of their additional power.

Still another reason why researchers prefer nonparametric tests concerns the level of measurement. Before exploring this, we must first examine three levels of measurement and related scales. The first level is called the *nominal* or *classificatory scale;* it is the weakest level of measurement. Categories of mental illness, numbers on a uniform, hair color, and telephone numbers are examples of nominal scales. Nominal measurement is, in effect, a means of naming or categorizing people, objects, events, or characteristics. A second level of measurement is called the *ordinal* or *ranking scale.* Ordinal scaling determines a relation between objects, events, people, or characteristics in terms of their being greater than, less than, or equal to one another on the basis of a selected criterion. Examples of ordinal scaling are socioeconomic status, ranks in military services, or ranks of college or university faculty. The third level of measurement is called the *interval scale.* The interval scale measures the same characteristics as the ordinal scale. In addition, the distances between any two numbers on the scale are of known size. In other words, the interval scale has a common and constant unit of measurement. Examples of interval measures are temperature (Fahrenheit and centigrade) scales and scores on an IQ test.

Why are the levels of measurement important to a researcher? A researcher should not use parametric statistical tests on nominal scale data; in such cases, nonparametric tests should be used. Some researchers feel that parametric procedures should be used only with interval level data, although journal articles often report the use of parametric tests with ordinal data. Excerpts 10.1 and 10.2 present statements from two journal articles in which the authors explain their choice of nonparametric statistical tests. In the first study the reason relates to the assumption of normality of the distribution ("Due to the skewness . . ."). The reason for using a nonparametric test for the second study in Excerpt 10.1 was ". . . because of the type of data."

A third possible reason for the use of nonparametric, rather than parametric, tests is the issue of number in the sample. Some researchers feel that parametric tests cannot be used with small sample sizes, but there are others who say that it does not make any difference. To summarize, the

---

EXCERPTS 10.1 and 10.2 REPORTED REASONS
FOR USING NONPARAMETRIC TESTS

Due to the skewness of the distributions, all data for the study were analyzed by means of nonparametric statistics.

SOURCE: R. D. Odom, R. M. Liebert, and J. H. Hill. The effects of modeling cues, reward, and attentional set on the production of grammatical and ungrammatical syntactic construction. *Journal of Experimental Child Psychology*, 1968, *6*, 131–140.

A nonparametric statistical technique was necessary because of the type of data collected.

SOURCE: G. A. Hayes. The integration of the mentally retarded and non-retarded in a day-camping program: demonstration project. *Mental Retardation*, 1969, *7*, 14–16.

---

procedure the researcher decides to use may be determined by the level of measurement, the researcher's feeling about sample size, or his attitude toward violations of the parametric assumptions.

Nonparametric statistical tests are similar to parametric procedures in many respects. Both (1) test a hypothesis, (2) involve a level of significance, (3) require a calculated value, (4) compare against a critical value, and (5) conclude with a decision about the hypothesis. They are different from one another in several respects. (1) There are many more tables of critical values for nonparametric tests than for the parametric procedures. (2) We will see that in order to reject the null hypothesis for some nonparametric tests, the calculated value must be less than the tabled critical value for a specific level of significance. For other nonparametric tests, as for most parametric tests, the calculated value must be greater than the tabled critical value at a given level in order to reject the null hypothesis. (3) Nominal scale data and, according to some researchers, even ordinal scale data cannot be used for parametric procedures, but there are nonparametric procedures for which all levels of measurement can be used.

Despite these differences, nonparametric statistics can be applied to research problems that might also be analyzed by parametric tests. For example, Table 10.1 shows some parametric statistics and the nonparametric procedures that can be applied to at least ordinal data. Two nonparametric procedures that can be applied to nominal data are discussed in the last two sections of this chapter.

SPEARMAN RANK-ORDER CORRELATION
COEFFICIENT (rho or r_s)

In Chapter 2, we discussed two correlational procedures that can be used to describe the relationship between two variables, the Pearson product-moment correlation and Spearman's rho. Spearman's rho is sometimes called the Spearman rank-order correlation coefficient (or rank correlation coefficient) and the symbol r_s is often used to distinguish it from r, the symbol for the Pearson product-moment correlation coefficient. There are four similarities between the Pearson r and Spearman's rho. (1) Both are reported as a two-digit number preceded by a decimal point ($+.95$, $.08$, $-.10$). (2) Both have a range of possible values from -1.00 to $+1.00$. As with r, the values for rho express the degree of relationship, that is, the nature and strength of the correlation (high positive, low positive, zero, low nega-

TABLE 10.1 PARAMETRIC STATISTICS AND ANALOGOUS NONPARAMETRIC PROCEDURES

PARAMETRIC	NONPARAMETRIC
Pearson product-moment correlation coefficient r	Spearman rank-order correlation coefficient (rho)
t test correlated samples	Sign test Wilcoxon matched-pairs signed-ranks test
t test independent samples	Median test Mann-Whitney U test
One-way ANOVA	Kruskal Wallis one-way ANOVA of ranks Median test
One-way ANOVA with repeated measures	Friedman two-way ANOVA of ranks
(No analogous parametric test)	Chi-square single-sample k independent samples

tive, high negative). (3) Both correlation coefficients can be tested for significance. For example, a researcher may wish to test the null hypothesis that the two variables under study are not associated in the population and that the calculated value of r_s differs from zero only by chance. To test whether the two variables are associated in the population, the subjects whose scores were used in computing r_s must be randomly drawn from some population. (4) Both r and r_s can be used in a correlation matrix (see, for example, Excerpts 2.13–2.15).

As you might expect, one of the major differences between the parametric Pearson procedure and the nonparametric Spearman procedure is the

TABLE 10.2 HYPOTHETICAL DATA FOR SPEARMAN'S RHO

DAYS	OBSERVER A	RANK A	OBSERVER B	RANK B
1	16	1	18	1
2	19	3	19	2
3	18	2	20	3
4	21	5	22	4
5	20	4	24	6
6	22	6	23	5

way the correlation coefficients are computed. Suppose a researcher wanted to determine the degree of relationship between two observers. Each observer recorded independently the frequency of some event for six days. The data might appear as in Table 10.2. If our researcher wanted to calculate the Pearson product-moment correlation, he would use the actual frequency scores (16, 19, 18, 21, 20, and 22 for Observer A and 18, 19, etc., for Observer B). To compute the Spearman rank correlation coefficient, the researcher would first rank the data for Observer A (assigning 1 to the lowest number) and then for Observer B. After ranking each set of data, the researcher would then use the two sets of ranks in a formula to determine the correlation coefficient. If the researcher wanted to test the significance of rho, he would refer to a table of critical values in a nonparametric statistics book.

The results of the Spearman rank correlation coefficient can be reported in many ways. Excerpts 10.3–10.5 are examples of the most frequently used

---

EXCERPT 10.3 SPEARMAN RANK CORRELATION COEFFICIENT MATRIX

CORRELATION (RHO) MATRIX OF THE
CONCEPTS AS RANKED SEPARATELY
BY EACH OF THE TRUSTEE GROUPS

		NECA	NVGA	ASCA	AVA
I	NECA		.86	.80	.79
II	NVGA			.78	.71
III	ASCA				.59
	Groups I, II, III Combined				.76

SOURCE: J. A. Bailey. Career development concepts: significance and utility. *Personnel and Guidance Journal*, 1968, 47, 24–28. Copyright 1968 by the American Personnel and Guidance Association.

---

variations. Excerpt 10.3 shows the use of a correlation matrix. Excerpts 10.4 and 10.5 illustrate how rho is reported in the body of the results section of the journal article. Notice that the author of the first example refers to the cor-

---

EXCERPTS 10.4 and 10.5 SPEARMAN'S RHO
REPORTED IN THE RESULTS SECTION

The Spearman rank-difference correlation coefficients (rho) between the boys' popularity choices and girls' popularity choices ranged from +.74 to +.24 for the different classes. . . .

SOURCE: G. S. Lesser. The relationship between various forms of aggression and popularity among lower-class children. *Journal of Educational Psychology*, 1959, *50*, 20–25. Copyright 1959 by the American Psychological Association. Reprinted by permission.

There was no relationship (rho = .11) between the presentation order of the items and the judges' values attached to the items. . . . There was a very slight tendency for the longer statements to be ranked higher in value (rho = −.23).

SOURCE: J. A. Bailey. Career development concepts: significance and utility. *Personnel and Guidance Journal*, 1968, *47*, 24–28. Copyright 1968 by the American Personnel and Guidance Association.

---

relation as the Spearman rank-difference, and the author of the second uses the term rho. Some journal articles might refer to the correlation coefficient as Spearman *r*, rather than the usual designation of rho or r_s. Excerpt 10.6 illustrates how Spearman's rho might be reported as a test of significance.

EXCERPT 10.6 SPEARMAN RANK CORRELATION
COEFFICIENT AS A TEST OF SIGNIFICANCE

Computing a Spearman rank correlation coefficient, a value of $r_s = .20$ was obtained. The correlation is not significant at the .05 level.

SOURCE: J. W. Slocum and R. H. Strawser. Racial differences in job attitudes. *Journal of Applied Psychology*, 1972, *56*, 28–32.

NONPARAMETRIC ANALOGS TO THE PARAMETRIC *t* TEST

Sign Test

The *sign test* can be applied to studies with two related samples when a researcher wants to establish that two conditions are different. The test uses plus and minus signs rather than quantitative data. As you can see from Table 10.1, the sign test's parametric equivalent is the correlated *t* test. However, the Wilcoxon test discussed in the next section is more like the correlated *t* test than the sign test, because the sign test uses only direction of score differences to determine whether the samples are significantly different from one another. The sign test is useful in ascertaining whether two conditions are different for related samples when the data for each pair can be ranked but quantitative measurement is not possible. The sets of observations or measurements must be related to each other in some meaningful way as was the case for the correlated *t* test discussed in Chapter 4. This relationship can be established by (1) using subjects who are carefully matched before being exposed to the treatment, (2) using subjects who are related, such as identical twins, or (3) using each subject as his own control in a pretest-posttest design. Also, the subjects must be a random sample from a larger population.

Let us examine a hypothetical example for which the sign test might be used. Suppose a researcher predicts that husbands consume more caffeine daily than their wives. The results of sampling 16 couples appear in Table 10.3. The first column contains the names of the couples sampled and the

TABLE 10.3 STUDY ILLUSTRATING
THE USE OF THE SIGN TEST

COUPLES (HUSBAND AND WIFE)	DIRECTION OF DIFFERENCE FOR EACH WIFE AND HUSBAND PAIR	SIGN
Ballou	H > W	+
Chapman	H > W	+
Decker	H > W	+
Friedman	H < W	−
Getty	H = W	0
Hileman	H > W	+
Jacobs	H < W	−
Lynch	H > W	+
McKinney	H = W	0
Neubert	H > W	+
Palumbo	H < W	−
Schultz	H > W	+
Stout	H > W	+
Thurmer	H = W	0
Watson	H > W	+
Yates	H > W	+

second column indicates the member of each wife and husband pair who consumes the greater amount of caffeine. In the third column a plus or minus sign is used to denote the direction of difference between the two members of the pair. A zero is used to indicate that no difference exists. The null hypothesis of the sign test is that there are an equal number of plus and minus signs. Accordingly, the researcher uses the number of plus and minus signs to obtain the calculated values. There are two calculated values, N and m. N is simply the total number of plus *and* minus signs; m is simply the number of minus signs. (For our example, in Table 10.3, $N = 13$ and $m = 3$.)

When the sign test is used with small samples, the researcher uses these two values to locate the critical value in an appendix table. (Such tables are found in most nonparametric statistics books.) The critical value is nothing more than a probability associated with the particular N and m values calculated by the researcher. Therefore, if the tabled critical value (i.e., the probability) is a smaller number than the researcher's level of significance, then the null hypothesis can be rejected. For example, if the level of significance is .05 and the critical value is .029, the researcher can reject.[1]

There are two popular formats for reporting the sign test in journal articles. One way is to present a table showing the sign ($+$, $-$, 0) for each pair. Excerpt 10.7 shows a table of paired median scores. The n columns in the table show the number for each group from which the medians were computed. In the last column, predicted direction of difference, the N (number of non-zero signs) used for the sign test is 7 (6 pluses and 1 minus).

EXCERPT 10.7 RESULTS OF THE SIGN TEST REPORTED IN A TABLE

MEDIAN LIKING SCORES OF SUBJECTS WHO SAID THEY FELT "VERY HAPPY" COMPARED WITH THOSE WHO DID NOT WHEN THEY OR THEIR PARTNERS WON A BINGO GAME

	Answers to Question 1				
SUBJECTS BY REWARD AND GROUP CONDITION	"VERY HAPPY" (VH)		"HAPPY" (H), "ALRIGHT" (A), "DON'T CARE" (DC)		PREDICTED DIRECTION OF DIFFERENCE: VH > H, A, DC
	n^a	Mdn	n^b	Mdn	
Player 4 (4-0-0)	17	13	3	−5	+
Helper 4 (4-0-0)	15	7	5	4	+
Player 3 (3-1-0)	14	−14	6	−4	−
Helper 3 (3-1-0)	18	2.5	2	−42	+
Player 2 (2-2-0)	31	3	9	−4	+
Helper 2 (2-2-0)	29	0	11	0	0
Player 1 (3-1-0)	11	10	9	0	+
Helper 1 (3-1-0)	10	19	10	−1.5	+

[a] Total $N = 145$.
[b] Total $N = 55$.

Evaluation of these data by the Sign Test (Siegel, 1956) indicates them to be very close to the acceptable level of significance ($p < .06$).

SOURCE: A. J. Lott, B. E. Lott, and G. M. Matthews. Interpersonal attraction among children as a function of vicarious reward. *Journal of Educational Psychology*, 1969, 60, 274–283. Copyright 1969 by the American Psychological Association. Reprinted by permission.

---

[1] The procedure described in this paragraph is appropriate for one-tailed null hypotheses. If the researcher is investigating a two-tailed null hypothesis, m must be defined as the number of plus *or* minus signs, whichever is smaller. Also, the tabled critical value (i.e., the probability) would be doubled before comparing it to the level of significance.

EXCERPT 10.8 RESULTS OF SIGN TEST REPORTED WITHOUT A TABLE

Moreover, the difference in frequency between role symmetry and self-interest, when compared alone, was also significant ($z = 2.02$, p < .03, sign test), as was the difference between self-interest and altruism responses ($z = 7.88$, p < .00001, sign test).

SOURCE: I. M. Lane and L. A. Messé. Equity and distribution of rewards. *Journal of Personality and Social Psychology*, 1971, *20*, 1–17. Copyright 1971 by the American Psychological Association. Reprinted by permission.

As stated in Excerpt 10.7, the authors found the critical value to be $p < .06$ (located in a table of values). Excerpt 10.8 illustrates the second format for reporting the sign test. An author may simply state the results of the tests in the results section of the article without presenting a table of the direction of differences. Note that the authors of Excerpt 10.8 report the results of the two sign tests as z values. If N is larger than 25, a researcher will use a different procedure to obtain the calculated value for the sign test (i.e., a procedure other than that described in the preceding paragraph). To interpret the calculated value, he then refers to a z table for the critical value. Thus, if a researcher reports z values, we know that he used a sign test for more than 25 pairs.

Wilcoxon Matched-Pairs Signed-Ranks Test

The nonparametric *Wilcoxon matched-pairs signed-ranks test* is equivalent to the parametric correlated or related samples t test. The Wilcoxon test takes into account the magnitude as well as the direction of the difference for each pair. It is a more powerful test than the sign test because more weight is given to a pair that shows a large difference. As with the sign test, subjects in the Wilcoxon test must (1) be carefully matched before being exposed to the treatment, (2) be related in some way (i.e., be identical twins), or (3) serve as their own control in a pretest-posttest design. Also, the subjects must be a random sample from a larger population.

Consider again the hypothetical example in which a researcher predicts that husbands drink more coffee daily than their wives. After sampling eight couples, he might record the results as in Table 10.5.

In the first column are the names of the couples. The next two columns list the number of cups of coffee consumed daily by each husband and wife. The fourth column, d, is the difference between the number of cups consumed by each husband and wife pair. The minus sign for the Jacobs and Palumbos indicates a difference which is opposite to the predicted direction (that is, in these two cases, the wives drink more cups of coffee daily than their husbands). The next-to-last column (rank of d) are the ranks of differences (without regard to the signs). The researcher assigns 1 to the lowest difference (the Jacobs), 2 to the second lowest difference (the Lynches),

TABLE 10.5 STUDY ILLUSTRATING THE
USE OF THE WILCOXON TEST

COUPLES	HUSBAND	WIFE	d	RANK OF d	RANK WITH LESS FREQUENT SIGN
Ballou	9	2	7	5	
Decker	11	1	10	6	
Hileman	6	0	6	4	
Jacobs	1	2	−1	−1	1
Lynch	5	2	3	2	
Palumbo	0	5	−5	−3	3
Stout	15	0	15	8	
Watson	12	1	11	7	

$$T = \overline{4}$$

etc. Then, the minuses, if there are any, are brought over to the rank-of-d column from the d column.

The last column, rank with less frequent sign, contains the ranks of the numbers of differences in the opposite direction (least frequent sign). The symbol T is the sum of the smaller like-signed ranks (for our hypothetical problem, $T = 4$). To test for significance, the researcher uses T as the calculated value and refers to a table of T critical values. He locates the N (8 in the above problem) and the predetermined level of significance. If the calculated value T is *less* than the critical value, the researcher rejects the null hypothesis at a specified level of significance (i.e., $p < .05$ or $p < .01$).

EXCERPT 10.9 RESULTS OF WILCOXON TEST REPORTED IN A TABLE

INDEX RATINGS OF INTELLECTUAL FUNCTIONING FOR MATCHED
PAIRS OF CHILDREN WITH DEAF OR HEARING PARENTS

RATING SCALE ITEM	NO. OF PAIRS	NO. WHERE CHILDREN WITH DEAF PARENTS RATED HIGHER	WILCOXON T VALUE	z
Intellectual ability	55	44 (80%)	255.5	4.31**
Use of intellectual ability	54	36 (67%)	326.0	3.58**
Works hard—strives to achieve	56	34 (61%)	619.0	1.46*

* $p = .07$.
** $p \leq .01$.
SOURCE: K. P. Meadow. Early manual communication in relation to the deaf child's intellectual, social, and communicative functioning. *American Annals of the Deaf*, 1968, *113*, 29–41.

There are typically two formats for reporting the results of the Wilcoxon matched-pairs signed-ranks test. Excerpt 10.9 illustrates the way the Wilcoxon test can be reported in a table. The number of matched pairs (55, 54, 56) for each Wilcoxon T value is greater than 25. As with the sign test, when N is larger than 25, the critical values found in a T table cannot be used. Instead, the author uses the T values to calculate z values and then uses the z table of critical values. The last column of the table reports the z values (4.31, 3.58, 1.46) for each Wilcoxon T value (255.5, 326.0, 619.0).

Excerpt 10.10 and 10.11 show the format used to report the Wilcoxon test in the context of the results section of an article. In the first excerpt the results of the first test were significant ($p < .005$) but the results of the second test were not ($p > .05$). The example in the second excerpt shows a slight variation of report format.

EXCERPT 10.10–10.11 RESULTS OF WILCOXON TEST REPORTED WITHOUT A TABLE

There was a statistically significant difference in responsiveness between the first and second quiet sleep epochs (Wilcoxon matched-pairs signed-ranks tests, $T = 0.00$ $p < .005$, $N = 14$). . . . Table 2 presents the means of behavioral response values during these sleep states. The difference was not statistically significant (Wilcoxon matched-pairs signed-ranks test, $T = 14.50$, $p > .05$, $N = 7$).

SOURCE: K. Schmidt and B. Burns. The behavioral arousal threshold in infant sleep as a function of time and sleep state. *Child Development*, 1971, *42*, 269–278.

A Wilcoxon matched-pairs signed-ranks test showed the difference between the two samples to be significant at greater than the .01 level of significance ($T = 0$, $N = 7$, $p < .01$).

SOURCE: S. Granowsky and W. J. Krossner. Kindergarten teachers as models for children's speech. *Journal of Experimental Education*, 1970, *38*, 23–28.

Median Test

In the last two sections of this chapter we have considered two nonparametric statistical tests (sign and Wilcoxon) that are analogous to the parametric correlated samples t test. In this section and the next, we present two nonparametric statistical tests that are analogous to the parametric independent samples t test.

The nonparametric *median test* is a procedure that a researcher can use to test whether two or more independent samples (groups) differ in central tendency. The median procedure tests the probability that two groups will have the same median. The null hypothesis of the median test is that the two groups are from populations with the same median. The median test is usually used when N is between 20 and 40.[2]

---

[2] There is also a requirement that no cell has an expected frequency of less than 5. For an explanation of expected cell frequency-requirement, refer to the section on chi-square on p. 219.

Suppose a researcher wanted to see if the frequency of a certain class of responses of 4-year-olds is a function of adult social approval and socioeconomic level. The results of our hypothetical experiment might appear as in Table 10.6. To use the median test, the researcher first determines the

TABLE 10.6 STUDY ILLUSTRATING USE OF MEDIAN TEST

	MIDDLE SOCIOECONOMIC LEVEL CHILDREN	LOW SOCIOECONOMIC LEVEL CHILDREN	TOTAL
ABOVE THE COMBINED MEDIAN	3	12	15
BELOW THE COMBINED MEDIAN	13	2	15
TOTAL	16	14	

combined median of responses for both groups of middle and low socioeconomic children. He would then construct a table showing the frequency of children in each group that are above and below the combined median. In Table 10.6, 3 children were above and 13 were below the combined median for the middle socioeconomic level children; 12 children were above and 2 were below the median for the low socioeconomic level children. Table 10.6 also shows the total number of children for each group (16 middle and 14 low socioeconomic).

To compute the calculated value for the median test, a researcher uses a chi-square (χ^2) formula. He then refers to a chi-square table of critical values which contains critical values for each number of degrees of freedom (1 to 30) at the predetermined level of significance (.001, .01, .02, .05, etc.). To find the number of degrees of freedom, a researcher subtracts one from the number of samples or groups. Our hypothetical example with two groups (high and low socioeconomic) has a df of 1 ($k - 1$). (The df will always be equal to 1 when the median test is used to compare two groups.) The researcher locates the critical value in the chi-square table at the intersection of the number of degrees of freedom and the predetermined level of significance. He would reject the null hypothesis if the calculated value is *greater* than the tabled critical value. Note that for the median test (or for any test in which a researcher uses a chi-square table) the calculated value must be greater than the critical value to reject the null hypothesis. The only table of critical values used for the median test is the chi-square table.

The formats for reporting the median test are the same as those we have considered in the previous sections. Excerpt 10.12 shows the use of a

---

EXCERPT 10.12 RESULTS OF MEDIAN TEST REPORTED IN A TABLE

TABLE 1 ACADEMIC AND SOCIAL SUCCESS IN RELATION TO SELF-EVALUATION

SELF-ESTEEM	*IOWA ACHIEVEMENT TEST*		*SOCIOGRAM*	
	ABOVE CLASS MEDIAN	BELOW CLASS MEDIAN	ABOVE CLASS MEDIAN	BELOW CLASS MEDIAN
Above Class Median	29	15	30	17
Below Class Median	18	25	13	27
	$\chi^2 = 5.1$; $p < .05$		$\chi^2 = 8.3$; $p < .01$	

SOURCE: S. Coopersmith. A method for determining types of self-esteem. *Journal of Abnormal and Social Psychology*, 1959, 59, 87–94. Copyright 1959 by the American Psychological Association. Reprinted by permission.

---

table to report the results of the median test. The samples or groups in this study are quite large. The median test is not generally used with samples this large. Excerpt 10.13 illustrates the format for reporting the results of the

---

EXCERPT 10.13 RESULTS OF MEDIAN TEST REPORTED WITHOUT A TABLE

A Median Test applied to the individual scores shows that those boys who received mean liking ratings which were above the median were more frequently chosen on the Treasure Hunt by 50% or more of their own cabinmates than were the boys who received average liking ratings below the median ($\chi^2 = 3.46$, $df = 1$, $p < .05$, one-tailed).

SOURCE: A. J. Lott and B. E. Lott. Some indirect measures of interpersonal attraction among children. *Journal of Educational Psychology*, 1970, 61, 124–135. Copyright 1970 by the American Psychological Association. Reprinted by permission.

---

median test in the results section of a journal article. In this excerpt the authors used the one-tailed test that we discussed in Chapter 3.[3]

---

[3] Recall the difference between a two-tailed test and one-tailed test. If the alternative hypothesis (alternative to the null hypothesis) is that the means or medians of two populations are different but the researcher is unsure about which one will be larger, the researcher will use a two-tailed test. If the alternative hypothesis is that the mean or median of one population is higher than the other, the researcher will use one-tailed test.

Mann-Whitney U Test

The *Mann-Whitney U test* is also analogous to the parametric independent samples t test. It tests whether there is a significant difference between two independent samples. Since the Mann-Whitney U test is a more powerful test than the median test, it is better as a nonparametric alternative to the t test. The null hypothesis for the Mann-Whitney U is that the two samples are from the same distribution. Like the independent samples t test, the Mann-Whitney U can be used with two samples of unequal number.

For a hypothetical example, suppose a researcher wanted to determine if there was a difference in the loss of hearing for factory workers at the beginning of an assembly line and workers at the end of an assembly line. All workers were measured for hearing loss annually and the numbers in Table 10.7 indicate the amount of hearing loss that occurred in one year (mea-

---

TABLE 10.7 STUDY ILLUSTRATING
THE USE OF MANN-WHITNEY U TEST

WORKERS NEAR THE BEGINNING OF THE ASSEMBLY LINE	RANK	WORKERS NEAR THE END OF THE ASSEMBLY LINE	RANK
10	6	12	8
15	9	3	1
9	5	11	7
5	2	7	4
6	3		
	$R_1 = 25$		$R_2 = 20$

---

sured in decibels). After collecting the data for each group, the researcher must rank the decibels across both groups from the lowest to the highest. Stated in terms of this process, the null hypothesis would be that the sums of the ranks for each group are equal, that is, that $R_1 = R_2$. From these sums of ranks (R) the researcher computes the calculated U value and refers to the appropriate table which gives critical values of U at different levels of significance. There is one table of probabilities for situations in which neither group is larger than 8, and another table for the cases in which one of the samples is between 9 and 20. As with the sign and Wilcoxon tests, the Mann-Whitney test requires finding a z value if one of the sampled groups is larger than 20.

Excerpts 10.14 and 10.15 show two ways of reporting the Mann-Whitney U. In Excerpt 10.14 results are reported in a table of z values because N for one group is larger than 20. In Excerpt 10.15 the results of the test are reported without a table. The authors of this excerpt refer to the null hypothesis in terms of the groups representing "the same population distribution."

**EXCERPT 10.4 RESULTS OF
MANN-WHITNEY U TEST REPORTED IN A TABLE**

	Attitude Measure			
	SUBJECT MATTER		*TEACHING MODE*	
STATISTIC	N GROUP	C GROUP	N GROUP	C GROUP
No. Cases	34	34	34	34
R	1009	1334	1043	1303
U		742°		708
Z_0		2.02		1.60

° $p < .05$.

SOURCE: E. C. Pack. The effects of testing upon attitude towards the method and content of instruction. Reprinted from the Summer 1972 issue of *Journal of Educational Measurement*, 9, No. 2. Copyright 1972 by the National Council on Measurement in Education, East-Lansing, Michigan.

**EXCERPT 10.15 RESULTS OF MANN-WHITNEY
U TEST REPORTED WITHOUT A TABLE**

The difference scores for the overpay group and the underpay group were compared separately against the difference scores for the control group by the Mann-Whitney U test (see Table 1). The value of U obtained for the overpay versus control comparison was 48; for the underpay versus control comparison, U was 47. A U of 26 or less would have been significant at the .05 level for each comparison. Since neither comparison was significant, the null hypothesis (that the control group represented the same population distribution on quantity as the overpay and underpay groups) was not rejected.

SOURCE: E. R. Valenzi and I. R. Andrews. Effects of hourly overpay and underpay inequity when tested with a new induction procedure. *Journal of Applied Psychology*, 1971, 55, 22–27. Copyright 1971 by the American Psychological Association. Reprinted by permission.

NONPARAMETRIC ANALOGS TO THE PARAMETRIC ONE-WAY ANALYSIS OF VARIANCE

Kruskal-Wallis One-Way Analysis of Variance of Ranks

The nonparametric Kruskal-Wallis one-way ANOVA is analogous to the parametric one-way ANOVA discussed in Chapter 4. The Kruskal-Wallis ANOVA (H) is used when a researcher wants to determine whether three or more independent samples come from the same population. It is not necessary to have an equal number of subjects or measurements for each sample. Again, when the researcher feels it is necessary to avoid the assumptions of the one-way ANOVA (F test), when the measurement is weaker than the interval level, or when the number in each independent sample is small (less than five), the Kruskal-Wallis H test might be a good alternative.

Let us examine a hypothetical example. Suppose a researcher selects three random samples of sophomore, junior, and senior students in a work-study program and administers an attitude scale to the three samples in order to assess their attitudes toward the program. A high score represents a positive attitude and a low score a negative feeling. The results of the attitude scores appear in Table 10.8. There are four sophomores, five juniors, and five seniors, or a total of 14 subjects. As with the Mann-Whitney U, the

TABLE 10.8 STUDY ILLUSTRATING THE USE OF THE KRUSKAL-WALLIS H

SOPHOMORES	RANK	JUNIORS	RANK	SENIORS	RANK
17	8	12	5	25	14
14	6	11	4	19	10
4	1	21	11	24	13
10	3	6	2	15	7
		18	9	23	12
$N_1 = 4$	$R_1 = 18$	$N_2 = 5$	$R_2 = 31$	$N_3 = 5$	$R_3 = 56$

researcher ranks the attitude scores without regard to the group in which the score belongs. A rank of 1 is assigned to the lowest score (4), the rank of 2 is assigned to the second lowest (6), and so forth. The researcher then applies a formula to obtain the calculated H value from these ranks and refers to the H table of critical values. If the calculated H value exceeds the critical value at the predetermined level of significance, the researcher rejects the null hypothesis. The H table is designed for three samples and for research problems in which no sample size is greater than five. If a researcher has more than three samples or has more than five scores or observations for any given sample, he has to use a χ^2 table of critical values with $k - 1$ degrees of freedom (k being equal to the number of samples). As with the H table, a researcher rejects the null hypothesis if the calculated χ^2 value is greater than the χ^2 tabled critical value at the predetermined level with $k - 1$ degrees of freedom.

Excerpts 10.16 and 10.17 illustrate the ways in which the Kruskal-Wallis one-way ANOVA is most frequently reported in journal articles. Excerpt 10.16 shows that the author used a table to present the results of the H between-groups for eight categories of written responses. Five of eight categories had a significant H at the .05 level; the last three in the bottom row were not significant at the required level. If the table caption did not indicate that there were four groups, the reader could determine the number of samples by adding one to the degrees of freedom (3 + 1). Excerpt 10.17 shows the way results are reported without a table. Again, the reader knows that there are three groups ($df = 2$) and that at least one of the groups has more than five scores because the calculated value is reported as a χ^2.

EXCERPT 10.16 RESULTS OF KRUSKAL-WALLIS REPORTED IN A TABLE

KRUSKAL-WALLIS H VALUES BETWEEN FOUR GROUPS
ON EIGHT CATEGORIES OF WRITTEN RESPONSES[a]

CATEGORY	THORNDIKE-LORGE LOW FREQUENCY HIGH FREQUENCY	PSYCHOLOG-ICAL TERMS TYPE	TYPE TOKEN	INDEFINITE SENTENCE
H between groups	21.29[a]	31.00[a]	13.34[a]	15.22[a]

	AMBIVALENCE CONSTRUCTIONS TOKEN	VERBS INTRANSITIVE TRANSITIVE	REFLEXIVES SENTENCES	RECOMMENDA-TIONS FOR TREATMENT SPECIFIC AMBIGUOUS
H between groups	17.02[a]	7.32	6.38	2.49

[a] Value required for significance at the .05 level, df = 3, is 7.82.
SOURCE: J. M. Zimmer. Content analysis of counselor and teacher responses. *Personnel and Guidance Journal*, 1968, 46, 456–461. Copyright 1968 by the American Personnel and Guidance Association.

EXCERPT 10.17 RESULTS OF KRUSKAL-WALLIS REPORTED WITHOUT A TABLE

A Kruskal-Wallis analysis of variance indicates a significant effect ($\chi^2 = 8.02$, $df = 2$, $p < .02$).

SOURCE: B. Weiner, H. Heckhausen, W-U. Meyer, and R. E. Cook. Causal ascription and achievement behavior: a conceptual analysis of effort and reanalysis of locus of control. *Journal of Personality and Social Psychology*, 1972, 21, 239–248. Copyright 1972 by the American Psychological Association. Reprinted by permission.

Friedman Two-Way Analysis of Variance of Ranks

The nonparametric Friedman test is called a two-way analysis of variance, but it is more equivalent to the parametric one-way ANOVA than to the parametric two-way ANOVA. The Friedman tests whether three or more samples of data come from the same population. The test is appropriate for situations in which the same subjects are measured repeatedly. For example, each child in an experiment might be exposed to three different kinds of adult social attention and two control conditions. Each child in such a study serves as his own control. The test can also be used for subjects matched on some characteristic and exposed to a number of different experimental conditions. In this situation a researcher would have several sets of subjects (each set containing so many matched subjects), and then one subject in each set would be randomly assigned to one of the experimental conditions. For example, children might be matched to four different age groups, with one child from each age group (set) randomly assigned to each experimental condition. If a researcher wants to avoid the assumptions of the one-way repeated measures ANOVA (F test), or for the other reasons mentioned in the introduction of this chapter, the Friedman test is a good alternative.

For a hypothetical example, suppose a researcher wanted to assess the effects of four experimental conditions on the responses of eight children. The study might be a behavior modification study that uses a reversal design—two baseline conditions, A_1 and A_2, and two treatment conditions, B_1 and B_2 (see Chapter 15 for a discussion of this design). The results of our hypothetical example appear in Table 10.9. The total number of responses

TABLE 10.9 STUDY ILLUSTRATING THE USE OF THE FRIEDMAN TWO-WAY ANALYSIS OF VARIANCE OF RANKS

	EXPERIMENTAL CONDITION				RANKED DATA			
	A_1	B_1	A_2	B_2	A_1	B_1	A_2	B_2
Charlotte	10	21	12	24	1	3	2	4
Emily	5	31	13	30	1	4	2	3
Jim	11	20	9	21	2	3	1	4
Joe	8	19	10	23	1	3	2	4
Ken	14	24	13	29	2	3	1	4
Linda	15	33	20	40	1	3	2	4
Mariana	11	23	18	28	1	3	2	4
Tom	7	18	12	17	1	4	2	3
					$R_1 = 10$	$R_2 = 26$	$R_3 = 14$	$R_4 = 30$

for each child is given under each experimental condition (A_1, B_1, A_2, B_2). Unlike previous tests, the data are not ranked over all groups. With the Friedman test the researcher ranks each set of data for each subject. For example, the researcher would rank Charlotte's scores assigning 1 to the score of 10, 2 to the score of 12, 3 to the 21, and 4 to the 24. Using the ranks for each subject, the researcher applies a formula to obtain a calculated value (χ_r^2) and refers to the χ_r^2 table of critical values which gives the exact probability associated with the calculated value. These exact probabilities can be directly compared to the researcher's level of significance. If the number of experimental conditions is greater than 4 or the number of subjects is greater than 9, the χ^2 table of critical values is used with the degrees of freedom equal to one less than the number of groups.

Excerpts 10.18 and 10.19 illustrate the two ways in which the Friedman test is most frequently reported. In Excerpt 10.18 the author used a table to present the results of the Friedman test applied to pupillary change as a function of five stimulus conditions for three age groups.

EXCERPT 10.18 RESULTS OF FRIEDMAN TEST REPORTED IN A TABLE

Table 2 contains rank orders for change in pupil size, based on both the amount and direction of pupillary change for each stimulus presented. The stimulus eliciting the greatest pupillary constriction received a rank of 1, and the stimulus elicit-ing the greatest pupillary dilation received a rank of 5. Friedman's nonparametric analysis of variance by ranks (Siegel, 1956) indicated that change in pupil size for the five stimuli differed across the three age groups.

RANK ORDERS OF STIMULI BASED ON AMOUNT AND DIRECTION OF PUPILLARY CHANGE[a]

			STIMULUS RANKS[b,c]		
GROUP	MOTHER	STRANGER	4-SQUARE CHECKERBOARD	144-SQUARE CHECKERBOARD	TRIANGLE
1m	5 (+.04)	1.5 (−.04)	3 (−.02)	4 (+.03)	1.5 (−.04)
2m	5 (+.11)	3 (+.02)	1 (−.04)	2 (+.004)	4 (+.04)
4m	3 (−.02)	5 (+.11)	4 (+.05)	1.5 (−.04)	1.5 (−.04)

[a] The mean differences on which the ranks are based are shown in parentheses.
[b] The stimulus receiving the largest pupillary constriction was ranked 1, and rank 5 was assigned to the stimulus producing the largest pupillary dilation.
[c] $\chi2 = 31.72$, $df = 4$, $p = < .001$.
SOURCE: H. E. Fitzgerald. Autonomic pupillary reflex activity during early infancy and its relation-ship to social and nonsocial visual stimuli. *Journal of Experimental Child Psychology*, 1968, *6*, 470–482.

Excerpt 10.19 shows the way the Friedman Test can be reported in the results section of a journal article. The authors used four Friedman tests to determine whether there was significant variation among the proportion of choices of various cues.

EXCERPT 10.19 RESULTS OF FRIEDMAN TEST REPORTED WITHOUT A TABLE

Differences were not significant for kindergarten subjects, either for the tri-grams ($\chi^2 = 1.05$, $df = 3$) or the quin-grams ($\chi^2 = 7.53$, $df = 5$). However, for first-graders, analysis of the trigram data showed variation significant at the .001 level ($\chi^2 = 21.62$, $df = 3$) and for the quingram data at the .005 level ($\chi^2 = 19.68$, $df = 5$).

SOURCE: J. B. Williams, E. L. Blumberg, and D. V. Williams. Cues used in visual word recognition. *Journal of Educational Psychology*, 1970, *61*, 310–315. Copyright 1970 by the American Psy-chological Association. Reprinted by permission.

Multiple Comparisons for the Kruskal-Wallis and Friedman ANOVA's

In Chapter 4 we discussed multiple comparison procedures that a researcher can use following a significant parametric F test. A researcher can also perform multiple comparisons among the groups (conditions or samples) following an overall test of significance for the two nonparametric tests discussed in the previous two sections. A significant Kruskal-Wallis H or Friedman χ_r^2 does not reveal which groups are significantly different, but several procedures can be used to make multiple comparisons if the re-

---

EXCERPT 10.20 USE OF THE MANN-WHITNEY
U TEST WITH THE KRUSKAL-WALLIS ANOVA

TABLE 2 RESULTS OF THE KRUSKAL-WALLIS ANALYSIS

SUBSCALE	H	H_o [*]
Understanding	181.800	181.844†
Probing	36.639	36.665†
Interpretive	46.099	46.157†
Supportive	108.919	109.944†
Evaluative	212.612	212.699†

[*] Values of H corrected for ties in ranks.
† Significant at .001 level ($df = 10$).

TABLE 3 RESULTS OF SELECTED INDIVIDUAL COMPARISONS USING THE MANN-WHITNEY U TEST

COMPARISON		
GROUPS	SUBSCALE	z [*]
Helping professions vs. all others	Understanding	15.45†
	Probing	3.25†
	Interpretive	−3.16†
	Supportive	−9.70†
	Evaluative	−13.68†
All counselors vs. ministers	Understanding	3.09†
	Probing	1.22
	Interpretive	−0.80
	Supportive	−2.37‡
	Evaluative	−2.70†
All counselors vs. nurses	Understanding	−2.25‡
	Probing	1.85
	Interpretive	−2.73†
	Supportive	1.98‡
	Evaluative	2.41‡
All counselors vs. lawyers	Understanding	6.65†
	Probing	−2.81†
	Interpretive	2.55‡
	Supportive	−5.30†
	Evaluative	−6.69†
All counselors vs. firemen	Understanding	10.80†
	Probing	4.74†
	Interpretive	−3.48†
	Supportive	−8.05†
	Evaluative	−11.15†

[*] Since the larger n is greater than 20 in all comparisons, U is distributed approximately normally (Siegel, 1956, pp. 120–121).
† Significant at .01 level.
‡ Significant at .05 level.

SOURCE: J. E. Jones. Helping-relationship response tendencies and occupational affiliation. *Personnel and Guidance Journal*, 1967, 45, 671–675. Copyright 1967 by the American Personnel and Guidance Association.

searcher wishes to determine which among all the k groups are significantly different from each other. The most frequently reported follow-up test for the Kruskal-Wallis one-way analysis of variance of ranks is the Mann-Whitney U test. A recommended procedure as a follow-up test to the Friedman ANOVA is the Wilcoxon matched-pairs signed-ranks test.

Most researchers will report the use of a follow-up test after a significant H or χ_r^2 is obtained. Excerpt 10.20 illustrates the use of the Kruskal-Wallis ANOVA with selected individual comparisons made with the Mann-Whitney U test. Notice in the first note to Table 2 that the values of H were corrected for ties in ranks, a procedure most researchers follow when there are many ties. Correction for ties increases the calculated value of H and tends to make the result more significant than it would have been if not corrected. In other words, when a researcher does not correct for ties, the test is considered more conservative. As we can see in Table 2, the calculated H value for understanding is 181.800 and the corrected calculated value, H_0, is increased to 181.844. Also, in Excerpt 10.20, the first note in Table 3 tells the reader that n was larger than 20; hence, the z table of critical values was used for the Mann-Whitney U.

Excerpt 10.21 illustrates how the results of multiple comparisons can be reported in the text of the results section. The multiple comparisons are for the Kruskal-Wallis analysis of variance illustrated in Excerpt 10.17.

EXCERPT 10.21 MULTIPLE COMPARISONS REPORTED WITHOUT A TABLE

Further nonparametric analyses with the Mann-Whitney U test yield significant differences between the high and low and middle and low groups ($z = 2.88$, $p < .005$; $z = 2.00$, $p < .05$, respectively).

SOURCE: B. Weiner, H. Heckhausen, W-U. Meyer, and R. E. Cook. Causal ascription and achievement behavior: A conceptual analysis of effort and reanalysis of locus of control. *Journal of Personality and Social Psychology*, 1972, *21*, 239–248. Copyright 1972 by the American Psychological Association. Reprinted by permission.

CHI-SQUARE

Single Sample Chi-Square Test

In the last eight sections we have considered nonparametric statistical procedures that can be used with at least ordinal level of measurement. In this section and the next we present two nonparametric procedures that can be applied to the nominal level of measurement.

The single sample chi-square (χ^2) test is one of the most frequently reported nonparametric tests in journal articles. The test is used when a researcher is interested in the number of responses, objects, or people that fall in two or more categories. This procedure is sometimes called a goodness-of-fit statistic. Goodness-of-fit refers to whether a significant difference exists between an observed number and an expected number of responses, people, or objects falling in each category designated by the researcher. The expected number is what the researcher expects by chance or according to some null hypothesis.

Suppose, for a hypothetical example, a college administrator wanted to determine if there is a significant difference between the number of students who actually sign up for six elective courses (the observed number) and the number expected by chance. The administrator might expect that by chance alone an equal number of students would enroll in each elective course. The null hypothesis for this study thus becomes that an equal number of students sign up for all six elective courses. Table 10.10 presents the hypothetical data.

TABLE 10.10 STUDY ILLUSTRATING THE USE OF SINGLE SAMPLE CHI-SQUARE TEST

	MYSTI-CISM	PROGRES-SIVE JAZZ	ZEN MEDIA-TION	CHANG-ING SEX ROLES	INTIMACY	SOCIAL CHANGE
Expected	20	20	20	20	20	20
Observed	35	12	21	22	11	19

After the observed number is obtained, the administrator uses a formula to compute the calculated χ^2 value. Then he refers to a chi-square table of critical values which contains critical values for each number of degrees of freedom (1 to 30) at the predetermined level of significance (.001, .01, .05, etc.). To find the number of degrees of freedom for the above study, the administrator would subtract one (1) from the number of categories (k). Thus, the number of degrees of freedom for the study in Table 10.10 is 5 ($k - 1$, or $6 - 1$). If the calculated χ^2 value is greater than the critical value at the previously established level of significance, the researcher would reject the null hypothesis.

As with the other nonparametric procedures, there are typically two formats used to report the results of the single sample chi-square test. Excerpts 10.22 and 10.23 illustrate how the results can be reported in a table or within the context of the results section. Notice in Excerpt 10.22 that the expected cell frequencies of E_1 and E_2 for both plural nouns and humans are 5. If a researcher uses the single sample chi-square test when any one expected cell (category) frequency is less than 5, he distorts and possibly inflates the χ^2 calculated value. The results become an overestimate and constitute a more liberal test of significance.[4] The expected cell frequencies in

---

[4] An adjustment procedure called the Yates correction for continuity can be used to prevent the chi-square test from being too liberal.

TABLE 3 OBSERVED AND EXPECTED FREQUENCIES OF INDIVIDUAL SUBJECTS
FALLING IN EACH OF FOUR POSSIBLE PATTERNS ACROSS PERIODS[a]

	N	A		B		C		D		CHI SQUARE
Plural nouns										
E_1	20	5	(5)[b]	5	(5)	4	(5)	6	(5)	.20
E_2	20	4	(5)	6	(5)	4	(5)	6	(5)	.20
Both Es	40	9	(10)	11	(10)	8	(10)	12	(10)	.50
Humans										
E_1	20	1	(5)	18	(5)	1	(5)	0	(5)	40.20**
E_2	20	5	(5)	13	(5)	0	(5)	2	(5)	16.60**
Both Es	40	6	(10)	31	(10)	1	(10)	2	(10)	56.10**

[a] The following four patterns for the three periods were considered the most probable:
A. Base–Up–Up
B. Base–Up–Down (apparent conditioners)
C. Base–Down–Up
D. Base–Down–Down
[b] Numbers enclosed in parentheses are expected values.
** $p < .01$.

SOURCE: J. D. Matarazzo, G. Saslow, and E. N. Pareis. Verbal conditioning of two response classes: some methodological considerations. *Journal of Abnormal and Social Psychology*, 1960, *61*, 190–206. Copyright 1960 by the American Psychological Association. Reprinted by permission.

Table 3 are large enough (5) so that the calculated chi-squares are not over-estimates. Of course, it is impossible for the reader to determine what the expected cell frequencies are in Excerpt 10.23.

A one-sample chi-square test was used to determine whether the frequency of correct test responses differed among the four treatment groups. The obtained $\chi^2 = 15.89$, $df = 3$, was significant at the .01 level.

SOURCE: M. R. Vitale, F. J. King, D. W. Shontz, and G. M. Huntley. Effect of sentence-combining exercises upon several restricted written composition tasks. *Journal of Educational Psychology*, 1971, *62*, 521–525. Copyright 1971 by the American Psychological Association. Reprinted by permission.

Independent Samples Chi-Square Test

The chi-square test may also be used for two or more independent samples. A researcher might be interested in determining whether or not the observations are significantly different from what might be expected by chance. Suppose a researcher wanted to see whether significant differences in favoring or opposing innovation exist between tenured and nontenured teachers. After collecting the answers to a questionnaire, the researcher might obtain the hypothetical observed cell frequencies that appear in Table 10.11. The table shows that 40 teachers without tenure favored innovation

TABLE 10.11 STUDY ILLUSTRATING THE USE
OF INDEPENDENT SAMPLES CHI-SQUARE TEST

	WITHOUT TENURE	WITH TENURE	
FAVORING INNOVATION	40	20	60
OPPOSED TO INNOVATION	10	30	40
	50	50	100

and 10 opposed innovation. For teachers with tenure, 20 were in favor of innovation and 30 were opposed. The marginal totals (50, 50, 60, 40) are used to compute the expected cell frequencies.[5] The observed cell frequencies and the computed expected frequencies are used in a χ^2 formula to obtain the calculated value. Then, the researcher looks up the critical value in the χ^2 table for the stated level of significance and degrees of freedom. The number of degrees of freedom is obtained by subtracting one from the number of rows and one from the number of columns, and then multiplying the remainders: $df = (r - 1)(c - 1) = (2 - 1)(2 - 1) = 1$.

The independent samples chi-square is frequently referred to as being based on a *contingency table*. We might refer to Table 10.11 as a 2 × 2 contingency table. The chi-square test can also be used for a larger number of rows or columns such as: 2 × 3, 3 × 4, 5 × 8, 10 × 10, etc. The same procedure would be used for computing the expected cell frequencies, the calculated value, the degrees of freedom, and finding the critical value. Again, the researcher rejects the null hypothesis if the calculated value exceeds the critical value. Also, the researcher should be aware of an inflated and distorted calculated χ^2 value if any of the expected cell frequencies is less than 5. The effect of having expected cell frequencies less than 5 makes the χ^2 test of significance extremely liberal, and again the Yates correction can be used. It is sometimes difficult for the reader to know what the actual computed expected cell frequencies are since these data are not usually reported in tables in the article. In Excerpts 10.24 and 10.25 the two formats for presenting independent samples chi-square test results are illustrated.

[5] The expected cell frequencies can be computed from Table 10.11. For example, if we wanted the expected frequency for the top left cell, we would use the marginal totals as: (50)(60)/100 = 30, for the lower right cell (50)(40)/100 = 20, etc.

EXCERPTS 10.24 and 10.25 TWO FORMATS FOR REPORTING
RESULTS OF INDEPENDENT SAMPLES CHI-SQUARE TEST

TABLE 2 2×2 CONTINGENCY TABLES FOR VALUES RANKED
FIRST (YES) AND NOT FIRST (NO) BY ADJUSTMENT GROUPS

	Yes	No	Yes	No
	Self-interpretation		Self-actualization	
Adjustment	45	38	13	70
Maladjustment	9	22	7	24
	$\chi^2 = 6.80, p < .05$		$\chi^2 = 0.38$	
	Relationship			
Adjustment	7	76		
Maladjustment	8	23		
	$\chi^2 = 4.53, p < .05$			

SOURCE: S. Katkin and E. Weisskopf-Jollson. Values and emotional adjustment. *Psychological Reports*, 1971, 28, 523–528. Reprinted by permission of author and publisher.

A 2×2 chi-square analysis was performed investigating the effects of Type of Training Problem and Dimension. The Ss were classified as having solved the original problem with or without help from E, and only the Dimension effect was reliable $(\chi^2_{(1)} = 8.53, \quad p < .01)$. More Ss required help when color was the relevant dimension.

SOURCE: J. C. Campione and V. L. Beaton. Transfer of training: some boundary conditions and initial theory. *Journal of Experimental Child Psychology*, 1972, 13, 94–114.

REVIEW TERMS

χ^2
chi square
χ_r^2
contingency table
correction for ties
direction of score differences
expected frequency
Friedman
H
independent samples
interval
k
Kruskal-Wallis
magnitude of change

Mann-Whitney U test
median test
nominal
observed frequency
ordinal
r_s
rho
sign test
single sample
Spearman rank-order
    correlation coefficient
T
Wilcoxon matched-pairs
    signed-ranks test

REVIEW QUESTIONS

1. What three factors may determine a researcher's decision to use a nonparametric statistical test rather than a parametric procedure?

2. Give an example of a nominal scale, an ordinal scale, and an interval scale.

3. What are five similarities between parametric and nonparametric procedures?

4. The Mann-Whitney U is an appropriate follow-up test for which one of the nonparametric tests?

5. List four similarities between the Pearson product-moment correlation coefficient and Spearman rank correlation coefficient.

6. Why is the Wilcoxon matched-pairs signed-ranks test considered to be more powerful than the sign test?

7. Which nonparametric test(s) can be used with nominal level data?

8. The Kruskal-Wallis is to the Mann-Whitney U as the Friedman is to the _____.

9. Which nonparametric test(s) is (are) analogous to the independent samples t test?

10. The parametric one-way ANOVA with repeated measures is similar to which nonparametric test?

RESEARCH DESIGN

In our discussion of inferential statistical analyses we have found that these statistical procedures are used by a researcher to compare the obtained results (dependent variable) with chance expectations. Also, we discovered that these procedures aid the researcher in making reliable inferences from the data or results. For example, a researcher can either reject or fail to reject a particular hypothesis. The variety of univariate, multivariate, and nonparametric inferential procedures that we presented in the last seven chapters can be used descriptively without a good experimental design. But statistical inference alone does not lead to a causal inference. Experimental design is the necessary basis for tests of statistical significance. In this part of the book, we present several types of experimental designs.[1]

An experimental or research design has two purposes: (1) to help a researcher answer a research question and (2) to control for possible rival hypotheses or extraneous variables

[1] We acknowledge a tremendous credit to Donald Campbell, Julian Stanley, Gene Glass and Glenn Bracht, for stimulating our thinking through their writings. They have much to do with the organization and content of the following chapters.

that might compete with the independent variable as an explanation for the cause-effect relationship. For example, a researcher might be interested in the effect of a new weight-reducing program (independent, experimental treatment, or intervention variable) on weight loss (dependent or outcome variable) of professional football players. In this example the independent variable (weight-reducing program) is presumed to effect a change on the dependent variable. But how can the researcher be sure that any observed weight loss is truly related to the administered program rather than to a different causal agent which is not identified as the independent variable? A poor experimental design does not eliminate or correct for extraneous variables or rival hypotheses (explanations) which might effect changes on the dependent variable in addition to the independent variable. As we can see, statistical tests are a researcher's tool for analysis, but they are limited in establishing causal relationships. The quality of the experimental design determines the degree to which a researcher can exercise experimental control, that is, the extent to which rival hypotheses or extraneous variables can be controlled or ruled out. Experimental control is associated with how much freedom or control a researcher has in the following: (1) the random assignment of individual subjects to comparison groups, (2) the extent to which the independent variable can be manipulated by the researcher, and (3) the time when the observations or measurements of the dependent variable occur, and (4) which groups are measured.

There are two criteria for evaluating the validity of any experimental design. The first criterion, *internal validity,* refers to casual relationships. Did the independent (experimental, treatment, or intervention) variable make a difference in the study, that is, can the researcher infer a cause-and-effect relationship? The second criterion, *external validity,* refers to representativeness or generalizability. To what extent can the results of the study be generalized to other populations, settings, treatments, or measurement variables? Of course, the ideal for the researcher is to select a design strong in both internal and external validity. However, this is not always possible, and the researcher should seek to have internal validity as the minimum requirement for the design he selects to use. Without internal validity the data from any study are very difficult to interpret. We will present the sources of internal and external threats as a guide for readers to evaluate research. Whether these threats apply depends on the particular circumstances of the study being evaluated.

In the following chapters we will discuss four types of experimental designs that have varying degrees of experimental control. The three pseudo- or preexperimental designs discussed in Chapter 11 have very limited experimental control. The three true experimental designs and their variations discussed in Chapters 12 and 13 can control a large number of possible rival explanations. The five quasi-experimental designs discussed in Chapter 14 can have more experimental control than the pseudoexperimental designs but not as much control as the true designs. There are many more quasi-experimental designs, but we feel that three of the five we present are among the most popular and appear most frequently in journal articles. In Chapter 15 we illustrate several operant or applied behavior analysis designs. Although they are really variations of the time-series designs presented in Chapter 14, we feel that they deserve separate consideration. As before, excerpts from journal articles are used to illustrate the method and analysis of results for each design. The reader should note that our discussion of example journal articles refers only to the design used and does not refer to the substantive significance of these studies.

CHAPTER 11

PSEUDOEXPERIMENTAL DESIGNS

In this chapter we present three pseudoexperimental or preexperimental designs: the one-shot case study, the one-group pretest-posttest design, and the static group comparison. These designs are called pseudoexperimental because they do not have built-in controls. In addition to the independent or treatment variable, there may be several other plausible explanations for the dependent variable changing or remaining the same. Researchers who use these designs have difficulty in assessing the effectiveness of the independent or treatment variable, and it is impossible to make completely valid inferences from the results of studies that use pseudoexperimental designs.

Because pseudoexperimental designs are used so frequently, we feel they are worthy of examination and discussion. There are more pseudoexperimental designs than the three we present in this chapter. However, examples of these designs should provide the reader with an awareness of their limitations. Our discussion of sample journal articles using these designs refer only to the design methodology and do not refer to the substantive significance of these studies. We do not intend to suggest that the problems or questions these studies seek to explore or answer do not make a possible contribution to their respective disciplines. However, the contributions of these studies would have been greater if the authors had used designs that control for the internal threats to valid inference. Also, the authors could have said more about the relationship between the independent and dependent variable. The three designs in this chapter do not provide satisfactory controls for the threats of internal validity. We present the seven internal threats to validity in this chapter; the external threats to validity are presented in Chapter 12, True Experimental Designs. The first design we consider has the most threats to valid inference.

ONE-SHOT CASE STUDY

The one-shot case study involves a group that has been exposed to some prior treatment (independent variable) or event. A diagram of the one-shot case study appears in Figure 11.1. X represents exposure of the group to

X	O
TREATMENT OR	OBSERVATION OR
INDEPENDENT	MEASUREMENT OF
VARIABLE	DEPENDENT VARIABLE

FIGURE 11.1. *Diagram of the one-shot case study.*

the independent variable; O refers to either measurement or observation of the dependent variable. The left-to-right dimension indicates the order in time of the X and O. In Figure 11.1 the independent variable X comes before the dependent variable O. As a hypothetical example, consider a researcher who wishes to see whether a film on sensitivity groups influences the level of apprehension before having a group experience. He would show the film first (X) and then measure (O) by a scale rating how the subjects felt after seeing the film.

This design can hardly be called an experimental study and perhaps should be more appropriately referred to as a descriptive study. The most obvious weakness of this design is the absence of control. All the researcher can do is describe the results of his observations; the design does not provide for a comparison of the results, so the researcher cannot compare the treatment results (measurements or observations) with the same group of subjects before treatment or with those of another group not exposed to X. The researcher in our hypothetical example does not have data from a pretest or another group with which he can compare his measures of the level of apprehension. The very minimum for gathering experimental information or evidence involves making at least one comparison.

The only comparisons or inferences a researcher can make from this design are based upon hypothetical data or common knowledge expectations of what might have happened to the dependent observations if X had not occurred. In other words, all the researcher or reader can do is make imagined conjectures about what the group was like before or without exposure to X. Thus, the researcher who uses this design knows little about the cause-and-effect relationship between the independent variable and the dependent variable. Changes in observations or measurement of the dependent variable cannot be attributed exclusively to exposure to the independent variable. Without some controlled comparison or control for internal threats to

---

EXCERPT 11.1 DESCRIPTION OF PURPOSE AND METHOD OF ONE-SHOT CASE STUDY

PURPOSE

The main purpose of the introductory statistics course described here was to arrange the conditions of learning so that a group of students with varied background and ability would master the course content. Performance and attitudinal data were collected to determine whether that purpose was realized and to evaluate the set of learning conditions employed.

METHOD

Contingencies

At the first class meeting, the instructor explained the rules of the course and gave a brief rationale for them. Students were told that they could work only in class; that the time allowed for work was a continuous 2.5 hour period each weekday; that they could arrive and leave whenever they chose, but the text book could not be taken out of class; that each unit of the book and the exercises following it had to be completed in writing before taking the exam for that unit; that each exam must be passed with fewer than three errors before beginning the next unit; that all errors on exams had to be corrected whether the exam was passed or not; that a grade of A would be given for passing all 24 unit exams and a grade of Incomplete for anything less; that individual tutoring could be had anytime by raising a hand or while an exam was being graded, but it had to be initiated by the student himself; that the instructor and teaching assistant were to be consid-

ered equals insofar as grading exams were concerned, and that the course was over when the last unit exam in the book was passed.

Unit exams varied in length and type of question depending on the nature of the material covered in the corresponding unit of the text. An effort was made, however, to include a number of questions requiring calculation from each unit. Students moved freely between the classroom and another room containing desk calculators.

Daily Study Sessions

A session began when the instructor unlocked the storeroom where the texts and exams were kept between sessions. Students found their own books and began work. When they were ready for an exam, they took it from a folder after showing the instructor or teaching assistant that the unit and the book exercises had been done. When they finished an exam, they had the option of bringing it either to the instructor or the teaching assistant for grading.

When they left for the day, books were left in the classroom, except in cases where the instructor gave the student the option of taking the book home for the weekend. This option was offered when a student needed more time to complete the work.

When asked on several occasions if they wanted lectures to supplement the text, only a handful said yes. Accordingly no lectures were given.

SOURCE: W. A. Myers. Operant learning principles applied to teaching introductory statistics. *Journal of Applied Behavior Analysis*, 1970, 3, 191–197. Copyright 1970 by the Society for Experimental Analysis of Behavior, Inc.

valid inferences, the rival explanations of differences in the dependent variable are numerous.

Example of Design

An illustration from the literature of the one-shot case study involves applying operant learning principles to teaching a college introductory statistics course. Excerpt 11.1 describes the method of instruction and the purpose of the study. The dependent variables were performance and responses to a questionnaire measuring attitudes. The performance measures were: number of days to finish the course, errors made on exams, and errors made in the programmed text. The attitude questionnaire had five parts, but we will only consider the first two parts, a description of which appears in Excerpt 11.2. If we diagram the one-shot case study as in Figure 11.1, we can show

---

EXCERPT 11.2 DESCRIPTION OF PARTS I AND II OF
ATTITUDE QUESTIONNAIRE OF ONE-SHOT CASE STUDY

PART I

In the first part were 15 items comparing the course as taught against what they imagined it would have been under the lecture method. Each item was rated on a seven-point scale whose mid-point was "no difference between the course as it was given and a lecture course" and whose end points were "a lot more by the course as it was given" and "a lot less by the course as it was given." Since they did not actually take the course by the lecture method (except for four students who had previously failed or dropped the course

when it was given by the lecture method), their judgments were necessarily between the course as it was taught and their impression of the lecture method from past experience.

PART II

In Part II, the students ranked a list of 13 features of the course in the order of "their importance to you." The fact that students have very different past histories of academic performance is reflected by the high variability of the rankings.

SOURCE: W. A. Myers. Operant learning principles applied to teaching introductory statistics. *Journal of Applied Behavior Analysis*, 1970, 3, 191–197. Copyright 1970 by the Society for the Experimental Analysis of Behavior, Inc.

---

the independent and dependent variables. Figure 11.2 might help you visualize the temporal sequence of the specific variables under study.

A description of the results of the performance and questionnaire measures appears in Excerpt 11.3. The stated purpose of the above study was to evaluate the set of learning conditions (method of instruction) by performance measures and student attitude to specific aspects of the course. Although two correlations are significant at the .05 level, the correlations between three measures of performance provide little or no help in evaluating the method of instruction. Also, the mean ranks for Parts I and II of the questionnaire are only descriptive data about student attitudes. Thus, the effects of this unique method of instruction on performance cannot be assessed unless the researcher can make some comparison or contrast. Com-

X

INDEPENDENT VARIABLE

Method of instruction

O

DEPENDENT VARIABLE

1. Performance
   a. Days to complete
   b. Errors on exam
   c. Errors in text
2. Attitude questionnaire
   a. Part I
   b. Part II

FIGURE 11.2. *Diagram of the example one-shot case study.*

EXCERPT 11.3 RESULTS OF PERFORMANCE AND
QUESTIONNAIRE MEASURES OF ONE-SHOT CASE STUDY

PERFORMANCE

SUMMARY TABLE. NUMBER OF STUDENTS AS A FUNCTION OF EACH
PERFORMANCE MEASURE. DATA FROM FIGURE 1 OF THE STUDY

	CLASS DAYS TO FINISH COURSE			TOTAL ERRORS ON UNIT EXAMS			ERRORS MADE ON PROGRAMMED TEXT		
	26–30	31–35	36–39	3–19	20–36	37–53	4–51	52–59	100–149
NUMBER OF STUDENTS	11	16	11	19	17	2	19	7	1

The Pearson product-moment correlation between number of days to finish the course and errors made on the exams was 0.414 ($p < 0.05$); between number of days to finish and errors made in the book, $r = 0.43$ ($p < 0.05$); and between errors in the book and errors on exams, $r = 0.04$ (n.s.).

QUESTIONNAIRE

Part I. Some of the judgments most favorable to the course were (1') "the amount of material fully understood" ($\bar{X} = 2.03$; the verbal equivalent lies between "a lot more" and "a moderate amount more" by the course as given); (2) confidence in solving statistical problems ($\bar{X} = 2.11$); (3) amount of time bored in class ($\bar{X} = 6.62$; the verbal equivalent lies between "a lot" and "a moderate" amount less by the course as given); (4) the extent to which natural differences of ability among the students were taken into account ($\bar{X} = 2.00$); (5) the extent to which the course failed to recognize you as an individual ($\bar{X} = 6.22$).

Part II. The item receiving the highest mean rank ($\bar{X} = 2.65$) and the lowest variability was "Being able to work at your own pace." The item receiving the next highest rank ($\bar{X} = 4.89$) was "Taking an exam on each set," followed by "Not being penalized for mistakes on exams by lowering the grade ($\bar{X} = 5.54$)."

SOURCE: W. A. Myers. Operant learning principles applied to teaching introductory statistics. *Journal of Applied Behavior Analysis*, 1970, 3, 191–197. Copyright 1970 by the Society for the Experimental Analysis of Behavior, Inc.

parisons in one-shot case studies are usually based on conjecture or imagination about the data before or without X. Usually the researcher makes these conjectures. However, in our example study the subjects were asked to call upon their imagination (of the lecture method) in rating the method of the course in statistics (see Excerpt 11.2). The same average ratings could have been obtained without exposure to X. The reader has no way of knowing whether X caused these ratings. Also, the correlations used in the study are only descriptive and do not provide statistical comparison.

To remedy the design, a comparison should have been made with another group of students who had the same course content presented in the traditional lecture manner. It is very possible that the course procedures described in Excerpt 11.1 *were* effective. However, isn't it also possible that the students responded favorably simply because the statistics course was set up differently from what the students were experiencing in their other courses? If so, might not a completely different instructional approach bring forth the same results? The major overriding weakness of the one-shot case study is that we cannot answer these questions: therefore, we are unable to tell for sure whether X was the cause of the "nice" results, whether some entirely different treatment could have produced exactly the same pattern of results, or whether the Ss would have responded in the same manner without any X at all.

Statistical Analysis

The results for the dependent variables in our example above appear in Excerpt 11.3. A summary table presents the number of students as a function of each performance measure. Also, the results of performance are presented in terms of correlation coefficients between the measures (errors on exams, errors in the text, and days to finish). The results of responses to Part I and II of the questionnaire are expressed in means. Note that the correlation coefficients were tested to see if they were significantly different from zero. As this example demonstrates, it is possible for a researcher to use both descriptive statistics and inferential statistics within the context of a one-shot case study. The use of inferential procedures, however, does not make up for an inferior design.

ONE-GROUP PRETEST-POSTTEST DESIGN

In the one-group pretest-posttest design, observations are made before and after the independent or treatment variable has been introduced to a group. Figure 11.3 shows the diagram for this design. O_1 is the pretest observation and O_2 the posttest observation. With the two observations the

$$O_1 \qquad\qquad X \qquad\qquad O_2$$
Pretest   Treatment   Posttest

FIGURE 11.3. *Diagram of the one-group pretest-posttest design.*

researcher is able to make one comparison or contrast. For our hypothetical example of the film on sensitivity groups, the researcher could have a one-group pretest-posttest design for the study by administering the level-of-apprehension rating scale both before and after the treatment.

This design is better than the one-shot case study, but there are still six possible uncontrolled extraneous variables (threats to internal validity) which also might explain the difference (or no difference) between O_1 and O_2. These uncontrolled variables—history, maturation, testing, instrumentation, statistical regression, and mortality—become confounded with the possible effect of the treatment variable. The variables become rival hypotheses for the difference between O_1 and O_2, and they are threats to the

---

TABLE 11.1 POSSIBLE SOURCES OF INTERNAL INVALIDITY FOR EXPERIMENTAL DESIGNS

1. History	5. Statistical regression
2. Maturation	6. Mortality
3. Testing	7. Selection
4. Instrumentation	

---

internal validity of the study. Therefore, it is impossible for the researcher to determine whether differences or no differences between O_1 and O_2 were caused by X or by one, or more, of the six extraneous variables. Discussion and analysis of these six threats to internal validity will perhaps be clearer if we examine a journal article using a one-group pretest-posttest design. The seventh threat in Table 11.1, mortality, will be discussed in the final section of this chapter.

Example of Design

In our example the researcher assessed the effects of counseling and tutoring on attitudes of teachers and achievement of seventh-grade under-achievers. The study started at the beginning of a spring semester. Eighteen seventh-graders were tutored for 10 weeks by academically gifted ninth-graders. Also, during the 10-week period of the study, the same counselor met with each seventh-grader once every other week in an individual counseling session. The counseling sessions were structured to permit expression of feelings concerning the tutoring program or any other specific concerns of the seventh-grader. Excerpt 11.4 describes the attitude and measurement procedures and the results of the t tests for the two dependent variables.

Using the diagram in Figure 11.3, we can also diagram the relationship of the independent and dependent variables. Such a diagram appears in Figure 11.4. Note that there are two dependent variables for the study,

---

**EXCERPT 11.4 PROCEDURE AND RESULTS
FOR ONE-GROUP PRETEST-POSTTEST DESIGN**

PROCEDURE

A 25-item attitude rating scale was constructed to measure attitude change. Both before and after the 10 weeks of tutoring and counseling, each of the tutee's 6 teachers rated his attitude toward school, self, authority, and his general outlook on life. Each item was rated on a 9-point scale, with 1 indicating extremely negative and 9 extremely positive. The results yielded six pre- and post-experimental raw scores for each individual tutee.

The *t*-test of the difference between means for correlated samples (Ferguson, 1966) was used in testing the difference between the pre- and post-experiment teacher ratings for each tutee. Therefore, 18 such *t*-tests, one for each tutee, were conducted.

The tutee grade point averages, as a group, were also tested for significance of difference between the pre- and post-tutoring and counseling by *t*-test.

RESULTS

Post-experiment ratings indicated a positive change in attitude for 16 of the 18 tutees. Two received lower ratings. Fifteen of the 18 had significantly higher teacher ratings (one-tailed at .05) following the 10-week experiment.

The second construct to which this study addressed itself considered the possibility of raising academic achievement through counseling combined with tutoring. The post-experiment data yielded a significant difference in mean grade point average for the group over a 10-week period. The *t* test described above was again utilized with the level of significance set at .05.

SOURCE: J. Wittmer. The effects of counseling and tutoring on the attitudes and achievement of seventh grade underachievers. *The School Counselor*, 1969, *16*, 287–289. Copyright 1969 by the American Personnel and Guidance Association.

---

teacher attitude ratings and grade point averages. Figure 11.4 shows that there are two independent variables—counseling and tutoring. However, because the researcher combined the two into one treatment, it is impossible to assess the separate effects of counseling or tutoring. Therefore, both counseling and tutoring must be considered as a single independent or treatment variable. Now, let us examine the six variables which might threaten the internal validity of our example study.

O_1	X	O_2
PRETEST-DEPENDENT VARIABLES	INDEPENDENT VARIABLE	POSTTEST-DEPENDENT VARIABLES
1. 25-item attitude rating scale; completed by teachers	Counseling and tutoring	1. 25-item attitude rating scale; completed by teachers
2. Seventh-graders' (tutee) grade-point averages		2. Seventh-graders' (tutee) grade-point averages

FIGURE 11.4. *Diagram of the independent and dependent variables for a one-group pretest-posttest design.*

Threats to Internal Validity

History. The first uncontrolled variable which may confound the effects of the independent variable is *history*. In addition to X (treatment variable), some other event may occur either in or out of the experimental setting which may have an effect on the dependent variable. In the study of the effects of counseling and tutoring such an event could have occurred in the school during the 10-week period between O_1 and O_2. The event might have been the election of student government officers or spring vacation or a nonscheduled and unanticipated occurrence such as the basketball team's getting to play in the regional tournament. The event could also be something that occurred outside the school and is not typically associated with the school, such as a regional or national event.

Another factor associated with history is the length of time between O_1 and O_2. The greater the time between O_1 and O_2 the more plausible the hypothesis of history becomes.

Experiments occurring in laboratory settings, with brief periods of time between O_1 and O_2, decrease the influence of the effects of history by limiting the possibilities for the intrusion of extraneous variables. It is difficult to know whether the 10-week period between O_1 and O_2 for the above example is adequate or too long. Also, it is difficult for the reader or author to know precisely what events either in or out of school might have influenced significant changes in the dependent variable. Our inability to identify these events certainly does not exclude history as a rival hypothesis; it should, in fact, make us more aware of the threat history poses to internal validity. The point is that the one-group pretest-posttest design does not control for history or what happens between O_1 and O_2. In addition to the independent variable X, history can be another explanation or source of what caused the differences between O_1 and O_2.

Maturation. The second threat to internal validity in the one-group pretest-posttest design is *maturation*. This variable refers to biological or psychological processes which occur with the passage of time and are independent of any external events. Changes within the subjects of an experiment may affect the observations or dependent variable.

How susceptible to biological changes are seventh-grade adolescents? How much change can occur within a 10-week period as a consequence of developmental processes? Of course, these questions cannot be answered by the above example study. However, they cast serious doubt as to the effectiveness of X and may provide an alternative explanation for the cause of significant differences between O_1 and O_2 on both dependent variables.

Another characteristic associated with maturation, which may be a plausible explanation for differences, is the psychological process. In our sample study, the seventh-graders may have performed better on the posttest because they were less anxious or more interested. Again, these

psychological processes are part of growing up or maturation. In other experiments this process could account for subjects performing not as well on
the posttest as the pretest because they were more fatigued, less interested,
more anxious, or more bored. The psychological process associated with
maturation becomes a possible extraneous source of differences in addition
to the treatment. The control of this variable is important for research involving psychotherapy, social psychology, child development, and educational and social remediation.

As with history, maturational variables become more crucial as the
time between O_1 and O_2 increases. The mere passage of time over a 10-week
period might have produced significant differences between O_1 and O_2
without the counseling and tutoring. Unless maturation is controlled by the
researcher, its adequacy as an explanation for differences is just as satisfactory and plausible as the treatment variable. The one-group pretest and
posttest design does not control for its effect.

Testing. A third source of variation that could explain the difference
between O_1 and O_2 in addition to X is the effect of *testing*. People usually
score higher when they take an achievement or an intelligence test the
second time or when they take an alternative form of a previously administered test. A similar effect is generally true for personality tests or adjustment
inventories. The scores on the second test usually reflect better adjustment,
although the opposite may sometimes be found. Also, on the second administration of an attitude scale, the ratings might be more positive or more
negative than the first time. In the above example study, significant changes
between the pretest ratings and the posttest ratings on the teacher attitude
scale might be the result of exposure to the pretest. The researcher does not
know if changes in teachers' attitudes are a result of the counseling and
tutoring or a result of the pretest.

Tests, inventories, and rating scales are referred to as *reactive measurements*. These measures or observations are called reactive because they may
change the event which the researcher is attempting to measure. Any measure is reactive if it has the potential for modifying the variables under
study, if it may focus attention on the experiment, if it is not a part of the
normal environment, or if it exercises the process under study. For example,
data which are collected by means of video or audio recorders or field observers might be reactive. These measures have the potential for changing
what is being measured even if X were not present. On the other hand,
there are *nonreactive* or *passive measures* which are a regular part of the
environment. Such nonreactive measures are school enrollment records,
traffic accident rates, birth and death rates, court records, voting trends, and
possibly participant observers. Whenever possible, researchers should attempt
to use nonreactive measures. However, if reactive measures have to be used,

a researcher must control for testing as a possible threat to the validity of the experiment because the reactive measure may itself be a stimulus for change. For the above study, we cannot say with any degree of confidence that pretesting of teacher attitudes about the students did not have an effect on the second rating (O_2). Thus testing becomes a third rival hypothesis or explanation for the significant difference between O_1 and O_2.

Instrumentation. A fourth uncontrolled rival hypothesis is the effect of *instrumentation,* that is, the effect of any change in the observational technique or measurement instrument which might account for an O_1–O_2 difference. Obviously, any mechanical device may deteriorate, malfunction, or become unreliable. The problem of instrumentation also becomes crucial when human beings are used as judges, observers, raters, coders, interviewers, or graders. The difference between O_1 and O_2 may be a result of the fact that the raters (1) are more experienced, (2) are more fatigued, (3) have learned about the purposes of the experiment, (4) have undergone maturation, or (5) have more relaxed or more stringent standards at O_2. Differences between O_1 and O_2 may also appear if the researcher uses different observers, interviewers, or coders for pretesting and posttesting.

The significant differences between O_1 and O_2 for grades and attitudes in the example study may be a result of instrumentation rather than counseling and tutoring. It is quite possible that the teachers' grading standards were relaxed at O_2 or that the teachers learned about the purposes of the experiment. Also, their attitudes might have changed significantly because the teachers' expectancies about the tutees changed. Thus, the one-group pretest-posttest does not control for instrument deterioration as possible explanation of change. Frequently, researchers provide measures of inter- or intrarater or judge reliability. However, this may not be a satisfactory substitute for control built into the design.

Statistical Regression. A fifth extraneous factor is *statistical regression.* This factor refers to the effect of the use of extreme scores at O_1 in a study. Usually there will be a shift of the mean of the extreme scores at O_1 to a higher or lower level at O_2, whichever is closer to the mean for the whole distribution of O_2 scores. Such shifts are prevalent in remedial research, where extreme scores are frequently used. Let us consider a hypothetical example of a new instructional program for gifted and low-achieving high school students. The students are selected on the basis of their extreme scores on a standardized achievement test. Figure 11.5 illustrates how the scores for both groups of students might appear for the school district. The pretest mean is 120 for the gifted, 100 for the entire school district, and 80 for the low-achieving students. After the new instructional programs have been completed, the posttest mean score shifts from 80 to 85 for the low achieving students and from 120 to 115 for the gifted students; both extreme scores have shifted toward 100, the mean for the entire distribution of O_2 scores.

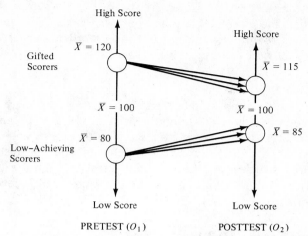

FIGURE 11.5. *Illustration of shift in extreme scores from pretest to post-test.*

Why does statistical regression occur? One reason is that there is al-ways some degree of imperfection in the measuring instrument which causes fluctuations in the test scores. For example, the subjects' scores on the pretest and posttest will not be the same and will usually vary within a certain range. The range in which scores vary for a test is referred to as the *error of measurement,* which is considered to be an index of imperfection (unreliability) for the measuring instrument. Thus, if extreme scores are used in a pretest-posttest design, they have on the average a greater prob-ability of shifting toward the mean for a distribution because the scores are at the extreme of the distribution.

In our example of the one-group pretest-posttest design, the seventh-grade underachievers were perhaps selected on the basis of low grades. In addition to affecting standardized test scores, statistical regression could account for significant changes in the mean grade-point averages. Failure to control for statistical regression presents a problem in interpreting the effects of the independent variable on the dependent variable. This is particularly true in remedial research in which subjects are selected on the basis of ex-treme scores.

Mortality. Another possible source of internal invalidity is *mortality,* the loss of subjects between the pretest and the posttest. If the subjects who drop out of an experiment are not similar to those who remain, the mean posttest score could differ from the mean pretest score simply because some of the subjects are not measured at O_2. Mortality is not a problem for the example journal article but could be a source of invalidity for other studies. For our hypothetical experiment on level of apprehension after viewing a film on sensitivity groups, subjects who might drop out could influence the mean ratings on the posttest. Unless aware of this, the reader could get the impression that the treatment was responsible for any changes that might appear between the pretest and posttest ratings when, in fact, the subjects who dropped out were the source of change.

Statistical Analysis

There are three appropriate statistical analyses that may be used for the one-group pretest-posttest design in order to compare O_1 and O_2. These are the parametric t test for correlated samples, the nonparametric sign test, or the nonparametric Wilcoxon matched-pairs signed-ranks test. The above example journal article used the t test for correlated samples (see Excerpt 11.4). Notice that there was a significant difference (at .05 level) between O_1 and O_2 for both dependent variables.

STATIC-GROUP COMPARISON DESIGN

The static-group comparison design has two groups. Comparisons are made between one group which is exposed to the experimental or treatment variable and one group which is not. Figure 11.6 shows the diagram for this design. Notice that the symbols (X and O_1) above the dashes represent the one-shot case study discussed earlier in this chapter. The dashes in Figure 11.6 indicate that the comparison groups are not equated by random assignment, that is, the subjects are not randomly assigned to the group that re-

$$X \qquad\qquad\qquad O_1$$
$$\text{-- -- -- -- -- -- -- --}$$
$$O_2$$

FIGURE 11.6. *Diagram of the static-group comparison design.*

ceived the treatment and the group that was not exposed to the treatment. Generally, the groups used in this design are intact or static. O_1 and O_2 are exactly vertical to one another, which means that the observations or measurements occur at the same time. The static group comparison design could be applied to our level of apprehension experiment. The researcher would: (1) find two intact groups, (2) show the film to one group (X) but not the other, and (3) measure the level of apprehension of all subjects in both groups at the same time after the one group has seen the film (O_1, O_2).

Example of Design

A description of the collection of data and the results of an example study using the static-group comparison design appears in Excerpt 11.5 and the study is diagramed in Figure 11.7. As we see from this figure, the treat-

---

EXCERPT 11.5 DESCRIPTION OF STATIC-GROUP COMPARISON DESIGN

STATEMENT OF PROBLEM

It was the purpose of this study to determine whether a significant difference existed between the freshmen who attended Pre-College Counseling at Auburn University and those who did not attend. More specifically, the study was to determine whether a statistically significant difference existed between cumulative grade point averages, curriculum changes, and dropout rates between two matched pairs of freshmen; one of the pairs attended Pre-College Counseling, and one pair did not attend.

COLLECTION OF DATA

A list of entering freshmen for the fall quarter of 1966 who attended Pre-College Counseling was obtained from the High School Relations Office at Auburn University. Also, a list of all entering freshmen for the fall quarter of 1966 was obtained from the Registrar's Office. These were used to divide the freshmen into two groups—those who attended Pre-College Counseling and those who did not attend. The students who did not attend Pre-College Counseling were matched according to the matched-pair method with those from the group who attended Pre-College Counseling. Age, sex, and American College Testing Program (ACT) scores were used as variables in matching the two groups. As a result of matching, 50 matched pairs were used in the analysis. Information for the comparison of grade point averages, dropout rates, and curriculum changes during their freshmen year was obtained from the Registrar's Office.

RESULTS

A *t* test was computed to determine whether there was a significant difference between the cumulative grade point averages of 50 matched pairs for the freshman year. The .05 level of confidence was used to determine the value of the *t*. The data concerning the results of the *t* test on Table 1 show that there is a mean difference of .31 between those who attended Pre-College Counseling and those who did not. The *t* was significant at the .05 level of confidence.

Chi-square tests were computed to determine whether there were significant differences between dropout rate and curriculum changes during the year for the matched pairs. The test for dropout rates resulted in a chi-square of 3.32, which is not significant at the .05 level. No analysis was made of the causes of dropout in this study; therefore, dropout in this study referred to withdrawal from the university for academic suspension, academic difficulty, personal reasons, or transfer to other schools. Nevertheless, there is no significant difference in dropout rates between the two matched pairs during the freshman year.

The test for curriculum changes resulted in a chi-square of .078, which also is not significant at the .05 level. Therefore, there is no significant difference in curriculum changes during the freshmen year between the two groups.

TABLE 1 A *t* TEST COMPARISON OF GRADE AVERAGES

VARIABLE COMPARED	ATTENDERS		NON-ATTENDERS		
	MEAN	SD	MEAN	SD	t
Cumulative GPA[a] (3 quarters)	1.39	.50	1.08	.59	2.8

[a] Based on the system in which A = 3.00, B = 2.00, etc.

SOURCE: M. H. Griffin and H. Donnan. Effect of a summer pre-college counseling program. *The Journal of College Student Personnel*, 1970, *11*, 71–72. Copyright 1970 by the American Personnel and Guidance Association.

ment variable (X) was a precollege counseling session, and the dependent measures were (1) number of dropouts, (2) cumulative GPA for three quarters, and (3) number of changes in the curriculum.

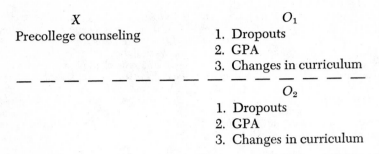

FIGURE 11.7. *Diagram of the independent variables and dependent measures for the example static-group comparison design.*

Threats to Internal Validity

Another source of invalidity for a study is *selection* of subjects. This source could not possibly be a threat to the first two pseudoexperimental designs presented in this chapter because only one group of subjects was used. However, selection is a problem for the static-group comparison design, which has two groups of subjects. The researcher cannot be certain that the two groups are equivalent unless the subjects are randomly assigned to the treatment and control groups, thus giving each subject an equal chance of being in either group. Selection becomes a problem whenever subjects who seek exposure to the treatment are compared with subjects who do not seek exposure. Significant differences resulting from a study could be the consequence of nonequivalent groups rather than the treatment or independent variable.

In our example article, matching on age, sex, and ACT scores does not insure equivalent groups. Also, if selection is a threat to a study because of nonequivalent groups, history and maturation will also become possible sources of invalidity because the researcher cannot be certain that the groups are exposed to the same events and that they have the same maturational processes. Selection may also involve the threat of regression if extreme scores are used to select Ss for the experimental group but not the control group (as would be the case in studies dealing with remediation, psychotherapy, or counseling). In other words, regression may be a threat to a study if the comparison groups are not equal. In summary, selection, history, maturation, and regression are all possible sources of invalidity when the static-group comparison design is used.

Testing and instrumentation are not problems for the static-group comparison design because the subjects are not measured twice. The last confounding source of validity for this design is experimental mortality. As you recall, this refers to the differences in groups because some subjects drop out of the experiment. The O_1–O_2 difference might be caused by dropouts rather than by the treatment variable. Mortality does not appear to be a problem for our example study. However, the static-group comparison does not control for experimental mortality.

We should be reminded that the above seven sources of internal threats to the validity of a study are only guides for evaluating research. Whether the above internal threats apply depends on the nature and particular circumstances of the study.

Statistical Analysis

There are four statistical tests that a researcher might use for the static-group comparison design. The parametric t test discussed in Chapter 4 could be used, or the researcher might decide to use one of the corresponding nonparametric tests, the Mann-Whitney U test or the median test. In addition, a chi-square test might be used. The authors of our sample journal article used the t test and the chi-square test (see Excerpt 11.5).

We do not intend to disparage all research that fails to conform to the tenets of controlled experimentation. However, if a researcher is investing a great deal of effort to collect data, it might be more valuable to select a design that provides at least one comparison and that reduces threats to internal validity. True experimental designs which control for most of the threats to internal validity are presented in the following chapters.

REVIEW TERMS

external validity
history
instrumentation
internal validity
maturation
mortality
O
one-group pretest-
   posttest design

one-shot case study
preexperimental design
pseudoexperimental design
reactive measurements
selection
static-group comparison design
statistical regression
testing
X

REVIEW QUESTIONS

1. What type of validity answers the question, Did the independent or treatment variable make a difference in the study or can the researcher infer a cause-and-effect relationship?

2. The process associated with the cumulative effects of learning and coping with environmental pressures of everyday experiences can become what type of rival hypothesis?

3. What do the dashes represent in a diagram of a static-group comparison design?

4. What do the symbols O and X represent in the diagrams used in this chapter?

5. List four examples of nonreactive measures.

6. Representativeness or generalizability refers to which type of validity?

7. Assuming that the comparison groups are relatively equal, how many sources of internal invalidity does the static-group comparison design control?

8. What threat to internal validity is most likely to be associated with remedial research?

9. The length of time between O_1 and O_2 can be a source of what type of internal invalidity?

10. Changes in the measurement or observation procedures can be a source of what type of internal invalidity?

CHAPTER 12

TRUE EXPERIMENTAL DESIGNS

We have learned that with a preexperimental design it is impossible to be certain that the treatment was the true cause of any observed differences in the data. Each of the preexperimental designs has uncontrolled extraneous variables that threaten the internal validity of the experiment. In this chapter we will examine three designs that have built-in controls for the threats to internal validity. If the author reports using a *true experimental design* and there is a difference in the data, then we can be relatively confident that the independent (treatment) variable was responsible for the observed differences, and that differences are not due to instrumentation, statistical regression, history, or other possible causes. The three true experimental designs that we will examine are (1) the pretest-posttest control group design, (2) the posttest-only control group design, and (3) the Solomon four-group design. In the last section of this chapter, we will examine sources of external invalidity of experiments which restrict the experimenters from generalizing their findings to other subjects, situations, or sets of conditions.

The true experimental designs were so designated by Campbell and Stanley to distinguish them from preexperimental and quasi-experimental designs.[1] The true experimental design has many variations other than the three considered in this chapter. Some of these variations will be presented in Chapter 13. From a historical perspective the three designs discussed in this chapter are the classic and, for some researchers, the recommended experimental designs.

We will use the same design symbols as used in Chapter 11. As you recall, O denotes the collection of data (dependent variable) from a group of subjects and X represents the exposure of the group to the independent or treatment variable. The left-to-right dimension in a design diagram indicates the order in time of O and X. Also, different groups of individuals are distinguished by a different row of symbols for each group. A new symbol, R, will be used in this chapter. This symbol denotes the random assignment of subjects to the various groups within a study.[2]

Randomization is an important process within the field of research design. Large groups which have been formed by means of random assignment of subjects are likely to be equal to one another (or nearly equal) on all conceivable variables. When groups are formed by means of *matching,* we can only be sure that they are equivalent in terms of the variable(s) used to do the matching. For example, a researcher may report that groups were equated by placing each member of a pair of subjects who had the same score on an achievement test in either the experimental or the control group. This kind of matching is not a satisfactory substitute for randomization, which is a more effective method for achieving equality among groups. However, randomization and matching can be combined. For example, a researcher may carefully match into pairs subjects from the total population to which he wishes to generalize and then randomly assign members of these pairs to the experimental or control groups. The combination of first matching and then random assignment will perhaps yield greater design precision than would randomization alone. Excerpt 12.1 describes a study in which the authors matched and then randomly assigned to treatment groups.

EXCERPT 12.1 COMBINATION OF MATCHING AND RANDOM ASSIGNMENT

The experimental design was a pretest-treatment-posttest design with random assignment of matched pairs to the two treatment groups. . . .

SOURCE: H. J. Sullivan, M. Okada, and F. R. Niedermeyer. Learning and transfer under two methods of word attack instruction. *American Educational Research Journal,* 1971, *8,* 227–239. Copyright 1971 by the American Educational Research Association, Washington, D.C.

---

[1] D. T. Campbell and J. C. Stanley. Experimental and quasi-experimental designs for research on teaching. In N. L. Gage (Ed.), *Handbook of Research on Teaching.* Chicago: Rand McNally, 1963, pp. 171–246.

[2] R can be used as the symbol for range (see Chapter 2), for the coefficient of multiple correlation (see Chapter 8), and also for randomization. The reader should be able to determine the correct meaning in a study from the context of the article.

PRETEST-POSTTEST CONTROL GROUP DESIGN

The pretest-posttest control group design has two groups of subjects which are compared with respect to measurement or observation on the dependent variable. Both groups are measured or observed twice; the first measurement serves as the pretest and the second as the posttest. To form the two groups, the researcher randomly assigns half of the subjects to one group and the other half to the second group. The two groups being compared are similar, or are assumed to be similar, to each other in terms of their characteristics at the beginning of the study because of random assignment. Measurements or observations of the dependent variable are collected at the same time for both groups.

The two groups differ in one very important manner—what happens to the subjects between the pretest and the posttest. The pretest-posttest control group design can be used by a researcher in two ways. In one variation of this design one of the two groups is exposed to the experimental treatment (independent variable) and the other group does not receive the treatment. The nature of the treatment depends, of course, on the area of the study and may consist of a new training program, social reinforcement, differentiated problem-solving strategies, milieu therapy for schizophrenia, an innovative counseling technique, a film on race relations, a weight-reducing program, a new method of teaching reading, etc. The group that receives the treatment is called the *experimental* or *treatment group;* the group which does not receive the treatment is called the *control group.*

In a second variation of the design each group is exposed to a different form of the experimental treatment, such as two kinds of problem-solving strategies, two training programs, two methods of teaching reading. There is no control group as it is typically defined, but this design allows the researcher to compare the differential effects of both forms of the experimental treatment. The researcher might label the groups experimental group 1 and 2, visual feedback group and auditory feedback group, or social reinforcement group and token reinforcement group.

The diagram for the pretest-posttest control group design is presented in Figure 12.1. The diagram shows that: (1) two groups are involved in the design (two rows of symbols), (2) each group is measured or observed at the same time before the treatment is applied to one group (the first column of O's), (3) each group is measured at the same time after the treatment has been applied (the second column of O's), (4) the subjects are randomly assigned to the two groups (R in each row), and (5) the first group receives

$$R \qquad O \qquad X \qquad O$$
$$R \qquad O \qquad\qquad O$$

FIGURE 12.1. *Diagram of the pretest-posttest control group design.*

the experimental treatment whereas the second group does not (X in the first row and a blank space in the second). For a study in which the researcher compares two forms of a particular treatment, instead of treatment versus control, we could revise the diagram by placing a Y in the blank space of the second row. For example, a company may wish to examine the effects of two industrial training techniques for factory workers, computer assisted instruction and printed programmed instruction. One training procedure could be represented by X in the first row (administered to the first group) and the other by Y in the second row (administered to the second group).

Threats to Internal Validity

The pretest-posttest control group design controls for most of the threats to internal validity. Before examining the relationship of this design to internal threats, let us look at a hypothetical example of the design. Suppose a researcher wants to investigate the effects of a new weight-reducing program for football players. The players are considered overweight based upon some specified criterion for playing weight. The researcher might have 40 players who are considered overweight. After deciding to use a pretest-posttest control design, the researcher could randomly assign 20 players to the experimental group and 20 players to the control group. Before treatment is applied to the experimental group, all players for both groups are weighed at the same time. Upon completion of the weight-reducing program for the experimental group, all 40 players are weighed again at the same time.

Since all 40 subjects are assigned at random to the two groups, we can assume that the groups are equal at the beginning of the experiment. Thus, there is no threat of selection biases. Since the groups are equal at the beginning of the experiment, history and maturation should affect both groups equally. If the players are selected on the basis of extreme weight, the phenomenon of statistical regression to the mean may affect the results. However, this is not a threat to internal validity since statistical regression should be present in one group as much as in the other group. For the same reason, testing will not be the cause of observed differences between the groups. It is unlikely that the effects of the first weighing-in could be responsible for the differences between the groups. Instrumentation should not be a problem in the pretest-posttest control group design if the same procedures and instruments are used for each of the two groups at the pretest and posttest time periods. In our hypothetical weight-reducing program the effect of instrumentation will be controlled if both groups are measured at the same time and on the same scale. If there is any defect or deterioration of the scale, both groups would be equally affected.

The only threat to internal validity that is not controlled by the pretest-posttest control group design is mortality. Since the subjects in one group are similar to those in the other group (subjects are randomly assigned to groups), we should expect an equal dropout rate for the two groups between the pretest and posttest. But exposure to the treatment may cause more subjects to drop out of, or remain in, the experimental group than the control group. This may be a crucial source of internal invalidity. If the subjects who drop out of a group are not similar to those who remain, the mean posttest score could differ from the mean pretest score simply because some of the subjects are not measured the second time. Football players exposed to our weight-reducing program may have had to drop out of the experiment because the program was too rigorous. The weight of those who dropped out could increase, or decrease, the mean score on the posttest because they were not weighed (posttest) the second time. As a result, the data analysis conveys the impression that the program was responsible (or not responsible) for any changes that might appear between the two groups on the posttest. For this reason, an author of a journal article should always report how many subjects from each group dropped out during the experiment.

Statistical Analysis

Two correct and two incorrect statistical analyses are often used with this design. One of the correct analyses utilizes *gain scores*. The scores are obtained by subtracting, for each subject, the pretest score from the posttest score. Mean gain scores are calculated for each of the two groups, and the parametric independent samples *t* test or the nonparametric Mann-Whitney *U* test or median test is used to compare the two groups. The second acceptable statistical procedure, considered by many researchers to be the better method, is the analysis of covariance. In this procedure the two posttest means are compared after having been adjusted for any differences between the two groups with respect to pretest means.

Researchers who use a pretest-posttest control group design sometimes employ an incorrect statistical analysis. It is inappropriate to use one *t* test to compare pretest and posttest means for the experimental group and then a second *t* test to compare pretest and posttest means for the control group. This kind of analysis makes the incorrect assumption that a significant difference in the first case and a nonsignificant differences in the second case substantiates the effectiveness of the treatment. Another incorrect procedure is to use one *t* test to compare the two pretest means and then a second *t* test to compare the two posttest means. This analysis wrongly assumes that a nonsignificant difference between the pretest means and a significant difference between the posttest means proves the effects of treatment.[3]

---

[3] For a discussion of why these *t* test comparisons are inappropriate, see J. P. Guilford, *Fundamental Statistics in Psychology and Education.* New York: McGraw-Hill, 1965, pp. 194–197.

Example of Design

Let us now examine an actual journal article in which a pretest-posttest control group design was used by a researcher. In this study 16 counselor trainees were randomly assigned to two groups. Excerpt 12.2 describes the method for this study. Subjects in the experimental group met for one and a half hours three times a week, while subjects in the control group were not exposed to the training program. All subjects were tape recorded in a coun-

---

EXCERPT 12.2 DESCRIPTION OF METHOD FOR PRETEST-POSTTEST CONTROL GROUP DESIGN

Eleven female nurses and five female occupational therapists working in the psychiatric division of a large city hospital were randomly assigned (eight per group) to either a control group which did not meet or to a training group which met for one and one-half hours, three times a week, for four weeks, for a total of 18 hours. Each nurse and occupational therapist was cast in the helping role as a counselor before and after counseling with a standard client given a mental set to discuss any personal problems or experiences which she might have which she could share with the counselor. Each session was tape recorded and three, three-minute excerpts were randomly selected from the tapes, and the excerpts were rated by two experienced raters in order to determine the levels of functioning before and after training. All of the scales, counselor empathy (E), respect (R), genuineness (G), concreteness or specificity of expression (C), self-disclosure (SD), and client depth of self-exploration (EX) ranged from the lowest level of functioning, level 1, to the highest level of functioning, level 5 (Carkhuff, in press). Pearson r intra-rater reliabilities on the counselor offered dimensions were as follows: E .99, .97; R .89, .96; G .97, .94; C .95, .88; SD .97, .82; inter-rater reliabilities on these dimensions were as follows: E .87; R .86; G .88; C .83; SD .83.

In addition, pre-post measures were taken on each nurse's and occupational therapist's values as measured by the Allport-Vernon-Lindsey Study of Values (AVLSV), administered in the usual way, and by a Moral Values Questionnaire (MVQ), adapted from Rosenthal's Moral Values Q Samples (Rosenthal, 1955). Based upon previous research, the trainer was functioning at minimally facilitative levels (level 3 or above). The trainer completed the AVLSV and the MVQ before training, but these indices were not readministered to him after training, since it was felt that they would not show any predictable change.

SOURCE: D. W. Kratochvil. Changes in values and interpersonal functioning of counselor candidates. *Counselor Education and Supervision*, 1969, 8, 104–107. Copyright 1969 by the American Personnel and Guidance Association.

---

seling session prior to and following the training program. Two experienced raters rated randomly selected excerpts of the pre- and posttest tapes for each counselor trainee (in both groups) in terms of several different scales of level of functioning. Ratings on these scales, plus scores on the AVLSV and the MVQ, were the dependent variables. Note that the author reports the Pearson r for intra- and interrater reliabilities for each scale (see p. 9 for an explanation of the importance of reliability). A diagram for this particular study is presented in Figure 12.2.

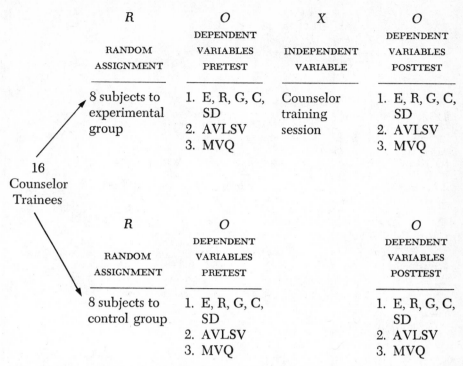

FIGURE 12.2. *Diagram of the independent and dependent variables for a pretest-posttest control group design.*

The author used *t* tests to make the following comparisons: (1) the two groups with respect to pretest performance on the various scales, (2) the pretest and posttest scores of the experimental group, (3) the two groups with regard to posttest scores, and (4) the pretest and posttest scores of the control group (see Excerpt 12.3 for a description of the results). Even with all of these *t* tests the researcher did not analyze the data correctly. The researcher should have used, for each dependent variable, either one *t* test to compare the gain scores of the experimental group with the gain scores of the control group or the analysis of covariance. Or better yet, a multivariate analysis of covariance could have been employed.

POSTTEST-ONLY CONTROL GROUP DESIGN

The posttest-only control group design is identical to the pretest-posttest control group design except that the pretest is not administered to either of the two groups. The diagram for this design appears in Figure 12.3. Notice that the design is similar to the diagram of the pretest-posttest control group design presented in Figure 12.1, except that there is no *O* for either group

---

EXCERPT 12.3 DESCRIPTION OF RESULTS
FOR PRETEST-POSTTEST CONTROL GROUP

RESULTS

Between the training and the control group there was no significant difference in over-all level of functioning ($t = .80$, n.s.) or on individual dimensions ($t(E) = .44$; $t(R) = 1.16$; $t(G) = .83$; $t(C) = .37$; $t(SD) = .00$; all n.s.) before training. After training the training group was functioning, over-all, at levels significantly higher than its pre-training levels ($t = 3.74$, $p < .005$) and the post-training levels of the control group ($t = 5.42$, $p < .005$). While the post-training levels of functioning, on the individual dimensions, of the training group were significantly greater at the .05 level than those of the control group ($t(E) = 2.33$; $t(R) = 2.64$; $t(G) = 2.38$; $t(C) = 2.28$; $t(SD) = 2.14$), they were significantly greater, at only the .10 level, than the pre-training levels of functioning of the training group ($t(E) = 1.81$; $t(R) = 1.82$; $t(G) = 1.41$; $t(C) = 1.79$; $t(SD) = 1.68$). The over-all post-training level of functioning of the control group was significantly lower than the control group's over-all pre-training level of functioning ($t = 3.16$, $p < .05$); however, a breakdown to the individual dimensions yielded a significant deterioration on only one dimension, concreteness ($t = 2.60$, $p < .05$). On both the AVLSV and the MVQ neither the average total change ($t = .06$ and $t = .36$, respectively, n.s.) nor the average movement toward the values of the trainer ($t = .56$ and $t = .71$ respectively, n.s.) was significantly greater in the training group than in the control group.

SOURCE: D. W. Kratochvil. Changes in values and interpersonal functioning of counselor candidates. *Counselor Education and Supervision*, 1969, 8, 104–107. Copyright 1969 by the American Personnel and Guidance Association.

---

before the treatment. As with the previous design, if the two groups are exposed to different forms of the same treatment, we would place Y in the space below X and refer to the two groups as experimental group 1 and experimental group 2 rather than experimental group and control group.

$$
\begin{array}{ccc}
R & X & O \\
R & & O
\end{array}
$$

FIGURE 12.3. *Diagram of the posttest-only control group design.*

As a hypothetical example, let us use again a study in which a researcher wishes to investigate whether a film on sensitivity groups influences the level of apprehension, as measured by a rating scale, that people might feel prior to their first group experience. Using the posttest-only control group design, the researcher would (1) randomly assign the subjects to two groups, (2) show the film to one group but not the other, and (3) measure the level of apprehension of all subjects. Or, consider our hypothetical weight-reducing program. Neither group would be weighed before the weight program was introduced to the experimental group. However, both groups would be weighed at the same time upon completion of the reducing program for the experimental group.

Threats to Internal Validity

The posttest-only control group design, through the random assignment of subjects to the two groups, controls selection, history, maturation, and statistical regression. Furthermore, the threats of testing and instrumentation do not exist since none of the Ss is measured twice. However, the

---

EXCERPT 12.4 DESCRIPTION OF METHOD FOR POSTTEST-ONLY CONTROL GROUP DESIGN

METHOD

The subjects were students who earned a maximum midterm grade point average (GPA) in the 1967 spring quarter of .66 (2 D's and 1 F). Fifty-nine students met this requirement. From this pool, 20 subjects were randomly selected. Twenty additional subjects, matched with the original 20 on midterm GPA, were then selected. Each subject in the resulting 20 matched pairs was randomly assigned to a treatment or control group. Seven subjects in each group were females; 5 of the treatment subjects and 3 of the control subjects were sophomores. The mean predicted freshman average grade (based on a combination of high school average and Scholastic Aptitude Test scores) was 1.59 for the control group and 1.58 for the treatment group (A = 4, B = 3, etc.). This suggests that although the probability of academic success (i.e., graduation) among these subjects was not high, they were performing appreciably below their predicted levels at midterm.

The subjects in the treatment group were telephoned by the receptionist at the counseling service and were asked to make an appointment with the director of testing and counseling as soon as possible. All subjects did so, and appointments were scheduled for the following three days. In fact, 18 of the 20 subjects scheduled and participated in a counseling interview within one week after mid-term (the 5th week of a 10-week quarter), and the remaining 2 subjects received counseling during the next week. The 20 students in the control group were not contacted by the counseling service, and none of them voluntarily sought counseling before the end of the spring quarter.

The counselor, who was the director of testing and counseling at SGC, had a master's degree in counseling and student personnel services as well as additional course work in counseling psychology. He also had had approximately three years' experience in college counseling.

The first counseling interview lasted from 30 to 50 minutes. It focused upon exploring the subject's academic problems and actively seeking concrete ways in which he could improve his grades by the end of the quarter. An attempt was made to verbally reinforce steps that the subject was taking or planning to take in an effort to improve his grades. At the end of the first interview, each subject was encouraged to schedule a second appointment for the next week, and 19 did so. During the second session, continued efforts were made to encourage behaviors that would lead to improved grades, and verbal reinforcement was again given to the subject for steps he was taking to improve his performance. On a subjective basis, it appeared that the large majority of subjects had taken such steps, and most of them expressed the belief that their situation was improving. In sum, 19 of the 20 treatment subjects participated in at least two counseling sessions, and 8 of them had between three and six interviews.

The effectiveness of counseling was assessed through comparing the treatment and control groups on the following indices: (a) final mean GPA for the spring quarter; (b) mean GPA for the four quarters following spring; (c) percentage of subjects receiving one or more suspensions during the four quarters following spring; and (d) percentage of subjects who graduated from SGC during the four quarters following spring.

SOURCE: C. J. Gelso and B. Thompson. Effects of emergency academic counseling. *Journal of College Student Personnel,* 1970, *11,* 276–278. Copyright 1970 by the American Personnel and Guidance Association.

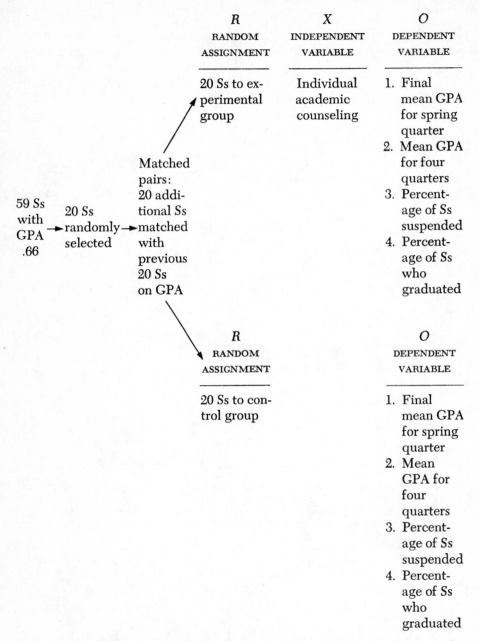

FIGURE 12.4. *Diagram of the example posttest-only control group design.*

threat to internal validity of mortality could be a problem and the researcher should always consider and report how many subjects from each group dropped out during the experiment. Also, as indicated earlier, randomization of Ss to groups is an extremely effective technique for achieving equality among groups. For this reason, there is general agreement that, unless there is some question as to the genuine randomness of the assignment, the post-test-only design is as good as, if not better than, the pretest-posttest design.

Statistical Analysis

The statistical analysis for this design is much less complex than the analysis for the pretest-posttest control group design. Most researchers simply use a t test to compare the two groups with respect to their posttest means. The nonparametric Mann-Whitney U test or median test may also be used. Sometimes, however, the analysis of covariance is used when covariate data are available for each subject. Of course, this covariate variable would have to be highly correlated with the dependent variable (see Chapter 7 for a discussion of covariance).

Example of Design

The journal article we have selected to illustrate the use of posttest-only control group design deals with emergency academic counseling. The description of the method for this study is presented in Excerpt 12.4 and Figure 12.4 diagrams the independent and dependent variables. Note that the 20 subjects, randomly selected from the population of 59, were matched on GPA with 20 other subjects from the same population. Then, each subject in the resulting 20 matched pairs was randomly assigned to the treatment or control group. There were four dependent variables.

Since the subjects for this experiment were chosen because of their low midterm GPA's, there might be a tendency to classify the design as a pretest-posttest control group design rather than a posttest-only design. The reason for deciding upon the latter classification is that the initial GPA scores (i.e., those used to select the subjects for the study) were not included in the statistical analysis. If the authors had used a t test to compare the treatment and control groups with respect to GPA gain scores, or if the initial GPA scores were used within the context of an analysis of covariance, then the design would have been a pretest-posttest control group design. Excerpt 12.5 describes the results of the study.

SOLOMON FOUR-GROUP DESIGN

One of the threats to external validity (generalizability of results) is the interaction between the pretest and the treatment. External validity will be limited to the extent that the pretest *sensitizes* the subjects to the treatment (see p. 265 for a more detailed discussion). If such an interaction is present, the results of a study cannot be generalized to nonparticipants of

EXCERPT 12.5 DESCRIPTION OF RESULTS FOR
POSTTEST-ONLY CONTROL GROUP DESIGN

RESULTS

The mean midterm GPA of both the treatment and control groups was .45 ($F = 0$, $D = 1$, etc.) and the SD was .22. Table 1 presents the final mean GPA and SD of both groups for the spring quarter and for the four quarters which immediately followed. To obtain this latter measure, the mean GPA of the 17 subjects in the control group who matriculated during one or more of the four quarters following spring was compared with the mean GPA of the 13 treatment subjects who did likewise. It can be seen that the mean GPA's of both groups for both spring quarter and the four following quarters are almost identical: t-tests of the difference between their means did not approach significance ($p > .05$).

In the two remaining comparisons, it was found that (a) 54 percent of the treatment subjects, as compared to 53 percent of the controls who matriculated during one or more of the four quarters following spring, earned at least one academic suspension; and (b) 31 percent of the treatment versus 11 percent of the control subjects who matriculated during these quarters graduated from SGC. Neither of these two differences attained the .05 level of confidence.

TABLE 1 FINAL GPA IN SPRING AND FOUR FOLLOWING QUARTERS

GROUP	SPRING QTR. GPA MEAN	SD	FOLLOWING QTRS. GPA MEAN	SD
Treatment	.69	.35	1.41	.68
Control	.71	.51	1.42	.64

SOURCE: C. J. Gelso and B. Thompson. Effects of emergency academic counseling. *Journal of College Student Personnel*, 1970, *11*, 276–278. Copyright 1970 by the American Personnel and Guidance Association.

the experiment unless they also receive the pretest. Since the purpose of almost every study is to generate results which apply to people not involved in the study, the existence of an interaction between pretest and the treatment could severely limit the generalization of the research. The Solomon four-group design was developed to identify those research situations in which the pretest does, in fact, sensitize individuals to the treatment.

The diagram for this design is presented in Figure 12.5. We can see that (1) subjects are randomly assigned to four different groups, (2) two

$$
\begin{array}{cccc}
R & O_1 & X & O_2 \\
R & O_3 &   & O_4 \\
R &   & X & O_5 \\
R &   &   & O_6 \\
\end{array}
$$

FIGURE 12.5. *Diagram of the Solomon four-group design.*

of the groups receive the treatment, (3) only one of the experimental groups is administered the pretest, (4) two of the groups do not receive the treatment, (5) only one of the control groups is administered the pretest, and (6) all four groups are administered the posttest. The Solomon four-group design is a combination of the pretest-posttest and posttest-only control group designs. The first two groups in Figure 12.5 represent the pretest-post-test control group design and the last two groups represent the posttest-only control group design. How could a researcher use the Solomon four-group design for our hypothetical weight-reducing experiment? Instead of randomly assigning the football players to two groups, the researcher would randomly assign the 40 subjects to four groups. The Solomon four-group design controls for all of the threats to internal validity that we have considered.

Statistical Analysis

A two-way analysis of variance (see Chapter 5) should be used to compare the four groups of posttest scores. Pretest scores will not be a part of the statistical analysis. Figure 12.6 shows the data set-up for this two-way

TREATMENT

		YES	NO
	YES	O_2	O_4
PRETEST			
	NO	O_5	O_6

FIGURE 12.6. *Two-way ANOVA to compare four groups of posttest scores.*

ANOVA. A significant F for the interaction source of variation indicates that the effectiveness of the treatment varies according to whether or not the subjects have been pretested. In such a case the researcher might wish to examine his data for simple main effects. To do this, he would first compare the treatment and control groups which received the pretest (O_2 vs. O_4); then, he would compare the two groups which did not receive the pretest (O_5 vs. O_6). If the first comparison leads to a significant difference and the second does not, then the researcher cannot generalize his findings (e.g., a significant main effect for the treatment-no treatment comparison) to nonpretested individuals.

Example of Design

Although the Solomon four-group design does not appear very often in the literature, we can present an example to demonstrate the use of the design. The author of the article was interested in finding out whether an interaction existed between an ethnocentrism attitude inventory (the pretest) and a subsequent film on ethnic prejudice (the treatment). Excerpt 12.6 describes the method used in this study.

EXCERPT 12.6 DESCRIPTION OF METHOD
FOR SOLOMON FOUR-GROUP DESIGN

METHOD

Two hundred and twenty-four students in four introductory classes at The American University served as Ss in the experiment. These groups were randomly assigned to four treatment conditions presented in Table 1. Two of these groups received a modified form of the California Ethnocentrism Scale consisting of 20 Likert-type items as the pretest attitude questionnaire. A high score represented high ethnocentrism. One of these two groups viewed the mental health film (b) on ethnic prejudice, "High Wall," 12 days after taking the pretest. After treatment, this group (Group I) was immediately posttested with the same questionnaire that served as the pretest. The other group (Group IV) was simply posttested 12 days later. Group II viewed the film and was posttested immediately afterward without having been pretested. Group III answered the questionnaire once. Two other groups, which were not included in the experimental design and which had a total N of 100, were simply pretested in order to examine the comparability of Ss initial attitudes on ethnocentrism. For an examination of the interaction hypothesis a factorial analysis of variance was used with .05 as the acceptable level of significance.

SOURCE: R. E. Lana. A further investigation of the pretest-treatment interaction effect. *Journal of Applied Psychology*, 1959, 43, 421–422. Copyright 1959 by the American Psychological Association. Reprinted by permission.

Table 1 of Excerpt 12.7 is a diagram of the Solomon four-group design. The results of the statistical analysis (a two-way factorial ANOVA) are presented in Table 3. The F corresponding to the T × P interaction was found to be less than 1, and this result was not significant at the .05 level of significance. In his interpretation of the findings, the author reports that "the pretest and treatment did not interact, which implies that the pretest did not sensitize Ss" to the treatment.

EXCERPT 12.7 DESCRIPTION OF THE RESULTS
FOR SOLOMON FOUR-GROUP DESIGN

RESULTS

The four groups of pretest scores, including Groups I and IV of the experimental design and the two groups not part of this design, were submitted to a Bartlett's Test and found to be homogeneous with respect to variance. A simple analysis of variance was then performed on the four pretest means. The resulting F ratio was not significant at the .05 level. The Ss were judged to have the same initial ethnocentric attitudes in each of the groups on the strength of this evidence.

TABLE 1 EXPERIMENTAL DESIGN

	GROUPS			
	I	II	III	IV
Conditions	Pretest 12 days Treatment Posttest	Treatment Posttest	Posttest	Pretest 12 days Posttest

The variances of the two sets of pretest scores of the experimental groups receiving a pretest were examined with Bartlett's Test and found to be homogeneous. A t test between the means of these two sets of pretest scores was not significant at the .05 level. The conclusion is drawn that the groups receiving the pretest in the experiment were initially homogeneous with respect to attitude toward ethnic groups.

Means and standard deviations of the posttest scores appear below. A Bartlett's Test was then applied to the four posttest results and the resulting chi square was not significant. A factorial analysis of variance was then performed on the posttest means for the four groups. A summary of the analysis of variance results appears in Table 3. The F ratio for the treatment effect was significant at the .05 level which implies that the film was successful

in changing opinion about ethnocentrism. The interaction effect between questionnaire and film was not significant. Consequently, the pretest and treatment did not interact, which implies that the pretest did not sensitize Ss to the communication even though the topic involved was of relative importance to them.

TABLE 2 SUMMARY OF MEANS AND STANDARD DEVIATIONS OF POSTTEST SCORES

GROUPS	M	SD	N
I (Pretest and communication)	46.04	15.6	52
II (No pretest and communication)	45.70	13.5	57
III (No pretest and no communication)	51.74	13.9	58
IV (Pretest and no communication)	55.65	14.4	57

TABLE 3 ANALYSIS OF VARIANCE ON POSTTEST SCORES

SOURCE	df	SS	MS	F	P
Treatment	1	61.23	61.23	16.69	<.05
Pretest	1	4.52	4.52	1.20	>.05
T × P (Interaction)	1	3.18	3.18	<1	>.05
Error[a]	221		3.68		

[a] Error term was computed by the Walker and Lev simple approximation method for unequal Ns.

SOURCE: R. E. Lana. A further investigation of the pretest-treatment interaction effect. *Journal of Applied Psychology*, 1959, *43*, 421–422. Copyright 1959 by the American Psychological Association. Reprinted by permission.

As we indicated at the beginning of this chapter, few experimental designs reported in professional literature of the behavioral sciences conform exactly to the three true experimental designs presented. In fact, most experimental studies use many variations of these designs. However, occasionally authors actually report the use of one of the true experimental designs. Excerpt 12.8 provides two illustrations.

EXCERPT 12.8 REPORTED USE OF TRUE EXPERIMENTAL DESIGNS

Design. The design was a Pretest-Posttest Control Group Design (Campbell and Stanley, 1963) with the 12 children randomly separated into an experimental and a control group.

SOURCE: R. O. Blackwood. The operant conditioning of verbally mediated self-control in the classroom. *Journal of School Psychology*, 1970, *8*, 251–258.

Experimental Design

The Solomon four-group design was chosen for the experiment (Campbell and Stanley, 1966). The design required the random assignment of subjects to four treatment groups such that Groups E_1 and C_1 would receive the pretest and posttest with E_1 being taught, and Groups E_2 and C_2 would receive the posttest with E_2 being taught.

SOURCE: J. Khatena. Teaching disadvantaged preschool children to think creatively with pictures. *Journal of Educational Psychology*, 1971, 62, 384–386. Copyright 1971 by the American Psychological Association. Reprinted by permission.

EXTERNAL SOURCES OF INVALIDITY

The true experimental designs and their variations do have built-in controls for the threats to internal validity. However, these designs do not have built-in controls for all of the sources of external validity.[4] External validity refers to the extent to which the results of an experiment can be generalized to subjects of other populations and to other environmental conditions (for example, other situations, treatments, experimenters, dependent variables). In other words, with what subjects and under what environmental conditions can the same results be expected? External validity can be classified into two broad types: population validity and ecological validity. *Population validity* concerns the generalization of the results to other subjects; *ecological validity* concerns the generalization of the results to other settings or environmental conditions similar to the experimental setting or condition. There are two possible sources of threats to external population validity and nine possible sources of threats to external ecological validity.

It is difficult to consider each source of external invalidity in relationship to the true experimental designs as we did for internal invalidity. A true experimental design controls for the seven possible sources of internal invalidity but may or may not control for all of the external sources of invalidity. Some of the sources of external invalidity are not a function of experimental design; instead, they are a consequence of inadequate description of the independent or the dependent variable in the procedure section of the journal article. If a study is replicated with another sample from the same population by another researcher, the results may not be the same as for the first study (that is, the results of the first will not generalize). Of course, there could be many reasons why generalization did not occur. One reason might be that the replicated study did not follow the procedures exactly or measure the dependent variable in the same way as the first study because the procedures or measures were inadquately described. Therefore, in order to evaluate many sources of external invalidity, a reader must examine the description of the independent and dependent variables for a specific study reported in a journal. Thus, while it is possible to control for all sources of

---

[4] We acknowledge a great obligation to Glenn H. Bracht and Gene V. Glass (The external validity of experiments. *American Educational Research Journal*, 1968, 5, 437–474) for the comprehensive refinement, elaboration, and extension of the sources of external invalidity.

internal invalidity by experimental design, some sources of external invalidity cannot be controlled by the design alone. For this reason, we present a discussion of eleven sources of external invalidity (see Table 12.1) to serve as a checklist to aid the reader in evaluating these possible sources rather than dealing with each source in relationship to a specific true experimental design.

TABLE 12.1 POSSIBLE SOURCES OF INTERNAL AND EXTERNAL INVALIDITY FOR EXPERIMENTAL DESIGNS

INTERNAL	EXTERNAL
1. History	Population
2. Maturation	1. Experimentally accessible population vs. target population
3. Testing	2. Interaction of treatment effects and subject characteristics
4. Instrumentation	Ecological
5. Statistical regression	3. Describing the independent variable (X)
6. Selection	4. Describing and measuring the dependent variable (O)
7. Mortality	5. Multiple-treatment interference
	6. Interaction of history and treatment effects
	7. Interaction of time of measurement and treatment effects
	8. Pretest and posttest sensitization
	9. Hawthorne effect
	10. Novelty and disruption effect
	11. Experimenter or Rosenthal effect

Population Validity

Experimentally Accessible Population Versus Target Population. There are two kinds of randomization—one deals with the random selection of subjects and the second with the random assignment of subjects. *Random selection* is concerned with selecting subjects who are representative of a popula-

EXCERPT 12.9 IMPORTANCE OF REPRESENTATIVENESS

THE SUBJECT IN CONTEMPORARY PSYCHOLOGY

Bias in the Selection of Subjects

In reading our journals, one receives the distinct impression that the only kind of people of interest to psychologists are college students! If college students were truly representative samples of the population at large, there would be no problem in generalizing from the results of our studies. But (fortunately or unfortunately) they do differ in highly significant ways from the general population, and we cannot have a truly meaningful science of human behavior by studying such a restricted sample.

How biased in subject selection is our research? Smart (1966) examined the two largest journals of the American Psychological Association reporting research with human subjects: the *Journal of Abnormal and Social Psychology* (1962–1964, Volumes 64–67) and the *Journal of Experimental Psychology* (1963–1964, Volumes 65–68). The present author reviewed the same journals for the period 1966–1967 (the *Journal of Personality and Social Psychology,*° Volumes 3–7, and the *Journal of Experimental Psychology*, Volumes 71–75), and the data from both surveys is contained in Table 1. Both surveys dealt only with nonpsychiatric subject groups.

Inspection of Table 1 reveals a striking degree of similarity between the two sets of data with a heavy concentration of college students as subjects. In addition to the great reliance on college students, both surveys revealed an overrepresentation of male subjects. The extremely small percentage of studies sampling the general adult population was particularly disturb-

° One of the journals created by splitting JASP.

ing; none of the studies published in the *Journal of Experimental Psychology* during those years used a sample of the general population. Further, this author's survey found that in this journal, the nature of the subjects studied was not specified in 3.6% of the articles for the 2-year period. No mention could be found in these articles of where the subjects were obtained, who they were, or even if they were male or female. This certainly seems a serious omission in the reporting of research.

The fact that college students are our primary focus of research has a number of important and sobering consequences. For example, approximately 80% of our research is performed on the 3% of the population currently enrolled in college (United States Department of Commerce, 1967). Regardless of how much our college enrollments may increase, college students most likely will never be truly representative of the total adult population, in terms of level of intelligence alone. Further, this pronounced emphasis on college students means that most of our research is conducted with a very young group, primarily ages 18–24.

Such students are probably at the peak of their learning and intellectual abilities and this could mean that many findings in learning, especially verbal learning, could be special to the college student with limited applicability to other groups [Smart, 1966, p. 119].

There is also the problem of social class representation for, as Smart noted, the college student population contains more upper- and middle-class people and fewer lower-class people than the general population.

tion. However, the *experimentally accessible population*, that is, the population from which the experimenter can select his subjects, may or may not be the same as the *target population*, that is, the population to which he wishes to have the results generalize.[5] The researcher in our hypothetical weight-

[5] This is also referred to as the interaction of selection and the independent variable by D. T. Campbell and J. C. Stanley (*Experimental and Quasi-Experimental Designs for Research.* Chicago: Rand McNally, 1966).

TABLE 1 HUMAN SUBJECT SOURCES IN AMERICAN
PSYCHOLOGICAL ASSOCIATION JOURNALS

SUBJECTS	JEP[a] SMART	JEP SCHULTZ	JASP[b] SMART	JPSP[c] SCHULTZ
Introductory Psychology	42.2%	41.2%	32.2%	34.1%
Other college	43.5%	42.5%	40.9%	36.1%
Precollege	7.0%	7.1%	16.9%	18.5%
Special adult	7.3%	5.6%	9.4%	10.1%
General adult	0	0	0.6%	1.2%
All male	22.3%	19.3%	33.6%	26.7%
All female	6.0%	6.0%	10.8%	10.6%

[a] *Journal of Experimental Psychology.*
[b] *Journal of Abnormal and Social Psychology.*
[c] *Journal of Personality and Social Psychology.*

SOURCE: D. P. Schultz. The human subject in psychological research. *Psychological Bulletin*, 1969, 72, 214–228. Copyright 1969 by the American Psychological Association. Reprinted by permission.

reducing program may wish to generalize the results to all football players on professional and collegiate teams. Yet, he may have used only one team within one regional division of the national professional football league. The team receiving the treatment is the experimentally accessible population while all other collegiate and professional football players represent the target population. As we can readily see, one team is not a random sample and not representative of the target population to which the researcher would like to generalize the results. To have representativeness, the researcher must randomly sample football players from many more professional and collegiate teams.

The second kind of randomization deals with random assignment. *Random assignment* is concerned with equating experimental and control groups. The randomization is generally done within a limited population (the experimentally accessible population). Researchers often emphasize only random assignment, giving almost exclusive consideration to the demands of internal validity with very little consideration to representativeness for the purposes of generalizing to different subjects. In Excerpt 12.9 Schultz indicates the importance of representativeness. Table 1 in the excerpt shows the results of his survey of two journals reporting research with human subjects and the results of an earlier survey by Smart.[6] It is important that researchers use both random selection and random assignment in their studies.

Interaction Between the Treatment and Subject Characteristics. Another source of population invalidity is the possible interaction between the treatment and subject characteristics. For example, one program of instruction may be more effective for one level of ability, one attitude bias, one level of education, or one personality characteristic than for another. Pro-

[6] R. Smart. Subject selection bias in psychological research. *Canadian Psychologist*, 1966, *7a*, 115–121.

grammed instruction may be more effective for children with average levels of ability than for children with high ability. Unless a researcher is aware of the possibility of this kind of interaction, he may assume that one treatment can be prescribed for all levels of a subject characteristic. Consider our hypothetical weight-reducing program. It is quite possible that the program is more effective for football players who are 5 to 10 pounds overweight than for players 20 to 30 pounds overweight. A researcher might be tempted to generalize his results to all overweight football players when, in fact, the results are valid only for players 5 to 10 pounds overweight. The same interaction is possible for other characteristics. For example, the weight program might be more effective for players who have a particular kind of personality or physical structure or who play a specific position on the team (that is, it may be effective for aggressive, tall linemen, but not effective for aggressive, tall backfield players).

Ecological Validity

Describing the Independent Variable (*X*). The generalization of the results of an experiment can be restricted if the researcher has not described in detail the procedures, activities, or length of time used in the treatment. Recall that a criterion for a good method or procedure section of an article is that the author report the treatment in enough detail so that the experiment can be replicated. If the researcher is vague in describing the independent variable, the reader would have difficulty in determining specifically what is to be generalized.

The researcher in our hypothetical weight-reducing program must specify the exact procedures and activities if the results are to be generalized or replicated. The specifics of the program must be described sufficiently to permit the program to be used by another team or trainer. For example, the method section should describe the chronological sequence of each procedure in the program, the length of time for each sequence, and the overall time period of the entire program. Authors of journal articles sometimes neglect to describe the overall period of treatment and the length of time for specific aspects within the treatment. These descriptions are crucial if the results of an experiment are to be generalized to other settings or situations.

Describing and Measuring the Dependent Variable (*O*). Researchers must also satisfactorily describe the dependent variable and the procedure used for measurement. For example, if the dependent variable for an experiment is rate of learning, the researcher must provide an operational definition of learning and describe how the rate of learning was measured and computed.[7] If a new counseling or therapeutic technique is used in an ex-

---

[7] An *operational definition* is a description of a characteristic or phenomena in terms of measurable activity. If a researcher does a study in learning or counseling, he must describe these concepts in terms of what the learner or counselor does. For example, learning might be defined as swimming 50 yards or adding a column of figures correctly, and counseling might be defined by specific kinds of verbal responses by the counselor.

periment, the researcher must, again, give an operational definition of change and the measures of change that were used.

A second potential source of invalidity associated with description and measurement related to the dependent variable is the degree of reliability of standardized measurement or test instruments. If the measuring instrument is not reliable, a true treatment effect may well be concealed by the inconsistency or fluctuation of the instrument. In other words, a Type II error will occur more often when the researcher's instruments are unreliable.

A third source of invalidity is the inappropriate choice of the measuring instrument. The measuring instrument must be valid for the concept or construct being investigated. An exaggerated example of an invalid instrument is the use of a personality test to measure academic achievement.

A fourth possible threat to generalization is the degree of reliability of judges or observers used to measure the dependent variable. When observers are used, the researcher should report the reliability (agreement between raters) of the judges' or observers' ratings or measures of the dependent variable. If low interjudge or observer reliabilities are reported or if reliability scores between judges are not included, it is difficult for the reader to be sure whether change on the dependent variable was caused by the treatment or by systematic fluctuations in the ratings of the judges.

A last possible threat to generalization concerns fallibility in the analysis of the dependent measures or data. For example, researchers might inaccurately transform frequency data to percentages, use the mean for a skewed distribution instead of the median, or perform an incorrect statistical analysis on gain scores. In such cases it is sometimes difficult for a reader of a journal article to be sure that change on the dependent measures was caused by the treatment or by fallibility in the analysis or transformation of the data.

The dependent variable in our hypothetical weight-reducing experiment was the weight of each subject. In this experiment it is important for the researcher to obtain reliable measures of weight. The researcher might have had the athlete stand on the scale and two observers record independently the weight to the nearest ounce. Whatever method is selected for recording weight, the procedure should be described and the reliability of agreement reported for the recorded weights. If the procedure for recording the weights (dependent variable) is vague or the recordings are unreliable, the researcher would have difficulty in generalizing the results to other situations. Our weight problem example perhaps would not present as much of a problem with reliability as behavior modification research (see Chapter 15). In these studies raters should be trained to observe in a natural setting and to record the frequency of several categories of behavior.

Multiple-Treatment Interference. A journal may report the results of an experiment in which two or more treatments have been applied to the same subjects. In such experiments it may be possible to determine the effect of the first treatment, but it is nearly impossible to assess the effect of the subsequent treatments. The researcher cannot separate the effect of each treatment because the results of the second treatment are confounded with the results of the first. The researcher does not know how much influence the second treatment has on the results. Multiple-treatment interference may also restrict generalization when the same subjects have been used in several different experiments. Perhaps the situation is most typical in college environments where a great deal of research is conducted with students who may have been subjects for several experiments.

If the hypothetical experiment of our weight-reducing program had two treatments, we would not know which of these two treatments could be generalized to other situations. Suppose that the first treatment consisted of applying a specified diet and the second treatment involved a program of isometric exercises. The researcher can determine the effects of the diet by weighing the players upon completion of the dietary program. However, it is impossible to determine what effect the isometric exercise program alone had on the dependent variable. The subjects would have already had a history of being on a prescribed diet which would be confounded with the possible effect of the exercises.

Interaction of History and Treatment Effect. Historical events occurring at the time of the experiment may interact with the treatment effect. These historical events may affect the results in such a way that the same results would not be found on another occasion. Some historical events of long duration such as national economic or defense programs might be difficult for the researcher to identify or may not be immediately obvious. However, a researcher can usually determine whether or not some historical event of short duration might have invalidated the results of an experiment. Such an event might be a regional play-off for the football championship. The results of the weight-reducing program may not be typical if the research was conducted at another time. Without the regional play-off as an incentive, the weight-reducing program might not work at all.

Interaction of Time of Measurement and Treatment Effects. Another possible threat to the generalization of the results concerns the time after treatment at which the measurement of observation is taken. Usually the effects of treatment are measured or observed immediately after treatment. However, the effect of treatment may not occur until sometime later. If a researcher measures retention immediately following an innovative teaching technique, he might observe that the technique has no effect on immediate retention. At some later time, the researcher may take another measure and find that the technique has had a significant effect on delayed retention. The interaction of time of measurement and treatment effects may also have the opposite effect, that is, the treatment effect may last only for a short period of time. Experiments which measure the dependent variable at several times after treatment increase the ecological validity of the results.

We may find that our hypothetical weight-reducing program is effective for only a limited period of time after the treatment period. For example, the effect of the program may not be observed until three days or one week after the weight-reducing program has been concluded. It is also possible, and perhaps more likely, that the effect of the weight program may last only for a short period of time after treatment.

Pretest and Posttest Sensitization. The observation or measurement of the dependent variable can sensitize the subjects to the treatment. It is possible that a pretest or a posttest confounds the results of the treatment with sensitization. Pretest sensitization is most likely to occur when experiments use personality tests, attitude or opinion measures, or interview techniques. Also, if measures of academic achievement are used, pretest sensitization may confound the effects of the treatment. For example, a researcher may find that subjects who are given a pretest questionnaire or interview show less, or more, change in attitude than subjects who do not receive the questionnaire. As discussed above, the length of time between the pretest and treatment might also be related to the degree of sensitization. Also, the length of time between treatment and posttesting might be crucial to the degree of sensitization.

There is also a possibility that the effect of treatment will occur only if a posttest is administered. The posttest may sensitize the subject or call his attention to the treatment in a way which might not have occurred if he had not received the posttest. A posttest questionnaire or achievement test questions may provide subtle cues which facilitate recall of information, concepts, or events of the treatment. Consider, for example, the experiment designed to assess the effects of a film about sensitivity groups on level of apprehension that people might feel prior to their first group experience. The posttest might facilitate the subjects' recall of details or scenes in the film which would influence apprehensiveness about sensitivity groups.

Although testing sensitization primarily concerns the effects of attitude, personality, opinion, and possibly academic measures, it could influence the results of our hypothetical weight program. The pretest might cause the subjects to be more conscious of their weight. The subjects' concern about weight might influence weight reduction, but the weight change would be independent of any treatment effect. On the other hand, it is very unlikely that the posttest sensitization would have any effect on this experiment.

Hawthorne Effect. Another variable which may confound or account for the effects of the treatment is the subject's knowledge that he is participating in an experiment. Several factors possibly associated with this variable, which is known as the Hawthorne effect, are the perceived demand characteristics of the experimental situation, evaluation apprehension, social desirability, and the placebo effect. *Demand characteristics* refer to all the cues in the treatment situation which might convey the purpose of the study to the subject and become a significant influence on the subject's behavior.

Evaluation apprehension is the anxiety that might be generated by participation in an experiment, thereby influencing the participant's behavior. *Social desirability* is the subject's motivation to do the "right thing" or to perform well in a situation in which he knows his behavior is being evaluated. For example, in research involving a dyad of subject and experimenter, the perceived role of subject can affect a subject's impression of what is socially desirable. Finally, the *placebo effect* is the subject's tendency to believe in the effectiveness of the treatment simply because it is administered in an apparently scientific setting. Since these four factors modify the subject's behavior in the treatment situation, but are unrelated to the actual treatment administered, they are threats to ecological validity, especially in remediation, psychotherapy, counseling, and other behavioral science experiments. Can you explain how these factors associated with the Hawthorne effect might influence our weight reducing program?

Novelty and Disruption Effects. The novelty or innovative nature of the treatment compared with the routine of the environmental setting may be a variable which is more influential than the actual treatment and may therefore confound the results. The effects may not occur under conditions in which the treatment is traditional or not novel. These effects are perhaps prevalent in situations where new and innovative experimental programs have been inaugurated.

Another aspect of the novelty effect is the disruption effect. The experimenter or person using an innovative program might feel uncomfortable and have difficulty adapting to the new program. After the experimenter or subject gains experience and facility with the new program, the results of the treatment (the new program) may be superior to the results obtained during or after the initial tryout period.

The weight-reducing program may have a novel or disruptive effect on the football players. The diet or exercises may be a departure from the everyday routine training program. The uniqueness and novelty of the program could be responsible for any differences in weight. On the other hand, the experimenter or trainer may lack experience with the actual procedures of weight reduction which could be responsible for a difference, or no difference, in loss of weight.

Experimenter or Rosenthal Effect. The last possible source of external ecological invalidity is sometimes referred to as the Rosenthal effect.[8] In some experiments it is possible that the experimenter may unintentionally modify the subject's behavior through active effects (nonverbal or verbal behavioral cues) or passive effects (appearance, sex, race, dress). The experimenter or trainer could communicate subtle cues which might influence the weight loss for our hypothetical example.

---

[8] For a detailed analysis of approximately 18 sources of experimenter effects which are related to ecological validity, see Robert Rosenthal, *Experimental Effect in Behavioral Research.* New York: Appleton-Century-Crofts, 1966.

Theodore X. Barber has explored the pitfalls in research with respect to nine investigator and experimenter effects.[9] Barber makes a distinction between the role of the investigator and the role of the experimenter. The investigator is responsible for the experimental design, the training of experimenters, the logistics of conducting the study, the analysis and interpretation of results, and writing the research report. The experimenter is the person who administers the experimental procedures and collects the data from the subjects. Barber describes four effects associated with the investigator: (1) paradigm effect (basic assumptions and ways of conceptualizing the area of inquiry), (2) loose protocol effect (script of step-by-step details of how the experiment is to be conducted), (3) analysis effect, and (4) fudging effect. There are five experimenter effects: (1) attributes effect, (2) failure-to-follow-the-protocol effect, (3) misrecording effect, (4) fudging effect, and (5) unintentional effect.

As we indicated in the last chapter, a researcher should attempt to have internal validity as the minimum requirement for his design and should control as many sources of external invalidity as possible. The true experimental designs we have considered in this chapter provide effective controls against the threats to internal validity, but they may not control for all of the sources of external invalidity. Perhaps the best way to insure external validity of an experiment is for the researcher (or another researcher) to replicate the experiment. Another possible technique for decreasing the threats to external validity is for the experimenter to build into the experiment simultaneous replication either in another setting or by another experimenter. Also, the extensions of the true experimental designs considered in the next chapter and the quasi-experimental design discussed in Chapter 14 may provide additional controls for some of the sources of external invalidity.

REVIEW TERMS

control group
disruption effect
gain scores
ecological validity
experimental group
Hawthorne effect
interaction of history and
    treatment effect

interaction of time of
    measurement and treat-
    ment effect
matching
multiple-treatment inference
novelty effect
operational definition
placebo effect

[9] For a detailed discussion of these nine effects see T. X. Barber, "Pitfalls in research: nine investigator and experimenter effects." In R. M. W. Travers (Ed.), *Second Handbook of Research on Teaching.* Chicago: Rand McNally, 1973.

population validity   random selection of Ss
posttest-only control group design   Rosenthal effect
posttest sensitization   Solomon four-group design
pretest-posttest control group design   subject characteristics
pretest sensitization   target population
random assignment of Ss   true experimental design

REVIEW QUESTIONS

1. What does the symbol R represent in the context of a diagram for a true experimental design?
2. Suppose a researcher uses 25 volunteers as the experimental group. To form the control group, a matched pair is found for each of the Ss in the experimental group based on IQ, age, socioeconomic status, and height. Is this procedure for forming the two groups as good as random assignment?
3. Is it possible for a researcher to use two experimental groups (instead of one experimental and one control group) in a pretest-posttest control group design?
4. Suppose a researcher uses a pretest-posttest control group design with Ss who have been identified on the basis of their very low scores on the pretest. Which of the threats to internal validity is most likely to operate in this situation?
5. Why doesn't randomization control for mortality?
6. Is it appropriate for a researcher to analyze the data from a pretest-posttest control group design with two independent samples t tests? Would two correlated samples t tests be appropriate?
7. What is the main difference between a static-group comparison and a posttest-only control group design?
8. The Solomon four-group design can be used to determine whether there is an interaction between _____ and _____. In this design how many groups are pretested? How many are given the treatment? How many are posttested?
9. If a research study has high internal validity, will it necessarily have high external validity?
10. One of the threats to external validity is the difference between the experimentally accessible population and the target population. If a researcher randomly assigns his subjects to the relevant comparison groups, does he eliminate this threat to external validity?
11. Suppose the same group of Ss are used in two consecutive experiments. Would the problem of multiple-treatment interference, if it existed, invalidate the results of the first study, the second study, or both studies?
12. As the time interval between the pretest and posttest increases, does it

become more or less likely that history will interact with the treatment so as to limit generalizability?

13. When will posttest sensitization be more likely to occur: (a) when Ss are given a posttest immediately after being exposed to the treatment or (b) when a longer time interval exists between the treatment and the posttest?

14. If the Ss know that they are participating in a study, they may act differently than they do normally. This phenomenon, which may limit the external validity of a study, is referred to as the _____ effect.

15. The possibility that a researcher may unintentionally affect the subject's behavior is sometimes referred to as the _____.

CHAPTER 13

EXTENSIONS OF THE BASIC TRUE EXPERIMENTAL DESIGNS

In Chapter 12 we discussed the three basic true experimental designs: the pretest-posttest control group design, the posttest-only control group design, and the Solomon four-group design. Although these are the classic true experimental designs, many others are available to researchers which also guard against the various threats to internal validity. We will now direct our attention to several designs that can be thought of as extensions of the designs discussed in the previous chapter. Specifically, we shall look at (1) pretest-posttest designs that involve three or more groups, (2) posttest-only designs that involve three or more groups, (3) extensions of the Solomon four-group design, (4) factorial and repeated measures designs, and (5) special factorial designs that are set up so as to increase external validity.

MULTIGROUP PRETEST-POSTTEST DESIGNS

A multigroup pretest-posttest design consists of (1) random assignment of Ss to three or more groups, (2) collection of pretest data from each subject, (3) exposure of each group of Ss to a different treatment condition, and (4) collection of posttest data from each subject. Thus, the multigroup pretest-posttest design is identical to the pretest-posttest control group design discussed in Chapter 12 except for the number of groups involved in the study.

In Excerpt 13.1 we see a study in which a four-group pretest-posttest design was used. The treatment conditions were defined by distinct sequences of learning materials, and all Ss were randomly assigned to one of the four experimental groups, administered a pretest, and given a posttest.

EXCERPT 13.1 FOUR-GROUP PRETEST-POSTTEST DESIGN

Procedures

The Ss for this experiment were 208 second- and third-grade children in nine classrooms within a single suburban community of greater Hartford, Connecticut. There were four experimental groups of 52 Ss, each group receiving the levels of the learning program in a different order. Because Treatment A presents Ss with the maximum amount of ordered sequence, this treatment has been labeled HiHi; Treatment B presents Ss with less ordered sequence, and is labeled Hi; Treatment C, presenting even less, is Lo; and Treatment D, presenting the minimum amount of ordered sequence, is labeled LoLo.

The eight levels of the program are collapsed into four levels for purposes of arranging the treatment groups and data analysis. Levels I and II are combined to 1; Levels III and IV are combined to 2; Levels V and VI are combined to 3; and Levels VII and VIII are combined to 4. The four different sequences which comprise the four experimental treatments are the following:

A	1	2	3	4	HiHi
B	1	2	4	3	Hi
C	1	4	3	2	Lo
D	4	3	2	1	LoLo

Members of each class group were randomly assigned to all four treatments.

A pretest and a posttest were administered to all Ss. Prior to their work with the learning program, Ss were given the pretest. Containing material similar to that on all levels of the learning program, its purpose was to show what Ss already knew about the material in the program.

The pretest includes 40 questions consisting of material similar to that on all eight levels of the program itself. Five questions are devoted to measuring the understanding of materials on all of the eight levels. The 40 questions on the pretest are arranged in random order. The posttest and pretest are interchangeable forms of the same test. Form A was administered to half of the Ss as a pretest and half of the Ss as a posttest. The same was true of Form B.

SOURCE: B. W. Eustace. Learning a complex concept at different hierarchical levels. *Journal of Educational Psychology*, 1969, 60, 449–452. Copyright 1969 by the American Psychological Association. Reprinted by permission.

In diagraming a multigroup pretest-posttest design, there is one row of symbols for each of the different comparison groups. An R at the beginning of each row designates random assignment of Ss to groups. Immediately following each R, an O denotes the collection of pretest data. To designate the different treatment conditions, a researcher may use different letters (X, Y, Z, etc.) or the same letter with different subscripts (X_1, X_2, X_3, etc.). Finally, an O is used at the end of each row to indicate the collection of posttest data from the Ss in each group. In Figure 13.1 we see two alternative ways in which a three-group pretest-posttest design could be diagramed. Since the diagram on the right in Figure 13.1 can be more easily extended to accommodate four or more treatment conditions, the present authors prefer this format to the one on the left.

R	O	X	O	R	O	X_1	O
R	O	Y	O	R	O	X_2	O
R	O	Z	O	R	O	X_3	O

FIGURE 13.1. *Two ways of diagraming a multigroup pretest-posttest design with three treatments.*

Both of the diagrams in Figure 13.1 represent a situation in which there are three distinct treatment groups. If a researcher were conducting a study to compare three approaches to counseling, either of these diagrams would be appropriate. In many research investigations, however, the researcher includes a control group along with two or more treatment groups. Thus, if a specific study were being undertaken to compare the relative effectiveness of two different approaches to counseling as contrasted with a control group (that does not receive any counseling), the diagrams in Figure 13.1 would become appropriate if we removed the Z and the X_3 from the left and right diagrams, respectively.

There are several acceptable ways in which the data associated with a multigroup pretest-posttest design can be analyzed statistically.[1] First, the researcher could use the analysis of covariance to compare the various groups in terms of posttest means, after these means have been adjusted to account for any differences that existed among the groups on the pretest. Second, he might compute a difference (i.e., gain) score for each subject by subtracting his pretest score from his posttest score, with a one-way ANOVA applied to the gain scores to see whether the groups differ in terms of the mean gain from pre to post. Third, the researcher might use a Lindquist Type I repeated measures analysis of variance. As you remember from Chapter 6, this particular statistical analysis provides three F ratios: one for the main effect of groups, one for the main effect of trials, and one for the groups-by-trials interaction. Of these, the interaction F ratio would provide the best indication as to whether the treatments have a differential effect on the Ss.[2] Fourth, a nonparametric procedure might be applied by ob-

---

[1] Two approaches to the statistical treatment of data are *not* acceptable. The first involves two separate one-way ANOVA's, one for the pretest scores and the other for the posttest scores. The second unacceptable approach involves several t tests, each of which compares the pretest and posttest scores for a separate group. For an explanation as to why these two approaches are considered to be inferior, see J. P. Guilford, *Statistical Analysis in Education and Psychology*, New York: McGraw-Hill, 1965, pp. 194–197.

[2] For reasons that cannot be discussed in this text, the F for the main effect of the treatments from the Lindquist Type I ANOVA (when applied to this design) will be too small and potentially will cause the researcher to make a Type II Error. However, the interaction F is not an underestimate; in fact, the groups-by-trials F will be the same as what would be obtained by applying a one-way ANOVA to gain scores.

taining a pre-post difference score for each subject, and then applying a Kruskal-Wallis one-way analysis of variance of ranks to the difference scores. Of course, an overall significant finding from any of these statistical approaches would require a follow-up by an appropriate multiple comparison investigation to find out exactly which groups are significantly different from one another.

In Excerpt 13.2 we see a study in which the researcher used the analysis of covariance in conjunction with a three-group pretest-posttest design. The three treatments involved different self-instructional approaches to teaching diagnostic problem solving (troubleshooting). Two of the

---

EXCERPT 13.2 THREE-GROUP PRETEST-POSTTEST DESIGN WITH DATA SUBJECTED TO THE ANALYSIS OF COVARIANCE

Prior to actual administration of experimental treatments, a pre-experimental instructional phase was conducted. Initially, the Ss received a filmstrip and record presentation dealing with principles of the automotive ignition system. Immediately after this instruction was given, an ignition system knowledge examination was administered. This was followed by administration of the Otis Mental Ability Test.

Students from the entire sample were then randomly assigned to one of the three instructional treatments: equipment oriented ($N = 15$), textbook oriented ($N = 15$), and programmed instruction ($N = 15$). Content of the three treatments was similar in that each presented students with the same three practice troubles to locate. Additionally, a common troubleshooting strategy was taught in each of the respective treatments. In all cases, the instruction presented was individualized in nature and allowed each student to proceed through the treatment at his own pace. The equipment oriented instruction was conducted using operational equipment and a troubleshooting flow chart for each student. An instructor assisted the student only if he had questions about how to read the flow chart or had mechanical problems with the engines. In all cases, students were required to locate the troubles "on their own." In this manner, the instructor variable could be adequately controlled. The programmed instruction consisted of a booklet developed

by the investigator which utilized both linear and intrinsic programming techniques. The booklet allowed students to "find" three different troubles in each of three different hypothetical engines. The troubles were identical to those which were used in the equipment oriented instruction. The textbook oriented instruction also provided students with troubles to locate. Content comparable to the programmed instruction treatment was used without including any reinforcement principles. Students in the equipment group averaged 15 minutes to complete their treatment while program and text group students averaged 22 and 13 minutes, respectively.

After each student completed his instructional treatment he was asked to fill out an attitude inventory (4). The inventory included items such as "I would like more instruction presented in this way" (positive), and "I felt frustrated by the instructional situation" (negative). Following completion of this instrument, a multiple-choice troubleshooting knowledge examination was administered. After all class members had completed the knowledge examination, a performance examination was administered. The performance examination required each student to locate troubles which had been placed in otherwise operational automobile engines. The troubles included in the criterion measure were different from those used in the three instructional treatments.

SOURCE: C. R. Finch. The effectiveness of selected self-instructional approaches in teaching diagnostic problem solving. *Journal of Experimental Education*, 1972, 65, 219–223.

treatments attempted to simulate student-equipment interaction through programmed instruction or text instruction, while the third treatment utilized actual equipment. The two pretest scores collected from each subject (based upon a knowledge examination and the Otis Mental Ability Test) were used as covariates within the analysis of covariance. A separate analysis was conducted for each of the dependent variables—troubleshooting knowledge, troubleshooting performance, and student attitude toward the instruction received. (Note that this study shows how the pretest data in a multigroup pretest-posttest design can come from measuring instruments that are different from those used to collect posttest scores. When this is the case, however, the analysis of covariance is the only legitimate procedure that can be used to analyze the data.)

MULTIGROUP POSTTEST-ONLY DESIGNS

A multigroup posttest-only design consists of (1) random assignment of Ss to three or more groups, (2) exposure of each group of Ss to a different treatment condition, and (3) collection of posttest data from each subject. Thus, the multigroup posttest-only design is identical to the posttest-only control group design except for the number of groups involved in the study.

In diagraming a multigroup posttest-only design, there will be one row of symbols for each of the different comparison groups. Each row will contain three symbols: R to designate random assignment of Ss to groups, a symbol to designate the treatment condition (X, Y, Z, or X_1, X_2, X_3), and O to indicate the collection of posttest data. In Figure 13.2 we see two alternative ways in which a multigroup posttest-only design could be diagramed.

R	X	O	R	X_1	O
R	Y	O	R	X_2	O
R	Z	O	R	X_3	O

FIGURE 13.2. *Two ways of diagraming a multigroup posttest-only design with three treatments.*

Both of the diagrams in Figure 13.2 correspond to a situation in which there are three distinct treatment conditions. For a study with two treatment conditions and one control condition, the Z and the X_3 from the left and right diagrams, respectively, would be eliminated. Of course, the diagrams in Figure 13.2 should not mislead the reader into thinking that only three treatments (or two treatments and a control) can be included in a multigroup posttest-only design. This design can accommodate four, five, or more groups. Since the diagram on the right in Figure 13.2 can be extended more easily to handle more than three groups, the present authors prefer this format to the one on the left.

The most frequently used statistical procedure for analyzing the data of a multigroup posttest-only design is a one-way analysis of variance followed, if necessary, by a post hoc multiple comparison investigation. In Excerpt 13.3 we see a four-group posttest-only design in which a one-way ANOVA was used to compare groups in terms of posttest means. Altogether, a total of four separate analyses were conducted: one on the immediate retention data, one on the PRQ items of the delayed retention test, one on the NRQ items of the delayed retention test, and one on the data from the Extended Curiosity Test. Following significant F ratios, the Duncan multiple comparison procedure was used to identify which specific means were significantly different from one another.

Although the statistical procedure in Excerpt 13.3 is the most popular, there are other possible ways of analyzing the data. For example, some researchers collect data on a concomitant variable at the same time that they collect the posttest data, that is, after the administration of the treatments. Scores from the concomitant variable are then used as a covariate within the analysis of covariance. This approach to data analysis requires that the researcher select his concomitant variable with extreme caution, because a crucial assumption of a covariance analysis states that scores on the covariate cannot be influenced by the treatment variable. When covariate data are collected prior to the administration of the treatments (as in a multigroup pretest-posttest design), this assumption will definitely be met. However, the validity of this assumption is questionable when the covariate data are collected after Ss have been exposed to the treatments. The only concomitant variable that is likely to meet the assumption is one which represents a stable aspect of the subject (e.g., IQ score or weight), unless the treatment is designed to change that aspect (e.g., to increase IQ score or reduce the Ss weight).

In addition to the analysis of covariance, the data from a multigroup posttest-only design could be analyzed by means of a nonparametric statistical procedure. For example, the Kruskal-Wallis test applied to posttest ranks would indicate whether the various treatment groups differ in terms of posttest medians. Another possible nonparametric approach would be chi-square, which determines whether the treatment groups differ in terms of the frequency with which subjects do something. Excerpt 13.4 is an example of the use of chi-square to analyze the data of a three-group posttest-only design.

SOLOMON FOUR-GROUP EXTENSIONS

Like other true experimental designs, the Solomon four-group design can be extended to accommodate additional groups of subjects. An extension of this design may be used because the inclusion of more than four groups will enable the researcher to answer more research questions and/or lead to a more powerful analysis.

EXCERPT 13.3 FOUR-GROUP POSTTEST-ONLY DESIGN WITH
DATA SUBJECTED TO A ONE-WAY ANALYSIS OF VARIANCE

METHOD

Material and Design

A chapter from *Griekenland. Het heden van Hellas* (*Greece. Hellas Today*) by A. C. De Vooys (1962) was selected and slightly altered. The experimental passage was 3,000 words long, and dealt with the habits and customs of Greek rural life. Forty-five multiple-choice questions (three alternatives), which required the recall of specific factual information, were constructed.

Fifteen of these were used both as prequestions and as delayed retention questions (prequestion-retention questions or PRQs); of the remaining 30 questions, 15 were used for testing for immediate retention, and 15 for the delayed retention test (new retention questions or NRQs). In addition, a test for Extended Curiosity was constructed. This consisted of a list of 10 topics, taken from the reading passage.

Two experimental treatments and two control groups were used:

1. *PG* (*prequestions, guess*). Before starting to read the passage the subjects in this treatment were given the set of 15 prequestions and were instructed to guess the answer, that is, select one of the alternatives to each question. Time available for this: 4 minutes.

2. *PNG* (*prequestions, no guess*). Before starting to read the passage subjects in this group were given the set of 15 prequestions and instructed just to read the questions. Time available for this: 4 minutes.

3. *CER* (*no prequestions, extended reading time*). While subjects in the PG and PNG treatments dealt with the prequestions, subjects in the CER condition were allowed to read the learning passage; thus they got 4 minutes extra reading time.

4. *C* (*no prequestions, no extra reading time*).

Subjects

The subjects were 72 undergraduates in psychology of Utrecht University. The subjects signed for two experiments. In order to control for foreknowledge only those subjects who had never been to Greece were allowed to enlist.

Procedure

The subjects were randomly assigned to one of the four conditions. They were run in homogeneous groups of three at a time. They were seated at three tables and were instructed according to their condition.

PG and PNG. "I'm going to give you an article about Greece, which I want you to study. But first I shall give you a sheet with a number of questions about the passage."

PG. "Try to answer the questions even though they may all seem equally probable or improbable to you. If you really don't know, just guess. Is that clear? Begin."

PNG. "Read through the questions, there is no need to hurry." After 4 minutes the experimenter removed the sheets containing the prequestions.

The reading material, which consisted of seven sheets of paper was handed out to all subjects. The subjects were then given the following instruction:

I have given you an (PG and PNG: "the") article about Greece. You will be allowed to study it for 15 minutes (CER: "almost 20 minutes"). You must try to complete the entire article in that time. If you have any time left the best thing to do is to scan the text, because you probably won't have time to read it all over again. Afterwards we'll check how much you remember of the passage.

Immediately after the reading task the experimenter removed the reading material, handed out the sheets with 15 immediate retention test items and instructed the subjects to answer the questions. After completing this test subjects were reminded to be on time for the next experiment, a week later, and they were asked not to discuss the experiment with other students. They returned a week later at the same time ostensibly for a different experiment, but were then given the test for delayed retention, consisting of the 15 PRQs and 15 NRQs in mixed random order. After completing this test, the subjects in the experimental treatments were asked to indicate which 15 questions belonged to the prequestion series from the learning session the week before. Finally all subjects were given the test for Extended Curiosity and asked to check those subjects they would like to read more about.

RESULTS

The mean retention and Extended Curiosity scores for the experimental and control groups are shown in Table 1. A one-way analysis of variance was performed on these results, and differences between group means were subsequently analysed with Duncan's multiple-range test at the .05 level of significance. For the immediate retention, analysis of variance showed a significant difference between the groups ($F = 2.99$, $df = 3/68$, $p <$

.05). Further analysis with Duncan's multiple-range test revealed that only the difference between PNG and CER was significant. Our main interest however concerned the test for delayed retention taken 7 days after learning. The differences here were considerably more substantial. On both the PRQs and NRQs, analysis of variance showed a significant difference between the groups: $F = 18.81$, $df = 3/68$, $p < .01$ and $F = 5.21$, $df = 3/68$, $p < .01$, respectively. Further analysis by means of Duncan's multiple-range test showed for the PRQs: PG > PNG > CER > C, that is, mean scores of PG and PNG were not significantly different, but each differed significantly from CER and C mean scores, while the difference in group means between CER and C was also significant. For the NRQs this comparison yielded PG < PNG < C < CER.

SOURCE: J. Peeck. Effect of prequestioning on delayed retention of prose material. *Journal of Educational Psychology*, 1970, *61*, 241–246. Copyright 1970 by the American Psychological Association. Reprinted by permission.

It should be noted that the basic Solomon design, even though it is considered to be a prestigious design, is not used with great frequency by applied researchers. Thus, it is not likely that extensions of the Solomon design will appear very often in the literature. Nevertheless, the remainder of this section is devoted to a discussion of two different studies which demonstrate how the basic Solomon design can be extended.

EXCERPT 13.4 THREE-GROUP POSTTEST-ONLY DESIGN WITH CHI-SQUARE USED TO ANALYZE THE DATA

METHOD

The present study was an attempt to utilize the video-tape playback as a behavior-shaping technique and to compare video-tape role playing with programmed material and a base-rate comparison group.

Referrals to counseling psychology by the ward physician were randomly assigned to three groups of 25 clients each: (*a*) control group, (*b*) programmed-materials group, (*c*) video-tape group. The description of the treatment procedures for the three groups is as follows:

Control group. The 25 clients randomly assigned to this group were put through the normal counseling procedures utilized by the ward counselors in each case. Be-

fore assignment to a ward counselor they were asked to do two things: (*a*) fill out a Standard Form 57 Civil Service application blank as though applying for a job; (*b*) take a job interview conducted by the assistant chief of personnel of the hospital.

Programmed-materials group. The programmed-materials group went through exactly the same procedure as the control group except that before being given the application blank and interview they were asked to complete a set of programmed materials developed locally by Cahoon and Watson (1966, 1967). As reported elsewhere in the literature, these programmed materials are focused primarily on job-finding techniques and application-blank completion for the hospitalized neuropsychiatric patient. They have been found to be quite useful in communicating certain objective and immediate bits of knowledge but are still of unknown value with regard to their ability to change behavior beyond the confines of the immediate learning situation.

Video-tape group. The video-tape group received these programmed materials and in addition was given a role-played job interview utilizing a member of the United States Employment Service as the prospective employer. Following the job interview, the patient was given an opportunity to view his own interview and to view other interviews individually and in small groups. All viewing was done with a trained vocational counselor who attempted to focus on specific behavioral changes which might have more effectively presented the patient in the interview.

Criteria measurements. Two immediate criteria measurements were made for the three groups. All patients within the group filled out Standard Form 57 Civil Service application blanks which were then rated by a personnel officer as to the dichotomous judgment of interview versus no interview on the basis of the application blank alone. In other words, the personnel officer made the judgment as to whether he would grant an interview to the person filling out the application blank on the basis of the application blank alone. In addition, following a job interview, the

personnel officer made the judgment whether he would give the patient a job on the basis only of the job interview. All judgments were made blind and were independent as to groups but a single rater was used.

TABLE 1 ANALYSIS OF APPLICATION BLANK DATA FOR RECOMMENDING A FURTHER INTERVIEW

SOURCE	YES	NO
Control	21	4
Programmed	18	7
TV	18	7
Total	57	18

Note. $\chi^2 = 1.32$, $df = 2$, *ns.*

RESULTS AND DISCUSSION

The results of the frequency count and statistical analyses are shown in Tables 1 and 2. The chi square generated by the data in both the application blank and interview measures are nonsignificant at the .05 level and the null hypothesis cannot be rejected.

TABLE 2 ANALYSIS OF INTERVIEW RATING DATA FOR RECOMMENDING EMPLOYMENT

SOURCE	YES	NO
Control	18	7
Programmed	12	13
TV	13	12
Total	43	32

Note. $\chi^2 = 3.88$, $df = 2$, *ns.*

The diagram for the study in Excerpt 13.5 makes it seem as though the basic Solomon design was used. However, each of the four groups was subdivided into male and female subjects. Thus, there really were eight groups in the study, not four. Furthermore, the entire design was conducted at five different grade levels (2–6). In terms of the statistical treatment of data, a three-way ANOVA was conducted five times, once for each grade level, and the results of these analyses were combined into one summary table that was presented near the end of the article. In each of these analyses, there were (1) main effects for the treatment variable, the effect of pretesting, and sex, (2) three first-order interactions, and (3) one second-order interaction. If the pretest had sensitized Ss to the treatment, it would have shown up in the pretesting-by-treatment interaction; however, this interaction was not significant in any of the five analyses. The inclusion of the third variable (sex) would have provided some interesting results if it had interacted with the treatment variable, but in each analysis the treatment-by-sex interaction was also nonsignificant

---

EXCERPT 13.5 SOLOMON FOUR-GROUP DESIGN WITH SEX USED AS A THIRD FACTOR IN THE ANALYSIS OF VARIANCE OF POSTTEST SCORES

Factorial Design

The student teacher-class pairs were randomly grouped into the experimental framework of Solomon's Four-Group design (21), which Campbell and Stanley (3:194–195) have recommended as a "true experimental design."

Subjects were randomly assigned to one of four groups. Group A represented the experimental group where pupils were given pre- and posttests and instruction in two selected phonetic generalizations in the interval between testing occasions. Group B represented the first control group where pupils were given pre- and posttesting but no phonetic instruction. Groups C and D received only posttesting, but treatment group C was taught phonetic rules and their application prior to testing and control group D was not.

Therefore, the study's experimental design is as follows:

		PRETEST		POSTTEST
Experimental				
Group A		0_1	x	0_2
Control				
Group B	R→	0_3		0_4
Control				
Group C			x	0_5
Control				
Group D				0_6

Where R = randomization, 0 = test observations, and x = treatment, i.e., instruction in phonetic generalizations.

The posttest scores of pupils in the four groups were converted to class means, and analysis of variance was used to test the statistical null hypothesis: $0_2 = 0_4 = 0_5 = 0_6$. With two levels each for the independent variables of pupil's sex (X), pretesting (Y), and phonetic treatment (Z), a $2 \times 2 \times 2$ factorial analysis was conducted for each grade level. Simultaneous analysis of all grade levels together was not done, because the tests were tailored for each grade level and cannot be considered parallel.

SOURCE: A. Yee. Is the phonetic generalization hypothesis in spelling valid? *Journal of Experimental Education*, 1969, 37, 82–88.

In Excerpt 13.6 we see another example of how the basic Solomon four-group design can be extended. Each of the original four groups in this study was randomly subdivided into two groups, with one group receiving familiar material and the other receiving unfamiliar material. A $2 \times 2 \times 2$ analysis of covariance provided the central statistical analysis. The three main effects were treatment (CAI vs. READ), pretesting, and type of material (familiar vs. unfamiliar). Scores from a concomitant variable served as the covariate. Results indicated significant main effects for treatment and type of material, but all four interactions turned out to be nonsignificant. Thus, the pretest was shown not to sensitize Ss to the treatments used in the study.

EXCERPT 13.6 COMPLICATED TRUE EXPERIMENTAL DESIGN
BASED ON THE SOLOMON FOUR-GROUP APPROACH

Procedure and Design

All testing was done by conventional pencil-and-paper procedures. A modified Solomon four-group design (Campbell & Stanley, 1963) was used to assess the pre-testing's effect on students' posttest attitudes. Since females outnumbered males, students were randomly assigned in equal numbers within sex to one of four groups: Experimental 1, Experimental 2, Control 1, and Control 2 (see Table 1). Those in the experimental groups were given an average of 45 minutes of CAI by individual typewriter-like terminals (IBM 1050) in small booths. Control groups were given reading selections from a general psychology text other than the one used in the course. These selections covered the same topics as the CAI program, and were read alone in a quiet room for approximately 45 minutes. Experimental 1 and Control 1 groups were pretested, while Experimental 2 and Control 2 groups were not pretested.

Students were further randomly subgrouped into A and B conditions. The A subgroups received CAI or general psychology reading material which was discussed in class and on which they were to be tested that week. The B subgroups received CAI or reading material that had not been covered in class and on which they would not be tested for several weeks. Students were instructed that the material that they would study was relevant to some future examination in the class, but were not told that they were being given different material in the various subgroups.

All students were posttested immediately after their experience, with Form A of the Brown scale given to those students who did not receive CAI. Those 44 students not chosen for the experiment were assigned to discussion groups.

TABLE 1 DESIGN FOR STUDY

GROUPS	PRETESTS	FAMILIAR	UN-FAMILIAR	POST-TEST	n
		PSYCHOLOGY COURSES			
Experimental 1_A	yes	CAI		yes	8
Experimental 1_B	yes		CAI	yes	6
Experimental 2_A		CAI		yes	10
Experimental 2_B			CAI	yes	8
Control 1_A	yes	Read		yes	8
Control 1_B	yes		Read	yes	8
Control 2_A		Read		yes	8
Control 2_B			Read	yes	8

SOURCE: A. Mathis, T. Smith, and D. Hansen. College students' attitudes toward computer-assisted instruction. *Journal of Educational Psychology*, 1970, *61*, 46–51. Copyright 1970 by the American Psychological Association. Reprinted by permission.

The basic Solomon design can conceivably be extended in additional ways, and our discussion here is not meant to be exhaustive. However, the authors have not seen the application of Solomon extensions beyond the two included in this section.

TRUE EXPERIMENTAL DESIGNS BASED ON FACTORIAL AND REPEATED MEASURES ANALYSIS OF VARIANCE AND COVARIANCE

In Chapter 5 we defined the terms factor and level. We must now distinguish between two different types of factors—active and assigned. With an *active* (or *manipulated*) *factor,* the researcher has the power to assign Ss to the various levels that make up that factor. For example, the method-of-counseling factor is active since the researcher can determine which specific Ss will receive each of the various methods of counseling. With an *assigned* (or *organismic*) *factor,* the researcher does not have the power to assign Ss to the various levels that constitute the factor. Instead, the nature of the levels is such that the Ss are grouped automatically in terms of defined physical or psychological traits. If sex is a factor, the researcher cannot exert any control whatsoever over who is in the male and female levels of the factor. Intelligence, anxiety level, grade level, socioeconomic status are further examples of assigned factors.

In a factorial design there are always two or more factors, each made up of at least two levels. The factors of the design, considered as a group, can be all active factors, all assigned factors, or a mixture of active and assigned factors. In order for the factorial design to quality as a true experimental design, at least one of the factors must be active. Therefore, of the three possible situations mentioned above, the first and the third have the potential of being true experimental designs. In both of these situations, the researcher has the ability to randomly assign Ss to the levels of at least one factor.

When all factors of the study are assigned factors, the researcher will not be able to randomly assign any of the Ss to the levels of any factor. This does not make the study inferior; it just means that the investigation should be classified as descriptive rather than experimental. In Excerpt 13.7 we see an example of a factorial design in which both of the independent variables were assigned factors. The researcher did not manipulate either of the two factors, striving for advancement or job level. All Ss were placed into the high and low levels of these factors because of their characteristics, and it would have been impossible for the researcher to randomly assign them to the levels of either factor.

Randomized Blocks Designs

When there is at least one active factor in a study, the researcher will be able to randomly assign Ss to the levels of that manipulated treatment factor. However, the Ss must be assigned from within each separate level of the nonmanipulated factor(s). For example, if a researcher conducts a study in which the two factors are method of teaching and student experience, the available Ss might first be subdivided into groups having a lot of experience, average experience, and a little experience. Then the Ss in each of these groups would be randomly assigned to the various teaching methods. Thus, in this type of design there will be as many separate random assignments of groups of Ss as there are levels of the nonmanipulated factor(s). Because of the nature of the subject assignment, this type of design is often

EXCERPT 13.7 TWO-WAY FACTORIAL
DESIGN WITH TWO ASSIGNED FACTORS

Procedures

Ss were administered the Worcester Scale of Social Attainment (WSSA) in a group. In a later session individual inquiry into responses was conducted. The WSSA is a broad assessment of the degree to which an individual meets certain standards of accomplishment in society, and is discussed more fully by Phillips and Cowitz (1953). Two sections of this scale were selected for use in this study: Striving for Advancement and Job Level.

Striving for Advancement (WSSA, Item 7) taps the extent to which an individual actively seeks to better his employment position. Scoring on this item is based on answers to the following check list.

1. I would like to go to school to get a better job.
    (a) Do you have a particular school in mind?
    (b) Name of school?
2. I have visited plants looking for a better job.
3. I am interested in another job if it is offered.
4. I read the help-wanted ads every day.
5. I have asked my friends to look for jobs for me.
6. I am going, or have gone to school lately for job improvement.
7. Sometimes I think about looking for something better.
8. I have written letters to employers.
9. I am keeping my eyes open.
10. I have tried to get interviews.
11. I would like to have a better job.
12. I would like to learn a trade.
13. I expect to work for myself.
14. I would like to work for myself.

Responses to these questions permit classification of Ss into following four categories:

1. Desire and intense activity, e.g., letters written about jobs; interviews applied for; contacts for better position actively sought.
2. Desire and tentative activity, e.g., is interested in another job if offered; does not seek a job; does not write although may look through ads; has vague long range goals.
3. Desire but no activity, e.g., says he would like a change but does nothing.
4. No desire and no activity.

Ss were placed in one of these four categories by two independent judges, whose judgments correlated .98 (Spearman rank order correlation).

Behavioral manifestation of the achievement motive is typically understood as a kind of striving measure, or a measure of how much effort is involved in trying to achieve; it does not necessarily reflect actual achievement. At the same time, however, it seems possible that actual achievement may influence the amount of achievement fantasy produced by an individual. For this reason, S's occupational level was used as a control for actual achievement in the area in which his behavioral expression of achievement motivation was examined. This control measure was taken from the Job Level item of the WSSA (Item 1). This item assigns scores to each level of occupation, as defined by the DOT. Higher scores are assigned to higher job levels.

The fantasy measure of achievement motivation was taken from the Role-Taking Task (RTT) devised by Feffer (1959). This is a projective test, based on Shneidman's Make-a-Picture-Story (MAPS).

RESULTS

Achievement fantasy scores were analyzed in a two-by-two factorial analysis of variance. Independent variables used were: Striving for Advancement (WSSA, Item 7) and Job Level (WSSA, Item 1). Ss in the two higher striving categories were compared with those in the two lower categories. Job Level was divided at the median into high and low groups.

SOURCE: D. M. Broverman, E. J. Jordan, Jr., and L. Phillips. Achievement motivation in fantasy and behavior. *Journal of Abnormal and Social Psychology,* 1960, *60,* 374–378. Copyright 1960 by the American Psychological Association. Reprinted by permission.

EXCERPT 13.8 TRUE EXPERIMENTAL DESIGN WITH
ONE ACTIVE FACTOR AND ONE ASSIGNED FACTOR

METHOD

Subjects

The Ss were 72 college freshmen enrolled in a class entitled Introduction to Educational Psychology. They were divided into ability levels on the basis of their "Gamma IQ" on the Otis Quick-Scoring Mental Ability Test. The mean of the high-ability group was 130 and that of the low-ability group was 121. The Ss were randomly assigned to treatments within their ability levels.

Treatments

Knowledge of Results. The Ss were administered a multiple-choice achievement test in educational psychology. The following class period, the test was again distributed among Ss and the experimenter (*E*) read the items with the correct answers. No discussion of the items was permitted. The class was told there was not sufficient time available for discussion. The *E* did not give any indication that a constructed response test would be administered the following class period.

No Knowledge of Results. The Ss were administered the same multiple-choice test, but were not given access to the test again.

Control. The Ss did not take the multiple-choice test.

Tests Employed

The 40 items in the multiple-choice test covered a number of topics in educational psychology. Reliability of the test estimated by the Spearman-Brown prophecy formula was .84.

On the third class period, Ss in all treatment groups were administered the criterion test. This test was a constructed response test utilizing the same stem of the multiple-choice item, but requiring Ss to compose or recall the correct answer. Their score was the number of recall responses that were in essence the same as one of the three wrong alternatives in the multiple-choice test. The *E* scored these constructed responses "blind," in that he did not know to which treatment group individual Ss had been assigned.

When the criterion test was scored for correct answers, the reliability coefficient estimated by the Spearman-Brown prophecy formula was .78. The mean number of correct responses for the Knowledge of Results treatment was higher ($M = 34$, $SD = 3$) than the Control ($M = 28$, $SD = 7$) or the No Knowledge of Results ($M = 31$, $SD = 5$). In comparing these three means, only the Knowledge of Results and Control difference was significant at the .01 level ($t = 3.78$; $df = 46$; two-tailed test). The Knowledge of Results and No Knowledge of Results comparison yielded a t of 2.47, which was significant at the .05 level ($df = 46$; two-tailed test).

RESULTS

To test for differences among the groups, a 3 (treatments) × 2 (levels) factorial analysis of variance was used. An F of 8.93 ($df = 2/66$; $p < .001$) was obtained for treatments, and an F of 4.99 ($df = 1/66$; $p < .01$) was obtained for ability levels. The Treatment × Ability Levels interaction was not significant, as it gave an F of 1.49 ($df = 2/66$; $p > .05$).

The Duncan multiple-range test revealed the No Knowledge of Results treatment resulted in significantly more ($p < .01$) errors than the Knowledge of Results or Control treatments. The Knowledge of Results and Control treatments did not differ significantly. The low-ability group made significantly more ($p < .01$) errors than the high-ability group. The interaction was not significant.

SOURCE: R. J. Karraker. Knowledge of results and incorrect recall of plausible multiple-choice alternatives. *Journal of Educational Psychology*, 1967, 58, 11–14. Copyright 1967 by the American Psychological Association. Reprinted by permission.

---

referred to as a randomized blocks design, with the term blocks being synonymous with the levels of the nonmanipulated factor.[3]

[3] The phrase *treatments-by-levels* design is sometimes used to describe a design in which there is one active factor and one assigned factor.

In Excerpt 13.8 we see a study in which there was one active factor and one assigned factor. The active factor, knowledge of results, had three levels, and the assigned factor, ability, had two levels. As indicated in the first paragraph, the high-ability Ss were randomly assigned to the three treatment conditions. Then, the low-ability Ss were assigned to the three treatment conditions. If the researcher had simply assigned all of his Ss randomly to the treatments (thus eliminating ability as a factor in the study), the design would have been a three-group posttest-only design, and the results would have been analyzed by means of a simple one-way ANOVA.

There are two reasons for including an assigned factor in a design, as was done in Excerpt 13.8. First, the researcher may be interested in whether or not the relative effectiveness of the treatments varies according to the types of Ss in the study. The answer to this question is provided by the interaction F ratio. In Excerpt 13.8 the lack of a significant F for the treatment-by-ability interaction shows that the comparative effectiveness of the three treatments was about the same for both high- and low-ability Ss. Second, the researcher may include an assigned factor in the design for the sole purpose of increasing the power (sensitivity) of the statistical analysis to a possible treatment effect. If the main effect for the assigned factor (and/or the interaction) is significant, then the F ratio for the main effect of the treatment factor will be larger than it would have been had the assigned factor been eliminated from the study. Thus, the inclusion of ability levels in Excerpt 13.8 caused the statistical analysis to be more sensitive to differences among the three treatments. If a one-way ANOVA were used to analyze the criterion scores in Excerpt 13.8 with classification of Ss by ability disregarded, the F for the treatment main effect would be smaller than 8.93, and it possibly might be nonsignificant.

Of course, a randomized blocks design can have more than two factors. The only restriction is that at least one factor be a nonmanipulated, assigned factor. In Excerpt 13.9 we see an example of a three-factor design in which there was one active factor (treatments) and two assigned factors (sex and ability). For obvious reasons, the only factor that involved random assignment of Ss to levels was the treatment factor.

Completely Randomized Factorial Designs

There are two main distinctions between a factorial design having some active and some assigned factors and a factorial design having all active factors. First, the procedure for assigning Ss to the treatment conditions is different. Since a randomized blocks design always has one assigned factor, Ss must be assigned to the treatment conditions from within each level of the blocking factor. When all factors are active, the researcher can assign Ss at random to the levels of each factor. For this reason, the term *completely randomized factorial design* is sometimes used to describe a factorial design in which all factors are active.

EXCERPT 13.9 TRUE EXPERIMENTAL DESIGN WITH
ONE ACTIVE FACTOR AND TWO ASSIGNED FACTORS

The purpose of the present experimental study was to identify and describe certain mechanisms of structuring that occur with highly specific, meaningful printed passages through the use of advance organizers. Previous relevant studies mentioned above have only considered variations on the paragraph abstract type of "advance organizer." In this study, four different types of advance organizers were compared: (a) paragraph abstract, (b) an enumerated sentence outline, (c) a true-false pretest, and (d) a completion pretest. Each of these pre-passage structuring methods dealt with the same set of ten key points from the reading passage employed as the learning task in the study.

PROCEDURE

A highly detailed passage on the non-religious aspects of Amish customs and life was adapted from Hostetler (19). The passage consisted of about 2,600 words. The material, although filled with specific details, was easy to read and could be managed with a reading rate of about 200 words per minute. Maximum reading time of 15 minutes was thus allotted to the passage.

From this passage, ten broad, subsuming concepts were identified as the basis for each of four advance organization aids. Five twelfth-grade social studies classes from a suburban school district in Montgómery County, Pennsylvania, provided Ss. Each of the five classes were homogeneously grouped. The factor of classes, therefore, corresponded to ability levels. The five classes represented the whole range of ability in the twelfth grade in this school. Within each intact class, the four treatments were randomly assigned on the basis of a seating chart supplied by the teacher and administrator before the study began. The actual class sizes on the day of the experiment ranged

from seventeen to twenty-nine. A total of 124 students were involved for the five classes.

On the day of the experiment, an investigator went to each of the five classrooms and remained there for the 50-minute duration. The first 5 minutes of the period were used to explain the purpose of the study. The Ss were assured that all results would remain anonymous. Instructions and one of the four advance organizer aids were appended to the front of the passage to be read.

After receiving general instructions, Ss of each group were told to process the advance organizer material carefully for 5 minutes according to the exact instructions printed on that sheet of the booklet. They were then told to read the passage itself carefully up to a 15-minute limit. After time was called, the advance organization sheet and the passage booklet were collected and at the same time the criterion test and IBM 1230 answer sheet were distributed.

The criterion test was of a 20-item multiple-choice format; each item had four options. The position of the correct option was randomly assigned within each test item. The items were arranged on the test according to the order in which the corresponding content matter originally appeared in the passage itself. The students were allowed 15 minutes for the test. The test items covered eight concepts which had been stressed in each of the advance organization methods (general, subsuming concepts) and twelve not covered by them (specific, subsumed concepts) but discussed in the reading passage itself.

RESULTS

Two $4 \times 2 \times 5$ (treatments by sex by ability level) analyses were run: one on the eight pretested items and one on the twelve non-pretested items.

SOURCE: B. B. Proger and L. Mann. Conceptual prestructuring for detailed verbal passages. *Journal of Educational Research*, 1970, 64, 28–34.

The second distinction between a randomized blocks design and a completely randomized factorial design concerns the number of research questions that can be answered by the study. With a two-factor randomized

EXCERPT 13.10 2 × 2 COMPLETELY RANDOMIZED FACTORIAL DESIGN

METHOD

Every S was given a single trial on the same list of 35 nouns. Each noun was taken from a different category in the Cohen, Bousfield, and Whitmarsh (1957) norms with the mean category frequency being 29.17. The design was a 2 × 2 factorial in which the first variable was the presence or absence of storage cues and the second was the presence or absence of retrieval cues. The four conditions formed by these variables are designated as C-C, C-NC, NC-C, and NC-NC. Twelve undergraduates from the State University of New York at Binghamton were assigned to each condition. The Ss were run in small groups numbering from one to five. The groups were randomly assigned to the conditions subject to the restriction of achieving an equal number per condition.

The nouns were read at a 3-second rate. When storage cues were presented, Ss were told that a category name (the category label from the Cohen et al., 1957, norms) would be read before each noun (e.g., a bird—RAVEN). They were told that they would not have to recall the category names, but that the category names would help them recall the nouns. When storage cues were absent, only the nouns were read (e.g., RAVEN). Immediately following presentation, a 4-minute recall period was given. When retrieval cues were presented, the 35 category names were listed in the same order as they occurred during input with a blank space beside each one. Brief instructions at the top of the recall sheet told S that each noun presented could be described by one of the categories and that he should print each noun he could recall beside its appropriate category. When retrieval cues were absent, only the blank spaces were provided, and brief instructions at the top of the recall sheet told S he should print the nouns in these spaces.

TABLE 1 MEAN NUMBER OF CORRECT RESPONSES

STORAGE CUES	RETRIEVAL CUES	
	ABSENT (NC)	PRESENT (C)
Absent (NC)	13.33	14.33
Present (C)	12.25	21.17

The Ss in both retrieval cue conditions were told to recall the nouns in any order they desired. They were not informed of their condition of recall until the time of recall.

RESULTS

The number of nouns correctly recalled by each S was computed (Table 1). Recall was higher when storage cues were presented than when they were absent, $F(1, 44) = 5.54$, $p < .05$, and recall was higher when retrieval cues were present than when they were absent, $F(1, 44) = 16.49$, $p < .01$. The most important finding, however, was the significant interaction, $F(1, 44) = 10.51$, $p < .01$, which was further analyzed by individual F tests. It was found that Group C-C recalled significantly more nouns than each of the other groups, the $Fs(1, 44)$ being 26.67, 15.66, and 20.58, for the comparisons with Groups C-NC, NC-C, and NC-NC, respectively. None of the comparisons among the latter three groups approached significance, $Fs(1, 44) < 1.46$.

DISCUSSION

The significant interaction shows that retrieval cues produce greater recall than no retrieval cues when storage cues are presented, thus replicating the findings of Tulving and Pearlstone (1966), but have no effect on recall when storage cues are absent. Essentially this same finding also has been obtained by Wood (1967) in work reported after the completion of the present research.

SOURCE: J. H. Crouse. Storage and retrieval of words in free-recall learning. *Journal of Educational Psychology*, 1968, 59, 449–451. Copyright 1968 by the American Psychological Association. Reprinted by permission.

blocks design there are three F ratios but only two legitimate questions that can be answered: (1) Is there a significant main effect for the treatment factor? (2) Is there a significant interaction between the treatment factor and the blocking factor? The third F ratio (related to the main effect of the blocking factor) does not really answer an experimental research question, because there is no experimental manipulation of this factor. With a two factor completely randomized design, on the other hand, each of the three F ratios provides an answer to an experimental research question (since both main effects and the interaction involve manipulated treatment factors).

In Excerpt 13.10 we see a study in which both of the two factors (presence vs. absence of storage cues and presence vs. absence of retrieval cues) were active. There were 4 treatment conditions associated with the 2×2 design, and 12 of the 48 Ss were randomly assigned to each treatment condition. Since both factors were active, each of the three F ratios answered a separate experimental research question. However, the significant interaction provides the most important information, and, as we noted in Chapter 5, results of the main effect F ratios must be interpreted with caution in light of the significant interaction.

As was the case with the randomized blocks design, a completely randomized factorial design can be extended to involve three or more factors. However, to qualify as a completely randomized design, all of the factors must be manipulated active factors, with randomization of Ss to the levels of each factor.

Repeated Measures Designs

In Chapter 6 we discussed two frequently used repeated measures analyses of variance, the Lindquist Type I ANOVA and the Lindquist Type III ANOVA. As you remember, the Lindquist Type I ANOVA involves one between-subjects factor and one within-subjects factor, while the Lindquist Type III ANOVA involves two between-subjects factors and one within-subjects factor.

In order for a repeated measures study to qualify as a true experimental design, at least one factor in the study must be an active factor. If the within-subjects factor is nothing more than a series of measurements across a time dimension (i.e., trials), then only the between-subjects factors have the potential for being active factors. Therefore, a Lindquist Type I study with trials as the repeated measures factor can be a true experimental design only if the one between-subjects factor is active, with random assignment of Ss to the levels of this factor. A Lindquist Type III study involving trials can be a true experimental design with random assignment of Ss to the levels of either (or both) of the between-subjects factors.

It should be noted that the repeated measures factor of a Lindquist Type I or III design can be active. This would be the case if Ss are exposed to a series of different experimental conditions. As contrasted with trials, an active within-subjects factor can be identified by the possibility of presenting the levels of this factor in different sequences. In fact, the researcher should randomly arrange the order in which the levels of the active repeated measures factor are presented to the Ss to prevent a possible learning (or forgetting) effect from becoming confounded with the main effect associated with the active factor.

In Excerpt 13.11 we find a Lindquist Type III true experimental design. Of the two between-subjects factors in this study, type of reinforcement was a manipulated factor, for the researcher had the power to decide whether an S would receive positive reinforcement, negative reinforcement, or no reinforcement at all. The other between-subjects factor, sex, was, of course, an assigned factor. Trials represented the within-subjects (repeated measures) factor.

EXCERPT 13.11 LINDQUIST TYPE III TRUE EXPERIMENTAL DESIGN

METHOD

Subjects

The Ss were 60 male and 60 female kindergarten children enrolled in four public elementary schools in Greencastle, Indiana. They ranged in age from 5 years, 5 months to 6 years, 9 months and were predominantly from middle and upper-lower class families.

Procedure

The children were tested in pairs with each S being randomly assigned a partner of the same sex. One member of the pair was designated as the performer (P) and the other as the observer (O). The P was instructed to stand in front of the apparatus while O stood to his left so that the responses of P could be easily observed. The Ss were then given the following instructions:

Now we are going to play a marble game. _____ (name of P inserted) is going to play first and then it will be _____'s (name of O inserted) turn. Let me show you how to play. Take a marble from here and drop it in a hole (E demonstrated by taking a marble from the bin and dropping it in one of the holes in the tray). You can use any of the holes you want but be sure to take one marble at a time and use only one hand. Now you try (P was given an opportunity to make a response). We won't use all the marbles that are here in the game. I will tell you when to stop and then it will be _____'s (name of O inserted) turn.

There were three learning conditions: (a) regular positive social reinforcement (praise), (b) regular negative social reinforcement (reproof), and (c) no reinforcement. The learning task, which required 6 minutes, was divided into 1-minute intervals. In the positive and negative reinforcement conditions, no reinforcements were given during the first 1-minute interval. Beginning with the second minute the first response after the first and thirtieth second of each minute was reinforced, giving a total of ten reinforcements during the learning task. In the positive reinforcement condition E made supportive or encouraging statements about P's performance, while the reinforcements in the negative condition consisted of critical comments. Since the purpose of the study was a comparison of the effects of positive and negative reinforcement, the reinforcements were worded so that both reward and punishment would produce the same response tendency—increase in rate of response. This was accomplished by including in the reinforcing comments the implication that speed of response was important. For example, the positive reinforcements included statements such as "Good—that's fast," while the negative reinforcements included such comments as "That's slow—not too good" or "That's not very fast." A reference to speed was included in the first reinforcement in each 1-minute interval. In the control group P performed the task for 6 minutes without receiving any reinforcements.

When P had completed the task, Ss were instructed to change places and O

was allowed to play the game for 6 minutes. During the time O performed the task no reinforcements were given by E.

At the end of the experimental session both Ss were commended for their performance.

RESULTS

The number of responses (marbles dropped through the holes by S) per 1-minute interval was used as the criterion measure. The data for the groups receiving the direct and vicarious reinforcement were analyzed in two separate $2 \times 3 \times 6$ repeated measures analyses of variance (Winer, 1962).

An analysis of variance of the data for the groups receiving direct reinforcement showed a highly significant Reinforcement main effect ($F = 11.42$, $df = 2/54$, $p < .01$). The curves shown in Figure 1 indi-

cate that the group receiving negative reinforcement showed the highest overall response rate after the first minute, while the control group responded at the lowest rate. The Sex and Trials main effects were not significant, nor were there any significant interactions ($p > .05$).

The analysis of variance of the data for the vicarious reinforcement groups indicated that the Reinforcement main effect was highly significant ($F = 7.13$, $df = 2/54$, $p < .01$). Figure 2 indicates

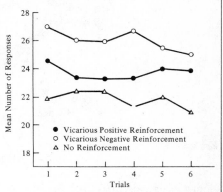

FIGURE 2. *Mean number of responses of groups receiving three types of vicarious reinforcement.*

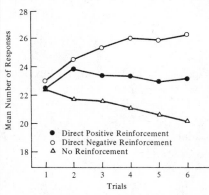

FIGURE 1. *Mean number of responses of groups receiving three types of direct reinforcement.*

that, as was the case in the direct reinforcement condition, the highest response rate occurred in the group which received negative reinforcement, followed by the positive reinforcement group, and then the control group. There were no other significant main effects or interactions ($p > .05$).

SOURCE: R. Kelly. Comparison of the effects of positive and negative vicarious reinforcement in an operant learning task. *Journal of Educational Psychology*, 1966, 57, 307–310. Copyright 1966 by the American Psychological Association. Reprinted by permission.

Diagraming Factorial and Repeated Measures Designs

In our discussion of preexperimental designs (Chapter 11) and true experimental designs (Chapter 12), we used X's and O's to construct a diagram for each design. Although some people have tried to extend this approach to the diagraming of factorial and repeated measures designs,[4] we

[4] B. W. Tuckman. *Conducting Educational Research.* New York: Harcourt Brace Jovanovich, 1972.

feel that the *X*-and-*O* technique is not well suited to these more complicated designs. Instead, the use of matrices with different cells is recommended. Although this approach to diagraming was previously introduced in Chapters 5 and 6, let us review how it would be applied to two of the excerpts that have been discussed in this chapter.

In Excerpt 13.8 a two-factor randomized blocks design was used in which there were three levels of a treatment factor and two levels of an assigned factor, ability. Figure 13.3 represents the diagram for this study.

TREATMENTS

		KNOWLEDGE OF RESULTS	NO KNOWLEDGE OF RESULTS	CONTROL
ABILITY	HIGH			
	LOW			

FIGURE 13.3. *Diagram of the randomized blocks design presented in Excerpt 13.8.*

There are six cells in the diagram. As indicated in the method section of the excerpt, 36 high-ability Ss were assigned to the cells in the top row of the diagram and 36 low-ability Ss were assigned to the cells in the bottom row of the diagram (12 Ss to a cell). Thus, for the purpose of the statistical analysis, there would be 12 pieces of data within each cell, and from these data the row, column, and cell means could then be computed.

In Excerpt 13.11 we looked at a Lindquist Type III true experimental design. The two between-subjects factors were type of reinforcement and sex, and the within-subjects factor was trials. The diagram for this design is presented in Figure 13.4. The third column of this diagram clearly indicates that there were a total of 60 Ss included in the study, with 10 Ss associated with each possible combination of the two between-subjects factors.

Diagrams similar to the two that we have just examined will not appear very often in journal articles. Sometimes, however, an author will present a table of means that resembles the diagram of the study. For example, Excerpt 13.10 contains a table of means that allows us to understand the set-up of the 2 × 2 completely randomized factorial design. In Excerpt 13.12 a table of means (and standard deviations) from a different article quickly informs the reader that a three-way factorial design was used, with one factor being active (type of reinforcement) and the other two (social class and grade level) being assigned. This particular table also provides information as to the number and names of levels within each of the three factors.

		TRIALS					
		1	2	3	4	5	6
POSITIVE REINFORCE-MENT	MALES	S_1 • • • S_{10}					
	FEMALES	S_{11} • • • S_{20}					
NEGATIVE REINFORCE-MENT	MALES	S_{21} • • • S_{30}					
	FEMALES	S_{31} • • • S_{40}					
NO REIN-FORCEMENT AT ALL	MALES	S_{41} • • • S_{50}					
	FEMALES	S_{51} • • • S_{60}					

FIGURE 13.4. *Diagram of the Lindquist Type III design presented in Excerpt 13.11.*

EXCERPT 13.12 DIAGRAM FOR A THREE-WAY FACTORIAL DESIGN
FORMED BY SPECIFICATION OF THE CELL MEANS AND
STANDARD DEVIATIONS

TABLE 1 MEANS AND STANDARD DEVIATIONS OF DIFFERENCE, SCORES
BY SOCIAL CLASS, GRADE LEVEL, AND TYPE OF REINFORCEMENT

| SOCIAL CLASS | Grade level | | | | | |
| | GRADE 2 | | | GRADE 6 | | |
	MATERIAL	PRAISE	SYMBOLIC	MATERIAL	PRAISE	SYMBOLIC
Middle						
M	5.66	6.64	6.58	5.75	8.25	9.66
SD	2.32	3.28	2.98	1.63	4.12	3.37
Lower						
M	8.41	5.41	5.25	6.75	7.00	6.33
SD	4.23	3.63	2.74	4.20	4.16	3.22

SOURCE: J. D. Cradler and D. L. Goodwin. Conditioning of verbal behavior as a function of age, social class, and type of reinforcement. *Journal of Educational Psychology,* 1971, *62,* 279–284. Copyright 1971 by the American Psychological Association. Reprinted by permission.

Statistical Analyses for Factorial and Repeated Measures Designs

The factorial and repeated measures ANOVA's that were discussed in Chapters 5 and 6 can be applied to studies in which all factors are assigned, all factors are active, or there is a mixture of active and assigned factors. In this chapter we have examined true experimental designs that fall into the last two categories, and the data from these designs are usually subjected to one of the analysis of variance procedures that were discussed in the earlier chapters. In other words, the most frequently used statistical procedure for

---

EXCERPT 13.13 ANALYSIS OF VARIANCE OF CHANGE SCORES
USED IN CONJUNCTION WITH A TWO-WAY FACTORIAL DESIGN

Students

Male students ($n = 203$) from university residence halls volunteered to complete the revised Byrne Repression-Sensitization Scale. . . . Subjects scoring 30 or less on this scale ($n = 41$) were considered repressors and subjects scoring 69 or above ($n = 40$) were considered sensitizers. Twelve repressors and 12 sensitizers were selected randomly as subjects for each of two experimental treatments: cognitive and affective. Four groups ($n = 12$) were defined: repressor-cognitive, repressor-affective, sensitizer-cognitive, and sensitizer-affective.

Maturity Ratings

An Attribute Rating Scale was developed to measure perceived maturity. Self and Counselor forms of the scale were developed.

Analysis of the Data

Pretest to posttest change scores were computed for self- and counselor-maturity ratings. A 2 × 2 factorial ANOVA (Defensive Orientation × Treatment) assessed differences between groups on self- and counselor-maturity ratings.

SOURCE: B. A. Baldwin and W. A. Cabianca. Defensive strategies of repressors and sensitizers in counseling. *Journal of Counseling Psychology,* 1972, *19,* 16–20. Copyright 1972 by the American Psychological Association. Reprinted by permission.

factorial and repeated measures designs is a regular analysis of variance of scores on the dependent variable. There are, however, other ways of collecting and analyzing data associated with these designs.

In Excerpt 13.13 we see a study in which a 2 × 2 factorial design was used. One of the factors was active (cognitive treatment vs. affective treatment) and the other factor was assigned (repressors vs. sensitizers). Prior to the exposure of any Ss to the treatment conditions, pretest scores were collected from each S on two dependent variables. Following the administration of the treatments, posttest scores were collected on the dependent variables. Change scores were then computed by subtracting each S's pretest

---

EXCERPT 13.14 TWO-FACTOR RANDOMIZED BLOCKS DESIGN WITH ANALYSIS OF COVARIANCE USING TWO CONCOMITANT VARIABLES

A 2 × 2 factorial design with a pretest and posttest was used for the study. There were two levels of instructional method—role-assumption and structured class; reinforcement control consisted of two levels, internal and external.

Procedure

Assigned Variables. The Ss were categorized as being externals or internals on the basis of their scores on the IAR scale (2). This scale was routinely administered to all boys entering the Advancement School as a part of the testing program during the first 2 days in residency. The median score on the IAR scale for those who elected to take the science course was used to establish categories of internals and externals (median for the total group was 25). Those above the grand median were considered to be internals (median on IAR for this group was 28), and those below the grand median were considered to be externals (median on IAR for this group was 21).

Manipulated Variables. The treatment which stressed learner-directed activities was called a role-assumption treatment. The sixteen students assigned to this treatment were told that they were to assume the role of a scientist and could spend their time during the semester studying anything they desired as long as it was related to science. The role of the teacher was one of helping the student to secure information and materials for experiments

or other explorations in which the student was interested. Students in this treatment performed two or more experiments during the semester. These experiments included such things as weather prediction, the effect of types of light upon plant growth, genetics, nutrition, and animal care.

The treatment which stressed teacher-directed activities was called the structured treatment. The sixteen students assigned to this treatment were given considerable direction as to the kinds of experiments likely to be of value in their learning. They were also given specific reading assignments for which they were held responsible. Except for grades, which were not given to students in either group, students in the structured group were, in every way possible, led to view the teacher as being in control of the learning situations and passing out reinforcements when they were justified.

The same two teachers taught both classes. Students were assigned to the treatments randomly, with eight students in each cell of the 4-cell design.

Criterion and Covariates. The criterion was achievement as measured by the Sequential Test of Educational Progress (STEP): Science (5).

Two control variables were used, control being exercised by means of the analysis of covariance. Intelligence (14) and the pretest STEP-Science achievement were used as covariates.

SOURCE: K. White and J. L. Howard. The relationship of achievement responsibility to instructional treatments. *Journal of Experimental Education*, 1970, 39, 78–82.

score from his posttest score, and a separate two-way ANOVA was used to analyze the change scores on each dependent variable.

Although scores on a pretest are sometimes used to obtain gain scores, they can also be used as a concomitant variable within an analysis of covariance. Thus, the statistical procedures that were examined in Chapter 7 will sometimes be used to analyze the data from true experimental studies that are based upon factorial and repeated measures designs. As an example, consider Excerpt 13.14. In this study the researcher used a two-way factorial design with one active factor and one assigned factor. Scores on the criterion were subjected to an analysis of covariance using a pretest and IQ scores as covariates. Assuming (1) that the correlation between each concomitant variable and the criterion variable was at least moderately high and (2) that the correlation between the two covariate variables was low, the covariance analysis provided the researchers with more powerful comparisons than a regular two-way ANOVA of the criterion scores would have permitted.

FIXED, MIXED, AND RANDOM MODELS FOR FACTORIAL AND REPEATED MEASURES DESIGNS

In both this and the previous chapter we have discussed several types of true experimental designs. Because these designs involve the inclusion of at least one active factor and random assignment of Ss to the levels of the active factor(s), the various threats to internal validity are controlled. In fact, high internal validity is the reason for the classification of these designs as true experimental designs.

However, high internal validity does not necessarily bring about high external validity. The former can exist without the latter, just as a measuring instrument can have high reliability without possessing high validity. For example, the authors of Excerpt 13.15 used a true experimental design, a multigroup posttest-only design, having high internal validity, but they acknowledge the potential limitation of their study in terms of external validity. Unfortunately, most authors are not this honest about possible threats to the

---

EXCERPT 13.15 STUDY WITH HIGH
INTERNAL AND LOW EXTERNAL VALIDITY

Experimental Design

A posttest-only control group was used in this experiment (Campbell & Stanley, 1963). Internal validity was controlled by random assignment of subjects to the four programs. Since the programs were administered to individual subjects the basic experimental unit was the subject. The n size for each treatment was 19, total $n =$ 76. External validity was a problem since the subjects were not randomly sampled from the universal population.

SOURCE: R. D. Tennyson, F. R. Woolley, and D. Merrill. Exemplar and nonexemplar variables which produce correct concept classification behavior and specified classification errors. *Journal of Educational Psychology*, 1972, 63, 144–152. Copyright 1972 by the American Psychological Association. Reprinted by permission.

external validity of their studies. Therefore, the reader of the professional literature must evaluate the degree to which the results can be generalized, even when a true experimental design has been used.

Regardless of the specific design being employed, there are certain things that a researcher can do to increase the external validity of his study. For example, the random selection of Ss from the population of interest helps to increase population validity. Or, the use of an unobtrusive measure as the posttest would eliminate posttest sensitization as a threat to ecological validity.[5] There are also certain research designs that can be used to help make sure that a study has high external validity. We have already looked at one of these—the Solomon four-group design. The pretest-by-treatment interaction from the two-way ANOVA of the posttest scores provides a direct test of pretest sensitization. If this interaction turns out to be nonsignificant, the researcher can generalize his results more confidently to other individuals who in the future will receive the treatment but not the pretest.

Another way to set up the research design so as to increase external validity is to include one or more random factors in the design. A *random factor* is a factor containing levels which are selected at random from a larger population of potential levels. For example, suppose a researcher is using a 3 × 3 factorial design in which one factor contains three treatment conditions and the other factor involves three grade levels, say first, fifth, and seventh. If the three grade levels are selected at random from the population of possible grade levels (K–8), then this factor would be random. However, if the first, fifth, and seventh grades are included in the study because the researcher has friends in these particular grades or because he got volunteers from these three grade levels, then the factor of grade level would be fixed, not random. A *fixed factor* is a factor containing levels which are not selected at random from a larger population of levels. A factor would also be fixed if all of the potential levels in the population of levels are included in the study. Thus, grade levels would be a fixed factor if all nine grades were included in the hypothetical study. For this reason, sex is always considered to be a fixed factor.

It is advantageous to have a random factor in an experimental study because the results of the statistical analysis can be generalized beyond the specific levels of the random factor to the population of levels. In the hypothetical 3 × 3 factorial design described above, a significant main effect for the treatment factor would allow the researcher to conclude that the three treatments have differential effects in the population of grade levels K–8. In a similar manner, a significant treatment-by-grade interaction would

---

[5]An *unobtrusive measure* is a measurement which can be taken without the Ss' being aware of the measuring process. For example, measuring the wear on the floor in an art museum would be one way of collecting unobtrusive measures regarding the popularity of different rooms in the museum.

For a more complete discussion of unobtrusive measurement, see E. Webb, D. T. Campbell, R. Schwartz, and L. Sechrest. *Unobtrusive Measures: Non-reactive Research in the Social Sciences.* Chicago: Rand McNally, 1966.

permit the researcher to conclude that the comparative effectiveness of the three treatments does not remain the same across the population of grade levels. However, if the factor of grade level had been fixed, rather than random, it would have been impossible to generalize the results of this study beyond the specific grade levels included in the research design.

If all of the factors in a design are fixed, the design is said to be based upon a *fixed model*. The term *random model* is used to describe a design in which all factors are random. And as you would guess, a *mixed model* is one that contains at least one fixed factor and at least one random factor. Although the external validity of experimental studies is increased due to the inclusion of one or more random variables, a review of the literature will show that most research designs are based upon the fixed model. In other words, you will not come across too many journal articles in which a random variable has been incorporated into the design. If you do, the design will probably represent a mixed model rather than a random model. The random model is used very infrequently in experimental research because the researcher is usually interested in specific levels of the manipulated treatment factor(s). Therefore, to randomly select levels of this factor from a larger population of levels would be absurd.

In Excerpt 13.16 we see a three-way factorial design in which there were two active factors, treatments and counselors, and one assigned factor, sex of the Ss. The factor of counselors was random, with the five counselors being selected at random from a population of 21 counselors. The other two factors were fixed. Thus, this design represents a mixed model.

Since the factor of counselors was random, the main effect and interaction results of the statistical analysis can be generalized beyond the specific counselors who were used in the study to the entire population of counselors. In other words, the researcher can feel relatively confident that his significant findings for the treatment main effect and the treatment-by-sex interaction would be found if the study were replicated using a different subgroup of counselors from the same population. Had the factor of counselors been fixed, however, the results of the replicated study might well be different from those of the original study.

It should be noted that the process for getting F ratios in the summary table varies according to whether the model underlying the design is fixed, mixed, or random. In discussing the analysis of variance and the analysis of covariance, we saw that the error or within-groups mean square (MS) was always divided into the other MSs to obtain the F ratios.[6] This simple rule applies only to those designs that are based upon a fixed model.[7]

[6] In a Lindquist Type I or Type III design, there were two MSs used to obtain F ratios, one for the between-subjects part of the summary table and one for the within-subject part.

[7] There is one exception to this guideline. If there is only one score per cell—possibly a group mean—in the data analysis, there will be no within-groups source of variation. In this situation, the most complicated (i.e., highest-order) interaction term is used as the denominator in forming all F ratios.

If the experimental design is based upon a mixed or random model, some of the F ratios will be formed by using the MS for error as the divisor, but other F ratios will be obtained by using the MS from one of the interactions as the divisor. While a discussion of the actual rule for determining how to obtain F ratios in mixed and random models is beyond the scope of this text, an examination of Table 2 in Excerpt 13.16 shows the application of the rule. The MS for error (294.58) was used to obtain the F ratios for the main effect of counselors and also the $A \times B$, the $B \times C$, and the $A \times B \times C$ interactions. If the model had been fixed, these particular Fs would have been found in the same way, and error MS would have been used to obtain the other Fs as well. Since the design was actually based on a mixed model, however, the Fs for the main effects of treatments and sex were found by dividing their MSs (587.24 and 149.51) by the MS for $A \times B$ (271.49) and

---

EXCERPT 13.16 THREE-WAY FACTORIAL
DESIGN BASED ON A MIXED MODEL

METHOD

Population

The sample consisted of 45 boys and 45 girls drawn randomly from a population of 302 eighth-grade students attending the junior high school in Chandler, Arizona. The Ss had no previous experience with school counseling. They were randomly assigned to one of three experimental groups of 15 boys and 15 girls each. The Ss in each group were then randomly assigned to five unknown counselors.

Counselors

The five male counselors used in the study were randomly drawn from the 21 male enrollees attending the 1966–67 National Defense Education Act Elementary School Counseling and Guidance Institute held at Arizona State University. At the time of the study, the counselors had been exposed to the same basic counseling theory and practicum experience. Each counselor saw a total of 18 Ss, 3 males and 3 females from each of three experimental groups.

Treatment

Previous to the study, a program was developed, featuring a model which was designed to orient Ss to counseling and to a style of participating in the counseling process. A script was written, acted out, and recorded on both video and audio tape. The program featured a male narrator, a male counselor, and a male eighth-grade client. The narrator introduced, analyzed, and concluded the counseling interaction which was simulated by the counselor and the client. The script portrayed how the client, the model, moved away from emitting verbal statements about others to emitting statements which contained more self-reference, for example, "I feel . . . , I think . . . ," rather than "they feel . . . , he thinks . . . ." The counselor verbally and nonverbally reinforced the client when he made statements which referred to self. The narrator explained the role of the counselor as an empathic listener and participator and further reinforced the concept that counseling is based on the client's talking about himself rather than others. A summary of the program and model procedures is shown in Table 1.

A video-tape recorder was used to record the simulated presentation by the narrator, client, and counselor. The presentation was approximately 14 minutes in duration. Following the video taping, which became the first treatment to be given to the first experimental group (T_1), an audio-tape recorder was used to convey the audio portion of the video tape to a standard audio tape. The audio tape became the second treatment and was presented to the second experimental group (T_2). The Ss who received the third treatment (T_3) neither saw nor heard the model and acted as a control group.

Experimental Procedures

1. Ninety initial counseling interviews were scheduled according to the experimental design. All Ss first reported to a reception area which was separate from the treatment and counseling areas. The Ss reported in groups of five and each group was composed of either T_1, T_2, or T_3 Ss in order to control for the variable of treatment interaction and a loss of validity.

2. The Ss received information from E regarding the nature of a counseling program and the name of the counselor they were to meet. This procedure alone constituted the treatment T_3.

3. The Ss who received T_3 then moved to the counseling area where they met with their assigned counselors in a 30-minute individual interview. The Ss who received either T_1 or T_2 moved to a treatment area after leaving the reception area and before counseling.

4. Counselors were instructed to behave as counselors from their own theoretical frame of reference and were not informed as to the nature of the investigation or treatments until the study was completed.

5. Following the interview, Ss reported immediately to another area, where they again met as a group of five and then returned to class. A master schedule was used which allowed for a synchronization of counselors, Ss, and treatments. The three basic experimental groups did not come into contact with one another during the investigation.

Criterion Instrument and Data Collection

Using the verbal response class of first person pronouns (FPP), it was possible objectively to gauge the amount of client self-reference by counting the number of FPP, "I" and "we," emitted during the counseling interview. The frequency of FPP was obtained from tape recordings made of the 90 interviews. The analyses

and coding process was simplified by sampling procedures. The standard 30-minute interview was divided into 14 2-minute segments, discounting the first and last minute of the interview. Five separate 2-minute segments were randomly drawn from these 14 segments and coded for each of the 90 interviews. The 90 10-minute samples were copied in random order onto master tapes and submitted to two coders for scoring. An analysis of variance was used to estimate the reliability of the coding of FPP and an $r = .99$ was obtained (Winer, 1962).

RESULTS

The statistical design in this study constituted a $3 \times 5 \times 2$ factorial experiment. Moreover, the factorial design was a mixed model in that two of the factors were fixed and one of the factors was random. The two fixed factors were the main effects of treatments and sex, while the main effect of counselors was the random factor. The design not only allowed an analysis of variance on the main effects, but permitted a testing of interaction between effects. F ratios were computed to test all hypotheses (Winer, 1962).

The analyses of variance used for FPP is presented in Table 2. In this table it may be observed that two of the main factors were found to have statistically significant differences between their levels and that there was an interaction effect attributable to a combination of two of the main factors.

TABLE 2 ANALYSIS OF VARIANCE—THREE-FACTOR EXPERIMENT USING FIRST PERSON PRONOUNS

SOURCE OF VARIATION	df	MS	F
Treatments (A)	2	587.24	2.16*
Counselors (B)	4	1191.02	4.06*
Sex (C)	1	149.51	.64
A × B	8	271.49	.92
A × C	2	481.91	1.48*
B × C	4	230.70	.78
A × B × C	8	323.52	1.10
Error	60	294.58	
Total	89		

* $p < .25$.

SOURCE: R. D. Myrick. Effect of a model on verbal behavior in counseling. *Journal of Counseling Psychology*, 1969, *16*, 185–190. Copyright 1969 by the American Psychological Association. Reprinted by permission.

the MS for B × C (230.70), respectively. The F for the A × C interaction was found by dividing the MS (481.91) by the MS for the A × B × C interaction (323.52). The main point here is that the inclusion of one or more random factors in a design causes the error MS to be used as the denominator in forming only a portion of the F ratios, rather than all of them as is the case with a fixed model.

REVIEW TERMS

active factor

assigned factor

completely randomized factorial
   design

fixed factor

fixed model

manipulated factor

mixed model

multigroup pretest-posttest design

multigroup posttest-only
   design

organismic factor

random factor

random model

randomized blocks design

repeated measures design

treatments-by-levels design

unobtrusive measure

REVIEW QUESTIONS

1. Would it be appropriate for a researcher to use several correlated samples t tests to analyze the data from a multigroup pretest-posttest design?

2. If the data from a multigroup pretest-posttest design are analyzed with a Lindquist Type I ANOVA, which of the resulting F ratios would indicate whether or not the treatments have a differential effect on the Ss?

3. In a multigroup posttest-only design, scores on a concomitant variable might be collected at the same time as scores on the dependent variable. If an analysis of covariance is then used to analyze the data, what important assumption might possibly be violated?

4. One reason for extending the basic Solomon four-group design to accommodate additional groups is to enable the researcher to answer more research questions than he could with the regular Solomon design. What is another reason for using an extension of the Solomon four-group design?

5. In the basic Solomon four-group design, there are two factors: (a) pretest vs. no pretest and (b) treatment vs. no treatment. If a third factor is added, say sex, how many groups will there be in the new extended design?

6. Can the analysis of covariance be used in conjunction with either the basic Solomon four-group design or an extension of this design?

7. Would occupational membership be considered active or assigned as a factor in an analysis of variance design?

8. In order for a factorial design to qualify as a true experimental design, how many of the factors must be active factors?

9. At a minimum, how many active factors (if any) and how many assigned factors (if any) must there be in a randomized blocks design?

10. Give two reasons researchers like to include assigned factors in their experimental designs.

11. How many legitimate research questions are answered by a 3×4 randomized blocks design? How many are answered by a 3×4 completely randomized factorial design?

12. Do the present authors diagram factorial designs by means of the symbols R, O, and X?

13. Can a study have high internal validity without having high external validity?

14. If a researcher includes a random factor in his design, does this mean that he selected the factor at random from a larger population of potential factors?

15. In a mixed or random model, will the MS for within-groups (error) be divided into all other MS values to obtain the F ratios?

CHAPTER 14

QUASI-EXPERIMENTAL DESIGNS

In the last two chapters we presented some true experimental designs of the kind frequently reported in journal articles. A researcher who uses these designs has almost total experimental control. For example, each subject can be randomly assigned to the experimental and the control comparison groups. Also, researchers can control completely when the treatment is applied and when the observations or measurements are made. In this chapter we present five *quasi-experimental designs*, in which a researcher has limited experimental control. Quasi-experimental designs can be used by researchers when true experimental designs are not possible or feasible. With a quasi-experimental design the researcher does not have total control, but he can control one or two of the following: when the observations are made, when the treatment or independent variable is applied, and which intact group receives the treatment.

Quasi-experimental and true experimental designs differ in the degree to which threats to internal and external validity are controlled. Generally, true experimental designs can have greater control over internal threats than over external threats to validity, while quasi-experimental designs can have greater control over external than internal threats to validity. For example, researchers who use a quasi-experimental design with naturally formed or intact groups will probably decrease *subject reactivity*. In contrast, subjects who are randomly assigned to comparison groups (as with true experimental designs) are likely to be more reactive and to be more aware of the purposes of the research. Thus, a researcher who uses a quasi-experimental design may gain greater control over the external threats to validity but may sacrifice control over some internal threats. As with the previous designs, we confine our discussion only to the internal threats to quasi-experimental designs as guides for evaluating research.

We will present five quasi-experimental designs. The first section of this chapter illustrates two types of pretest-posttest quasi-experimental designs, the nonequivalent control group and the separate-sample pretest-posttest design. The next section presents two time-series designs, the single group and the multiple group. The last section presents a correlational analysis quasi-experimental design.

PRETEST-POSTTEST DESIGNS

The nonequivalent control-group and separate-sample pretest-posttest designs presented in this section allow the researcher to control for the time when subjects are observed and/or which subjects are exposed to the treatment.

Nonequivalent Control Group Designs

The nonequivalent control group design has two groups that are compared on observations before and after the exposure of one group to the treatment. This design is similar to the pretest-posttest control group design, except that subjects in the nonequivalent control group design are not assigned randomly from a common population to the experimental and control groups. Thus, random assignment of subjects is what distinguishes the pretest-posttest control group design from the nonequivalent control group design. The latter design is appropriate for research conducted in natural or field settings. A nonequivalent group may be helpful to confirm or rule out some plausible rival hypothesis when use of a true experimental design is not possible. This design is better than the preexperimental one-group pretest-posttest design because it provides an additional comparison (a control group).

The diagram for the nonequivalent control group design is presented in Figure 14.1. The diagram shows that (1) two groups are used (two rows in the diagram); (2) each group is measured or observed (O) at the same time before the treatment (X) is applied to the experimental group; (3) each group is measured at the same time after the treatment has been applied to the experimental group; (4) the subjects are not randomly assigned to the two groups (no R's in the diagram); and (5) the experimental and control groups do not have pretreatment sampling equivalence, that is, the groups are nonequivalent (indicated by the horizontal dashes separating the two groups). This design can also be used as a *nonequivalent comparison group design* in which two treatments (X and Y) are used, and it also can be used with more than two groups.

$$O \qquad X \qquad O$$
$$\text{------------}$$
$$O \qquad\qquad O$$

FIGURE 14.1. *Diagram of the nonequivalent control group design.*

Although the two groups are nonequivalent, the researcher does attempt to have the experimental and control groups as similar as possible when using this design. There are two versions of this design based on the degree of similarity between the experimental and control groups. In an *intact nonequivalent design* the researcher uses intact or naturally assembled groups, such as two sections of a housing development, two military units, two departments in a company, or two classrooms. The researcher then randomly assigns the treatment to one of the groups. The other version of the design is the *self-selected experimental* group design. The experimental group in this version consists of subjects who volunteer, are self-selected, or seek to be exposed to the treatment, whereas the control group subjects do not seek exposure. Counseling, psychotherapy, weight reduction, physical fitness, reading, or any other self-improvement treatment are examples of the type of research in which subjects might seek exposure to the independent variable.

Consider the above versions of the nonequivalent control group design (intact and self-selected experimental group) applied to the following hypothetical example. A company might examine the effects of human relations training on the number of products sold for sales personnel in one department of a company and compare the results with the sales personnel in another department (intact groups). Or, a group of sales personnel might seek human relations training and the effects of exposure for this group (self-selected experimental group) might be compared with another group of sales people who do not have human relations training. As we can see, the intact nonequivalent design is a better approximation of the true experimental design because it may control for more threats to internal validity. The self-selected design is much weaker because the groups are likely to be highly dissimilar. However, the self-selected version of this design is better than the one-group pretest-posttest design.

Threats to Internal Validity. To evaluate what threats to internal validity are controlled by the nonequivalent control group design, we must examine the particular circumstances of a study. However, there are some general guides for evaluating research that uses a nonequivalent control design. Selection is always a problem for this design because a researcher can never be certain that the groups are equivalent unless the subjects are randomly assigned to comparison groups. Selection is more of a problem for the self-selected version of this design than for the intact version. Even with intact groups, a researcher should attempt to make the groups as equivalent as possible. For example, intact groups such as a freshman and a sophomore high school English class may have more selection bias than two English classes from the same year in high school. If selection is a threat, history and maturation are also threats. A researcher cannot be certain that both groups have been exposed to the same events and that they have the same maturational processes. A researcher cannot be sure that sales personnel who are given human relations training compared with sales personnel from another department have the same intrasession history or the same maturational process occurring between the pre- and posttest.

Regression and mortality can also be threats to the nonequivalent control group design. Statistical regression will be a threat to the validity of this design if either of the comparison groups has been selected on the basis of extreme scores. Consider what would happen if the Ss in a study were people who are very much overweight and seek exposure to a weight-reducing program or sales personnel who have low sales records and wish to improve their skills by seeking human relations training. In both examples, a difference in scores from the pretest to posttest between the two groups may be the result of statistical regression rather than the effect of the treatment. Also, if too many subjects drop out from one of the comparison groups, mortality can pose a threat to this design. For example, subjects in the experimental group who have been pretested and received treatment may drop out before being posttested. The degree of differences in scores from the pretest to the posttest between the two groups may be an artifact of mortality rather than treatment.

This design may control for instrumentation and testing. The nonequivalent control group design (as well as the time-series designs discussed in the next section) may have an advantage over the true experimental design because it can control for subject reactivity. The researcher may be more concerned about controlling subject reactivity (external validity) than internal validity.

Statistical Analysis. The discussion of statistical procedures for the pretest-posttest control group design (Chapter 12) is also relevant to the nonequivalent control group design. One correct statistical analysis uses gain scores. Mean gain scores are calculated for each comparison group and the independent samples t test, the nonparametric Mann-Whitney U test, or the median test may be used to compare the two groups. Application of the analysis of covariance to this design might be used. However, because of nonrandomization of subjects to comparison groups, assumptions associated with the analysis of covariance might be violated. Thus, the analysis of covariance is less desirable for this design than for the pretest-posttest control group design. It is somewhat incorrect to use two *t* tests to compare the pretest and posttest means for each comparison group. Nor should the researcher use one *t* test to compare the two pretest means and then a second *t* test to compare the two posttest means.

Example of Design. In the journal article example of the nonequivalent control group design, the authors examine the effects of training on the Goodenough Draw-A-Man test. A description of the method and the results

---

EXCERPT 14.1 NONEQUIVALENT CONTROL GROUP DESIGN

Procedure

The initial Draw-A-Man test was administered to both groups according to Goodenough's (2) instructions. With only a plain piece of paper and a pencil on their desks, the children were given these directions: "On these papers I want you to make a picture of a man. Make the best picture that you can. Take your time and work very carefully. I want to see whether the boys and girls in Charles Hay School can do as well as those in other schools. Try very hard and see what good pictures you can make." No suggestions or comments were made to the children while they worked.

The experimental Ss were taken individually from the classroom four different days (two days each week for a period of two weeks) following the initial Draw-A-Man test. They were asked to assemble a jigsaw puzzle of a male figure constructed out of three pieces of ⅛ inch plywood glued together with a piece of masonite of similar size as backing. The figure was approximately 17 inches in height and 7 inches across; it consisted of the following 14 separate pieces: hair, head, neck, trunk, shoulders, arms, hands, legs, and

feet. The following instructions were given for the training sessions: "You have five minutes to try to put this puzzle together. I would like you to work as quickly and quietly as you can without any help from me. If you finish before the five minutes are up, you may begin again." No help was given to the child as he worked and discussion was discouraged.

Three days after the fourth puzzle session, the Draw-A-Man test was readministered to the control and experimental groups, again following Goodenough's standard instructions. The tests were scored according to the directions given in the original manual (2).

RESULTS

The *t* test for correlated samples was computed between the means of the first and second tests for each group. The results are tabulated in Table 1. It is apparent that the hypothesis is strongly supported. While the control group showed no significant change from the first to the second test, a marked and significant increase in scores was evident for the experimental group.

TABLE 1 SHIFTS IN DRAW-A-MAN SCORES FOR TWO GROUPS

GROUP	N	INITIAL TEST MEAN	FINAL TEST MEAN	t	SIGNIF.
Experimental	20	20.5	24.7	10.38	.001
Control	14	16.2	15.7	.70	ns

SOURCE: G. R. Medinnus, D. Bobitt, and J. Hullett. Effects of training on the Draw-A-Man test. *The Journal of Experimental Education*, 1966, 35, 62–63.

for this study appear in Excerpt 14.1, and a diagram for this sample journal article is presented in Figure 14.2. Note that 14 kindergarteners were used as the control group and 20 first-graders comprised the experimental group. Also, notice that the experimental subjects were taken individually from the classroom during a period for two days each week. The dissimilarity between the two groups is revealed from the pretest means (Table 1 in Excerpt 14.1). For this study, perhaps it would have been better to have both the experimental and control groups from the same grade.

Excerpt 14.1 describes the results and presents a table showing the pretest and posttest means. The correlated samples t test of pretest and posttest means for each group is not the best statistical test of significance for this study (see the discussion of statistical procedures in Chapter 12). A better treatment of the data would have involved an independent samples t test on the gain scores for the two groups. The problem or question concerning the effects of training could have perhaps been more accurately answered if intact classes from the same grade had been used and if a more legitimate statistical approach had been employed.

	O DEPENDENT VARIABLE	X TREATMENT	O DEPENDENT VARIABLE
20 first-graders	Draw-A-Man test	Four training sessions (experience in assembling a jigsaw puzzle of a male figure)	Draw-A-Man test
14 kindergarteners	Draw-A-Man test		Draw-A-Man test

FIGURE 14.2. *Diagram of the example nonequivalent control group design.*

Separate-Sample Pretest-Posttest Designs

The separate-sample pretest-posttest design may be applied in a situation where the researcher cannot randomly separate subgroups for differential treatments. However, the researcher can exercise some experimental control by the random assignment of which subjects are to be observed before and after treatment. This design can be used for large or small populations, such as high school classes, factories, cities, or military units, and allows representative sampling of a population that has been specified prior to treatment. Figure 14.3 illustrates this design. The diagram shows that: (1) the two rows represent randomly selected equivalent subgroups (note that if the two groups were not equivalent or were intact there would be a horizontal dash line between them); (2) one sample is measured before the treatment is applied; (3) both samples receive the treatment, but the

$$R \qquad O \qquad (X)$$

$$R \qquad\qquad X \qquad O$$

FIGURE 14.3. *Diagram of the separate-sample pretest-posttest design.*

treatment received by the first sample is irrelevant to the design and to the results of the study (indicated by the parenthesis); and (4) the second subgroup is measured or observed after exposure to the treatment.

Let us examine a hypothetical example that uses this design. Suppose a superintendent of schools in a city wants to assess the attitudes of teachers on several aspects of in-service training. The school superintendent does not want to have all the teachers in the city fill out a lengthy questionnaire twice; nor does he want pretest sensitization to limit his study's external validity. To circumvent these problems, he randomly selects two subgroups of city teachers. One equivalent subgroup fills out the questionnaire several days before a particular innovative in-service training session and the second sample completes the questionnaire several days after the in-service program. By using this design the superintendent is able to assess whether this innovative in-service training program changed the attitudes of teachers about in-service training.

Threats to Internal Validity. The separate-sample pretest-posttest design does not control for the internal threats of history, mortality, and possibly maturation and instrumentation. For example, unique or extraneous events could account for observed changes. In our hypothetical example the city council's approval of a salary increase for teachers one day prior to the innovative in-service program could account for a favorable attitude toward in-service training instead of the actual program.

Also, seasonal cycles or trends could cause changes and compete as rival hypotheses. In our hypothetical example changes in teacher attitude could be influenced by having a vacation period or the end of the school year in close proximity to the in-service program. The posttest subgroup could complete the questionnaire the day before spring break and possibly reflect positive attitudes toward in-service training. Thus, the actual in-service program would compete with the spring break (seasonal fluctuations in attitude) as a rival hypothesis.

This design also fails to control for mortality. In our hypothetical example, if a considerable number of teachers drop out or are not posttested, the results of the questionnaire could be confounded with mortality. Mortality might not be a problem with our hypothetical example, but it could be a problem if subjects who receive the treatment drop out before being posttested.

If the time between testing is long for a study investigating techniques in psychotherapy or remedial education, maturation could become a rival hypothesis. Changes on the dependent variable may represent natural developmental processes which may be operating even if the treatment had not been applied.

Instrumentation could also represent a threat if it is possible for the measuring instrument to change between the pretest and posttest, as might be the case if data are collected by interviewers or testing procedures. These procedures could be unreliable from the pretest to the posttest period and change or shift in the way the interviews are conducted or tests are administered.

As with the nonequivalent control group design, the separate-sample pretest-posttest design is more likely to have greater external validity (generalizability) than the true experimental designs. Usually fewer demands are

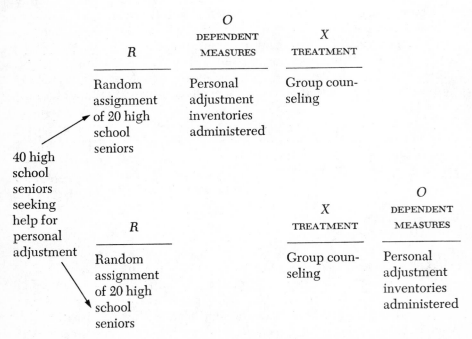

FIGURE 14.4. *Diagram of the hypothetical study using the separate-sample pretest-posttest design.*

made on the subjects in the field or natural settings than in the laboratory. However, researchers using this design must sacrifice internal validity in order to achieve greater generalizability or external validity.

Statistical Analysis. The appropriate statistical analysis for the separate-sample pretest-posttest design can be either the nonparametric Mann-Whitney *U* or median tests, or the parametric independent samples *t* test.

A Hypothetical Study. As a hypothetical study, a researcher might examine the effects of group counseling on personal adjustment of high school seniors. The researcher might have 40 seniors who seek help for some personal problems. The researcher might wish to use several personal adjustment inventories as dependent measures. Because the selected inventories take considerable time to complete, the researcher gives the inventories to one subgroup before the group counseling and administers these inventories to another subgroup after group counseling. By so doing, he would also protect against the possibility of pretest sensitization. The procedure for this hypothetical study might include meeting once a week for a 10-week period. An independent sample *t* test could be used to compare the pretest means with the posttest means for each subgroup. Figure 14.4 is the diagram for this hypothetical study.

TIME-SERIES DESIGNS

The time-series designs have repeated observations or measurements before and after treatment. These designs are excellent techniques for researchers wishing to evaluate the effects of a *planned treatment* or an *unplanned (ex post facto) treatment*. For example, unplanned research might assess the effects of a social reform program or a law that deals with legalized abortion on adolescent marriages, the implications of the equal rights amendment on male custody of children, or the effect of import duties on the sale of foreign automobiles. The only comparison a researcher can make with a single-group time-series design is before and after treatment. Researchers can make two or more comparisons with a multiple-group time-series design—the before-and-after series for the experimental group and the before-and-after series for the control group.

Single-Group Time-Series Designs

The typical pre and post designs that we have considered involved observations taken immediately before and after treatment. The single-group time-series designs use several measurements or observations before and after introducing the independent (treatment or intervention) variable. Some researchers feel that the time-series designs, like all quasi-experimental designs, are a good substitute for unfeasible true experimental designs and are very sensitive for investigating causal claims.

A. SINGLE GROUP WITH TEMPORARY SINGLE TREATMENT.

$$O_1\, O_2\, O_3\, O_4 \quad X \quad O_5\, O_6\, O_7\, O_8$$

B. SINGLE GROUP WITH CONTINUOUS SINGLE TREATMENT AND WITHDRAWAL.

$$O_1\, O_2\, O_3\, O_4 \quad X \quad \overline{O_5\, O_6}\, O_7\, O_8$$

C. SINGLE GROUP WITH CONTINUOUS SINGLE TREATMENT.

$$O_1\, O_2\, O_3\, O_4 \quad X \quad \overline{O_5\, O_6\, O_7\, O_8}$$

D. SINGLE GROUP WITH MULTIPLE TEMPORARY TREATMENTS.

$$O_1 O_2 O_3 O_4 \quad X \quad O_5 O_6 O_7 O_8 \quad Y \quad O_9\, O_{10}\, O_{11}\, O_{12}$$

FIGURE 14.5. *Diagrams of four variations of the single-group time-series design.*

Figure 14.5 shows four variations for the single-group time series. In diagram A four observations are made before the introduction of X (O_1 to O_4) and four after X (O_5 to O_8). A hypothetical example of the use of this design might be a study to determine the effects of human relations training for sales personnel on average number of sales during designated time periods. The weekly or monthly sales (dependent variable) would be recorded before the human relations training (independent variable X) and after training. The point to note about diagram A of Figure 14.5 is that the treatment variable (human relations training) is *temporary;* it does not continue during observations O_5 to O_8, but only for the period between observation O_4 and O_5.

Diagram B represents a second use of the single-group time-series design. Here the independent variable is applied for a brief period which continues through one or more observations and then is withdrawn. We have illustrated *continuous treatment* by a line drawn over the O's that are affected.[1] Consider a hypothetical example in which the treatment involves training cardiac patients to monitor and control their heart rates. Before training in biofeedback procedures are applied, the patients' heart rates (dependent variable) are recorded at O_1 through O_4. Biofeedback procedures (independent variable) are applied between O_4 and O_5. These procedures are applied during observations 5 and 6 and then withdrawn during observations 7 and 8, so the line is drawn over O_5 and O_6 but not over O_7 and O_8.

---

[1] Another way to illustrate continuous treatment is to have X precede each O to which it is applied, for example: $O_1\, O_2\, O_3\, O_4 \quad XO_5\, XO_6\, XO_7\, XO_8 \quad O_9\, O_{10}$. Treatment is applied from O_5 to O_8, but not applied to O_9 and O_{10}.

As another hypothetical research example of the use of the single-group time-series design, suppose a researcher wanted to investigate whether a city ordinance, legalizing liquor by the drink, decreased (or increased) the number of arrests (dependent variable) for drunken driving within the city limits. A researcher could use the time series by recording the number of arrests that occurred monthly for several months before the city ordinance and later record the monthly number of arrests after the city ordinance was passed. Diagram C in Figure 14.5 illustrates this type of time-series design. Notice that the independent variable (city ordinance) is applied continuously and throughout the observations O_5 to O_8, so the line is drawn over all observations after the introduction of the treatment.

The last variation of the single-group time-series design has two or more treatments applied successively to the group. A researcher could examine the effects on a group of sales personnel of a workshop on human relations training (X) and later a workshop in techniques for determining potential customers (Y). In diagram D in Figure 14.5 each O could represent the number of sales during a particular time period. In this hypothetical example the treatments are temporary and do not continue during O_5 to O_8 or O_9 to O_{12}. However, there can be situations in which the researcher applies the first treatment (X) continuously $(O_5$ to $O_8)$ and then applies the second treatment (Y) continuously $(O_9$ to $O_{12})$. The diagram for this example would have one line over O_5 to O_8 and a second line over O_9 to O_{12}. Or, a researcher might apply the first treatment and later add a second treatment without withdrawing the first. The diagram for this example would look like this:

$$O_1 \, O_2 \, O_3 \, O_4 \quad X \quad \overline{O_5 \, O_6 \, O_7 \, O_8} \quad \frac{X}{Y} \quad \overline{O_9 \, O_{10} \, O_{11} \, O_{12}}$$

There are two types of measurements or observations a researcher can use with single-group time-series designs. In our examples used to illustrate diagrams A, B, and D of Figure 14.5, the observation or measurement was *group repetition,* that is, the same intact group of subjects was observed at successive points throughout the entire series. In the example used to illustrate diagram C, the observation was *group replication,* that is, the observations are not of any specified group of subjects or individuals and the composition of the groups may vary at each O. Group repetition or group replication may be used with any of the diagrams in Figure 14.5.

The four types of time-series designs that we have considered may include more observations before and after the independent variable than just the four that we have used. By having many observations, the time-series designs allow the researcher to focus on the processes of change of the dependent variable. Also, the time-series designs allow the subjects to serve as their own control because measurements or observations are recorded for several periods of time before the introduction of the treatment variable.

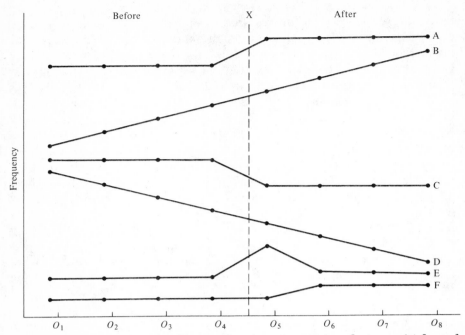

FIGURE 14.6. *Six possible results for time-series designs.* (*Adapted from D. T. Campbell and J. C. Stanley.* Experimental and Quasi-Experimental Designs for Research. *Chicago: Rand McNally, 1966.*)

Several possible results may be obtained in the time-series designs. If the results of treatment or intervention are effective, there should be an abrupt change in the level or direction of the time series when the treatment is applied. Six hypothetical results of time-series designs are illustrated in Figure 14.6. The results plotted for time series A and C appear to show a change in the dependent variable when the treatment variable (X) was introduced (between O_4 and O_5). The dependent variable increased for time series A and decreased for C. To interpret the effect of the treatment, a reader must know whether the treatment was temporary or continuous. Continuous treatment or intervention may result in a time series, like A and C, which indicates change from the point at which treatment is applied. If the treatment is temporary, the time series for the dependent variable may appear as illustrated in E, a temporary increase at O_5 and a return to the before level at O_6, O_7, and O_8. However, the reader should be alert to the possibility that the time series A and C could reflect the application of a temporary treatment if the dependent measures represent behavior that is irreversible, such as reading comprehension.

The results illustrated in the hypothetical time series B, D, and F must be interpreted with some caution. For example, in time series B, there appears to be an increasing trend that does not appear to be a function of the independent variable (changes in the dependent variable do not occur when the independent variable was introduced). Also, the decreasing trend of the dependent variable in time series D does not appear to be a function of the treatment variable. The change in time series F occurred after O_5 and may or may not be a function of the treatment variable. The reader should note that results of time-series studies that typically appear in journal articles usually show much more variability than the six hypothetical time-series results in Figure 4.6. As we shall discuss in the next section, a great deal of instability (variation) in the time-series data can be an internal threat to the validity of the research if the instability is unaccountable.

Threats to Internal Validity. Although single-group time-series designs can control for more threats than the one-group pretest-posttest design, there are still several possible threats to the internal validity of these designs. It is plausible that some coincident event (history) other than the treatment could be responsible for changes in the dependent variable. History is more likely to be a threat when the time-series design is an ex post facto analysis rather than a planned study. For example, it could be a problem in our hypothetical study of the effects of a city ordinance concerning liquor (diagram C of Figure 14.5) because it is likely that this would be an unplanned study. Also, the more infrequent the observations, the more plausible history becomes as a rival hypothesis. Thus, seasonal changes could influence the dependent variable and be confounded with the introduction of the treatment variable. The single-group time-series designs do not have a comparison group to control for the internal threat of history.

Mortality could be a possibly threat to the validity of these designs. For example, it might become a rival hypothesis in the hypothetical sales-training study if several sales personnel leave their jobs after the introduction of the human relations training. However, previous records (before and during the study) of resigning and replacement could insure that personnel changes do not provide a good rival hypothesis.

Maturation may not be a problem for these designs if the results are similar to the A, C, and possibly E time series illustrated in Figure 41.6, that is, if the change in the dependent variable does not occur until X is introduced. Maturation could account for the hypothetical results in the B and D time series of Figure 14.6.

Testing seems to be an implausible rival hypothesis for changes occurring between O_4 and O_5 for the time-series designs. However, it could be a problem if tests are used and reflect changes at points other than those at which the treatment is applied.

Instrumentation could be a plausible explanation for differences before O_4 and after O_5 if the measurement procedures are not constant or reliable throughout the entire time series. If in the hypothetical studies in Figure 14.5 different methods of recording sales (diagram A), different biofeedback equipment (diagram B), or different procedures for recorded arrests (diagram C) were used after introducing the independent variable, instrumentation may be mistaken for the treatment effect. Also, a low interobserver reliability might cause a misinterpretation of the results whereby it is assumed that the treatment is responsible for any change from before O_4 to after O_5. It is also possible that changes in the time series could be caused by observers' awareness of the experimental conditions of the research or knowledge of when the independent variable is applied. Reliability and awareness are important factors in time-series designs when observers are used (as we will see in the next chapter when we discuss applied behavior analysis designs). A researcher who uses observers in time-series designs should report in the journal article interobserver reliability, whether or not the observers were aware of the time periods or the experimental procedures of the study, and whether there were any subtle changes in the instrumentation or observational procedures.

Statistical regression as a threat to internal validity probably will not be a rival hypothesis for a before-and-after difference, unless extreme records or scores are obtained for several O's prior to treatment. For example, in our hypothetical study in diagram C of Figure 14.5 the treatment (city ordinance) may have been instigated at a particular time because the number of previous arrests was extremely high, so that statistical regression could account for any decrease (in number of arrests) from that point on.

Selection could be a problem in the time-series designs if the same subjects are not observed or measured at all O's. In the hypothetical studies in Figure 14.5, the same sales people and cardiac patients (diagrams A and B) are measured repeatedly before (O_1–O_4) and after (O_5–O_8) the introduction of the independent variable. However, in the hypothetical study of number of arrests (diagram C), the same people may not be represented at all O's (group replication) and therefore selection is likely to be a rival hypothesis. The more the intact group changes in composition from O to O the more likely selection becomes as a rival hypothesis. For example, people moving into and out of our hypothetical city in diagram C might represent an 85 percent annual change in the composition of population from which the measured sample is drawn. As the percentage of change in the composition of population increases, so does the threat of selection to internal validity. Although group-replication measurement may increase the problem of selection, it does reduce the likelihood of other threats to validity in that it is less prone to subject reactivity and practice effects than is group-repetition measurement. Readers should be aware of the repetition versus replication measurement distinction in judging the validity of a time-series study.

One last possible threat to the validity is the *instability* of the dependent measures. Instability refers to unaccountable variation which may be mistaken for the treatment or intervention effect. Although the internal threat of instability of dependent measures can be a problem for other designs that we have presented, it represents a very serious threat to time-series designs. The problem of instability can be handled by statistical analysis to determine whether or not there is a change before and after the introduction of treatment to the series.

Threats to the internal validity of time-series designs depend largely on the particular situation to which the design is applied. To judge the threats to a single-group time-series design, a reader of a journal article must assess the application of the design to the complexity and uniqueness associated with a particular problem or situation. For example, planned time-series studies may have less internal threats and may establish stronger causal inferences than ex post facto research. Our above discussion of threats to internal validity also applies to the multiple time-series designs discussed in the next section.

Statistical Analysis. The time-series designs employ a very sophisticated statistical procedure which is different from any that we have considered in previous chapters. The procedure is based on an *integrated moving average model* and weighs heavily on the observations closer to the point at which the independent variable is introduced. (A more detailed explanation of this statistical technique is beyond the scope of this book.) As with the other statistical procedures we discussed in previous chapters, the statistical procedure for time-series designs uses levels of significance and either one-tailed or two-tailed tests.

In Figure 14.6 we provided hypothetical results of time-series designs. Usually researchers who use a time-series design present the results in a similar graphic form. However, a line graph alone may not be satisfactory for determining whether there is a change after the introduction of treatment. When a researcher has highly variable data between observations, it is important to submit the data to statistical analysis and to test for significance. A frequency graph of highly variable data is not an adequate procedure to determine the probability of change after treatment.

Example of Design. The journal article selected to illustrate the single-group time-series design explored the effects of videotape self-confrontation in human relations training.[2] The author hypothesized that videotape feedback to the T-group would have a significant effect on moving the group toward the selected goal of establishing a democratic atmosphere in which each member could participate and would be valued for his personal contribution. Excerpt 14.2 describes the method, the subjects, the procedure, the two dependent variables, and the analysis of data.

The statistical analysis of the time-series data for each group indicated

---

[2] The complete study from which this excerpt was taken is actually a multiple-group time-series design. However, we have used the data and statistical analysis for one group (Group C) to illustrate a single-group time-series design.

EXCERPT 14.2 DESCRIPTION OF A
SINGLE-GROUP TIME-SERIES DESIGN

METHOD

The study was conducted as a time-series research design. A time-series design is a research procedure whereby periodic measurements are taken on some group or individual, an experimental change is introduced into this time series, and the results are indicated by a discontinuity in the measurements recorded in the time series (Campbell & Stanley, 1963, p. 37). A control group is not required for this type of research because the group under study serves as its own control. Studies which match experimental and control groups of 30 or more subjects may achieve a valid equivalence, at least theoretically. However, it is questionable if a small group of 8–10 people can ever be adequately matched, especially in terms of the complex prerequisites and personality characteristics which contribute to group process. Therefore, the own-control, time-series design is particularly appropriate for small-group research studies. In the current study, repeated measures of the dependent variables were collected over several separate but consecutive time periods. The independent variable under study, that is, videotape feedback, was introduced into the time series and the data examined to determine if such feedback resulted in significant group change.

Subjects

The present study employed three separate groups. One group would have been sufficient in terms of the research design but the experiment was repeated with two additional groups to assure that results were not limited to a single, unique group.

The subjects in the three T groups were matched according to age (overall mean age for all groups was 30.4 years), sex (male-female distribution for the three groups was 5–2, 4–3, and 5–2), and education (mean grade completed was Grade 10). The institution is government sponsored and designed to provide basic matriculation (Grade 12 diploma) for people who have been away from school for some time but who wish to complete their high school education. Students attend regular daytime classes and are compensated by the government so they are not employed during the school year (Souch, 1970).

Students of the type enrolled at this institution tended to have experienced difficulty and frustration in regard to obtaining satisfactory and rewarding employment. Most were married and had families, thus presenting certain pressures and added responsibilities not found among the usual youthful student. Their decision to return to school for retraining indicated considerable motivation for self-improvement and suggested that they could benefit from the T-group experience.

The study was presented to the subjects as an opportunity to learn more about themselves and to improve their methods of interacting in a group situation. All subjects were informed that the sessions would be videotaped and that they would be allowed to see some replays at some time in the future.

Trainers were chosen from a group of doctoral students studying in the general area of group approaches to behavior modification. Criteria for selection were interest in the project, willingness to participate, and availability. The three trainers chosen for the study were male graduate students completing the PhD in Counseling Psychology at the University of Alberta. One trainer was in his mid 30s; the other two were in their early 40s. All three trainers had previous experience with groups. They had no knowledge of the hypotheses or specific variables of the study and therefore could be considered impartial participants in the study.

Procedure

Each T group met twice a week for a 90-minute period. The groups attended 10 such meetings, which may be considered adequate for group change (Schein & Bennis, 1965, p. 76).

The time-series analysis required approximately 30 time periods, hence each session was arbitrarily divided into three 25-minute segments through a short coffee break and a leader-originated break. This resulted in 30 segments but some time was lost during the time allotted for videotape replays so a final total of 28 separate videotaped segments were accumulated. These 25-minute segments of group interaction are the measurement units applied to the time-series model.

All sessions were conducted in a small television studio. Group members were seated in comfortable chairs arranged in the form of a horseshoe with a videotape camera directed into the open end of the horseshoe. Recording equipment and procedures have been described elsewhere (Martin & Zingle, 1970).

A group met for five sessions (15 segments) before seeing any videotape replay. Just prior to the sixteenth segment, the group's attention was directed to a 21-inch playback monitor and a 10-minute section of videotape recording, selected randomly from the previous segment of group interaction, was shown to the group. The T-group leaders asked subjects to note their characteristic manner of group behavior and interaction, both verbal and nonverbal. Participants then observed the entire 10-minute section without interruption but were free to talk about the replay as they watched it.

A regular 25-minute segment of group interaction followed this first videotape feedback stimulation. The group then observed a second videotape replay, again randomly selected, of this just-completed segment of group interaction. The procedure was repeated a third time so that the group experienced three videotape feedback situations between regular periods of group interaction.

Dependent Variables

Two objective measures of group behavior were selected to evaluate the effects of the videotape stimuli. Dependent variables were chosen in relation to the previously mentioned T-group goal. Development of a democratic and sharing group atmosphere was assessed by measures of verbal output for group members. Verbal output was defined as verbal quantity for group members (i.e., how much each person spoke), and verbal frequency for group members (i.e., how often each person spoke).

A multichannel events recorder was employed to record individual speaking times calculated from eight 1-minute random samples of each 25-minute segment. The total time that a member spoke during a sample was the cumulation of the length of lines on the recorder tape corresponding to that person; the number of times that a member spoke during a sample was the number of separate lines recorded on his specific channel.

Final measures for these two variables were simple variances calculated for all group members during any one segment. Means were not employed because the stated T-group goal is related to variability among group members rather than to average group behavior. Thus, the eight 1-minute samples of verbal behavior in any given segment contributed to a single measure of variability for that segment and the total series generated 28 measures of group variance of verbal quantity (total time variance), and 28 measures of group variance of verbal frequency (frequency variance) for each of the T groups. The hypothesized increase in democratic, cooperative group interaction would be reflected by decrease in group variance for either of the verbal output variables.

Analysis of Time-Series Data

Campbell and Stanley (1963, p. 38) have shown that introduction of a treatment variable to a time-series sequence may cause change in the level and/or the slope of the series. Thus, if significant, meaningful results are to be obtained, both level and slope must be examined for change in any time-series design.

The recent integrated moving average model described by Maguire and Glass (1967) and Glass and Maguire (1968) is designed for such analysis and was applied to the data to determine if either level or slope of the dependent trend variables changed as a function of videotape feedback.

SOURCE: R. D. Martin. Videotape self-confrontation in human relations training. *Journal of Counseling Psychology*, 1971, *18*, 341–347. Copyright 1971 by the American Psychological Association. Reprinted by permission.

that the effects of videotape feedback were not necessarily predictable, that is, videotape feedback in relation to the T-group goal may be either beneficial or detrimental. Excerpt 14.3 illustrates the time-series graph for each

---

EXCERPT 14.3 DESCRIPTION AND GRAPH OF THE
RESULTS FOR SINGLE-GROUP TIME-SERIES DESIGN

Group C did not experience the hypothesized decrease in level of the dependent variables. Instead of immediate change in level, videotape feedback seemed to initiate a gradual but steady change in this group with a significant decrease in slope for both total time variance ($t = $ from -1.88 to -1.67; $p < .05$) and frequency variance ($t = $ from -3.02 to -2.69; $p < .01$). Graphical presentation of trends in verbal output for this group appear in Figure 3.

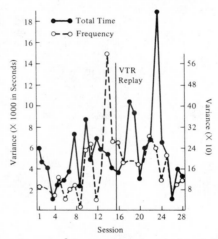

FIGURE 3. *Total time variance and frequency variance of verbal output for Group C.*

SOURCE: R. D. Martin. Videotape self-confrontation in human relations training. *Journal of Counseling Psychology,* 1971, *18,* 341–347. Copyright 1971 by the American Psychological Association. Reprinted by permission.

---

dependent variable for Group C before and after VTR replay. The results of the statistical analysis are also included in this excerpt. The immediate change in level after VTR replay was not statistically significant for either of the two dependent variables. However, the slope (the general trend of change) indicated a significant change. In other words, VTR feedback for Group C appeared to be the stimulus for a slow change in the pattern of group behavior and interaction (slope), but the change did not occur immediately after the introduction of VTR replay (level).

Multiple-Group Time-Series Designs

The multiple-group time-series design is similar to the single-group design discussed in the previous section, but it has two or more groups that are usually intact or naturally formed. It can be used with group-repetition or group-replication measures and with planned or ex post facto research. Some of the variations of the multiple-group design are similar to the non-equivalent control group design discussed at the beginning of this chapter. However, the multiple-group time-series design includes a series of observations or measurements before and after treatment, while the nonequivalent group design includes only one observation before and after treatment for the groups. The multiple-group time-series design is considered by some researchers to be one of the best quasi-experimental designs because the effect of the treatment variable can be compared once with a series of a second group and once with the O's of the experimental group before the treatment was introduced.

The four variations of the single-group time-series design illustrated in Figure 14.5 can be used as multiple-group time-series designs by adding another group. The diagram for these multiple variations would have a dash line (indicating nonequivalence) separating the time series for the different groups. The second group can be a comparison group with X or a control group without X. Consider how we might diagram the hypothetical examples related to Figure 14.5 if we add a control group. For diagram A, a group of sales personnel who do not receive human relations training is added. A group of cardiac patients who do not receive training in biofeedback to monitor and control their heart rates is added in diagram B of Figure 14.5. For diagram C another city with similar demographic features that does not have a city ordinance legalizing liquor by the drink can be added to make the design a multiple-group time-series. The single group with multiple temporary treatments can be used as a multiple group by adding a control group to diagram D. For each of the above hypothetical examples of multiple-group time-series, ideally treatment is randomly assigned to one of the groups.

Other variations of the multiple-group time-series design are illustrated in Figure 14.7. In diagram A two treatments are staggered. First X is applied to one group and later Y is applied to a second group. The researcher might decide that the application of two treatments at once is too costly. For example, human relations training (X) might be given to the sales personnel of one department in a large company and later training in techniques for determining potential customers (Y) given to another department's sales personnel. The treatment in this hypothetical example is temporary, but it can also be continuous.

Another variation of the multiple-group design has reversed and temporary multiple treatments (diagram B of Figure 14.7). This variation will control for more threats to external validity than the design illustrated in A. At the same time a researcher can apply human relations training (X) to one group and techniques for determining potential customer training (Y) to the second group. Later the training is reversed for each group.

A. MULTIPLE GROUP WITH STAGGERED AND TEMPORARY MULTIPLE TREATMENTS.

$$O \quad O \quad O \quad O \quad X \quad O \quad O \quad O \quad O$$
-- -- -- -- -- -- -- -- -- -- -- -- -- -- -- -- --
$$ O \quad O \quad O \quad O \quad Y \quad O \quad O \quad O \quad O$$

B. MULTIPLE GROUP WITH REVERSED AND TEMPORARY MULTIPLE TREATMENTS.

$$O \quad O \quad O \quad O \quad X \quad O \quad O \quad O \quad O \quad Y \quad O \quad O \quad O \quad O$$
-- -- -- -- -- -- -- -- -- -- -- -- -- -- -- --
$$O \quad O \quad O \quad O \quad Y \quad O \quad O \quad O \quad O \quad X \quad O \quad O \quad O \quad O$$

C. MULTIPLE GROUP WITH TEMPORARY SINGLE TREATMENT.

$$O \quad O \quad O \quad O \quad X \quad O \quad O \quad O \quad O$$
-- -- -- -- -- -- -- -- --
$$O \quad O \quad O \quad O \quad X \quad O \quad O \quad O \quad O$$
-- -- -- -- -- -- -- -- --
$$O \quad O \quad O \quad O \quad X \quad O \quad O \quad O \quad O$$

D. MULTIPLE GROUP WITH STAGGERED AND CONTINUOUS SINGLE TREATMENT.

$$O \quad O \quad O \quad O \quad X \quad \overline{O \quad O \quad O \quad O} \quad \overline{O \quad O \quad O \quad O} \quad \overline{O \quad O \quad O \quad O}$$
-- -- -- -- -- -- -- -- -- -- -- -- -- -- --
$$O \quad O \quad O \quad O \quad \quad O \quad O \quad O \quad O \quad X \quad \overline{O \quad O \quad O \quad O} \quad \overline{O \quad O \quad O \quad O}$$
-- -- -- -- -- -- -- -- -- -- -- -- -- -- --
$$O \quad O \quad O \quad O \quad \quad O \quad O \quad O \quad O \quad \quad O \quad O \quad O \quad O \quad X \quad \overline{O \quad O \quad O \quad O}$$

FIGURE 14.7. *Diagrams of four variations of the multiple-group time-series design.*

Thus, the researcher can compare the differential effects of two treatments and possibly control for multiple treatment interference.

The last two illustrated variations of the multiple-group time-series design have more than two groups. The multiple group with temporary single treatment is shown in diagram C of Figure 14.7. We have illustrated the design with temporary treatment, but it can also be used with continuous treatment. Consider our hypothetical example of a weight-reducing program for professional football players. The same program can be applied to three groups. If the weight training is continuous, we would have lines over all the O's following X for each group. Human relations training can be applied temporarily by a researcher to three groups, and the diagram for this research would appear as in C of Figure 14.7. There are two other variations of the multiple-group temporary single treatment design. A researcher can use a multiple group with multiple treatment design. The diagram would be similar to C, but we would replace the X's in the last two groups with Y and Z to denote different treatment conditions. For example, a researcher might expose three groups of overweight professional football players to three different treatments: diet control techniques (X), a running program (Y), and transcendental meditation (Z). The treatments for this hypothetical example would be continuous. The last variation of the design C time series involves applying the same treatment to three different groups. For example, a researcher might be interested in determining whether the weight-reducing program for professional football players is equally effective for professional baseball players, basketball players, or golfers. Or, another researcher might wish to assess the differential effects of one treatment or intervention technique on the responses of boys and girls or different age groups.

The last variation of the multiple-group time-series design is illustrated in diagram D of Figure 14.7. It is more likely that this design would be used for continuous treatment conditions, although it can be used for temporary treatments. In this variation a single treatment is staggered or applied successively to each group. For example, human relations training can be applied successively to the sales personnel of three departments in a large company. If the number of products sold increases at the point when treatment is introduced for each group, a researcher can make a fairly strong inference about the causal relationship between the treatment variable and dependent variable.

The discussion of the hypothetical results as illustrated in Figure 14.6 also applies to multiple-group time-series designs.

Threats to Internal Validity. The discussion of the sources of invalidity for the nonequivalent control group design is also relevant for the multiple-group time-series designs. Unless the researcher can be reasonably certain that the comparison groups are similar, differences in the dependent variable between groups may be caused by differences between the groups rather than the treatment. As we have indicated, the best technique for equating groups is random assignment of each subject to the groups, but a researcher cannot always do this. Consider our hypothetical example of the effects of human relations training on sales. The two intact groups of sales personnel might not be equivalent. This would be the case if a treatment is applied to junior and less experienced sales personnel while the control group contains senior sales personnel with more years of experience. If a researcher found no differences in sales at the end of the study, inferences about no treatment effectiveness might be a doubtful conclusion because of selection bias.

History and maturation also pose threats. Statistical regression is another source of invalidity if extreme scores are used for one group at the beginning of treatment. Differences between the experimental and control groups could be accounted for by mortality rather than the treatment if too many subjects drop out of one group. (Of course, it is difficult for a reader to know whether or not mortality is a problem for a study unless the author reports how many subjects drop out.) As we have said before, a discussion of internal validity can be only a guide for evaluating research. Threats to the validity of a study must be assessed in terms of the circumstances and design associated with a specific study.

Statistical Analysis. As with the single-group time-series, graphic presentation of data is not enough for a researcher to determine whether changes have occurred as result of treatment in the multiple-group time-series designs. The statistical procedures described for the single-group time-series are also appropriate for multiple-group time-series designs.

Example of Design. The journal article selected to illustrate the multiple-group time-series design studied the effects of a crackdown on speeding on highway fatalities. Excerpt 14.4 describes the background of the study.

EXCERPT 14.4 DESCRIPTION OF METHOD FOR
MULTIPLE-GROUP TIME-SERIES DESIGN

In 1955, 324 people were killed in automobile accidents on the highways of Connecticut. Deaths by motor vehicle accidents had reached a record high for the decade of the fifties as the usually hazardous Christmas holidays approached. Two days before Christmas, Governor Abraham Ribicoff of Connecticut initiated an unparalleled attempt to control traffic deaths by law enforcement, and announced his crackdown on speeders in that state.

Ribicoff believed, along with many safety specialists, that excess speed was the most common contributing factor in traffic deaths, and that control of speed would result in diminished fatalities. He believed that previous efforts to control speeding under the usual court procedures and by the existing "point system" had

been inadequate. In a study of three months' records of the police court in Hartford, it was noted that no more than half the persons originally charged with speeding were so prosecuted, the charge often being diminished to a less serious one. Ribicoff wanted to initiate a program with reliable procedures and strong sanctions as a means to control speeding and thus to reduce traffic deaths.

On December 23, 1955, Governor Ribicoff announced that in the future all persons convicted of speeding would have their licenses suspended for thirty days on the first offense. A second violation was to mean a sixty-day suspension, and a third conviction for speeding would result in indefinite suspension of the driver's license, subject to a hearing after ninety days.

SOURCE: D. T. Campbell and H. L. Ross. The Connecticut crackdown on speeding: time-series data in quasi-experimental analysis. *Law and Society Review*, 1968, 3, 33–53.

The dependent variable was the number of traffic fatalities per 100,000 population. The nonequivalent control consisted of adjacent and similar states—New York, New Jersey, Rhode Island, and Massachusetts. The traffic fatalities for these states were compared with those for Connecticut, the experimental state. Excerpt 14.5 shows graphic evidence of a steadily decreasing rate of traffic fatalities for Connecticut from 1956 to 1959; the rate of traffic fatalities for the control states did not decrease as rapidly. In other words the gap between Connecticut and the control states gets steadily larger in 1957, 1958, and 1959.[3]

[3] For a detailed description of the statistical procedures and analysis of these time-series data, see G. V. Glass, Analysis of data on the Connecticut speeding crackdown as a time-series quasi-experiment, *Law and Society Review*, 1968, 3, 55–76.

EXCERPT 14.5 GRAPH OF THE RESULTS
FOR MULTIPLE-GROUP TIME-SERIES DESIGN

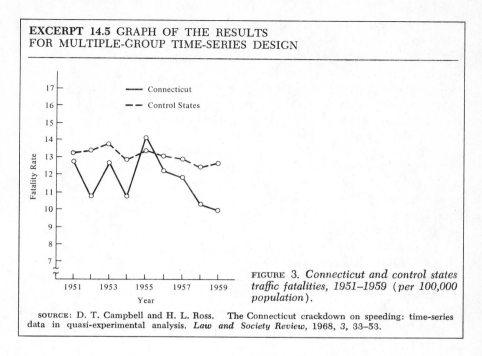

FIGURE 3. *Connecticut and control states traffic fatalities, 1951–1959 (per 100,000 population).*

SOURCE: D. T. Campbell and H. L. Ross.   The Connecticut crackdown on speeding: time-series data in quasi-experimental analysis. *Law and Society Review*, 1968, 3, 33–53.

CORRELATIONAL ANALYSIS DESIGN

We have discussed quasi-experimental designs in which the independent variable could be manipulated by the researcher. In the designs presented in the first section, the researcher can randomly apply the independent variable to one of two groups or subgroups. In the two designs presented in the next section, the researcher can control when the independent variable is introduced. In the multiple-group time-series designs he can control when to apply the treatment and to which group. We will now present a quasi-experimental design that can be used when neither manipulation of the independent variable nor randomization is possible. Natural sequences or seasonal events can be investigated using cross-lagged panel correlational analysis, and the technique can be applied to economic, political, demographic, and behavioral variables. A correlational analysis design cannot usually establish a causal relationship as can a carefully controlled true experimental design, but it can help in possibly providing some hint of causality.

Cross-Lagged Panel Correlational Analysis

Lacking the freedom to control and manipulate the events to be investigated, a researcher can rely upon correlational techniques to infer direction of probable causality. For example, when data are derived from surveys or other observations, a researcher can determine the possible direction of causation between two variables. Here, the researcher must rely to a great extent upon statistical treatment rather than upon direct manipulation of the independent variables. As we discovered in Chapter 2, a correlation coefficient, by itself, does not imply causation. A researcher cannot infer a causal relationship from a single correlation of two variables. However, with the recent development of the cross-lagged panel correlational technique, a technique which uses several correlation coefficients, it is possible for a researcher to differentiate the relative plausibilities of competing causal or directional interpretations between two variables.

The cross-lagged panel correlational technique is based on *time precedence*. The assumption of time precedence is that when an event consistently precedes the occurrence of another event (and the reverse does not occur), the researcher can consider two possibilities: (1) event or variable A (although there may be others) has an influence on event B; (2) both A and B are effects of some other general variable. Correlational techniques can be used to study the strength of a relationship between two variables, but a reliable estimate of causality cannot be made from one coefficient of correlation. However, if a researcher has coefficients of correlation relating two variables at more than one point in time, a causal influence may be inferred about the directional effect of one variable on another. Figure 14.8 presents the diagram for the cross-lagged panel correlational technique.

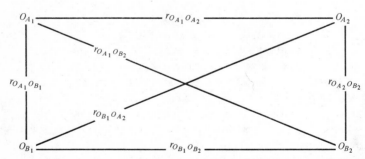

FIGURE 14.8. *Diagram of the cross-lagged panel correlational technique.*

Two variables are measured for one group of subjects on two different occasions. O_{A_1} refers to the first measurement of one variable and O_{B_1} refers to the first measurement of the second variable (notice in Figure 14.8 that the two symbols are on a vertical line, indicating that the observations or measurements occur at the same time). The second set of observations of the same variables, taken later, is designated as O_{A_2} and O_{B_2}. Repeated measures (or test-retest) of the same variable (designated by the subscripts 1 and 2) are referred to as *panel data*. Thus, in Figure 14.8 there are two panels (or vertical arrangements of symbols). The first panel or measure is indicated by the line O_{A_1}–O_{B_1} and the second panel by the line O_{A_2}–O_{B_2} (there could be another panel O_{A_3}–O_{B_3}).

There are six correlation coefficients (r) in Figure 14.8. The $r_{O_{A_1}O_{B_1}}$ is the symbol for the correlation coefficient for the first panel of the two variables O_{A_1} and O_{B_1}; $r_{O_{A_2}O_{B_2}}$ is the correlation coefficient for the second panel or measure of the two variables. Measures within the same panel (vertical lines) are referred to as *unlagged measures*. (Lag refers to a period of time between measures.) The $r_{O_{A_1}O_{A_2}}$ is the correlation coefficient for the first and second panel of the O_A variable, and $r_{O_{B_1}O_{B_2}}$ is the coefficient for the first and second panel, the O_B variable. These correlations for two panels of one variable (horizontal lines) are called *lagged measures,* and they reflect the consistency in each variable over time. The remaining two correlation coefficients, $r_{O_{A_1}O_{B_2}}$ and $r_{O_{B_1}O_{A_2}}$ (on the two diagonals), are *cross-lagged correlations,* that is, they are correlations of one variable in the first panel with the second variable in the second panel.

A researcher may use the two cross-lagged correlations in relationship to the other four correlations to infer a possible causal influence or source. In other words, a change in one variable consistently followed by a change (either an increase or decrease) in another variable may satisfy the notion of causality. Four hypotheses are in competition when a researcher uses the cross-lagged panel correlation technique: (1) an increase in variable A will cause an increase in variable B, and a decrease in A will cause a decrease in B; (2) an increase in A will cause a decrease in B, and a decrease in A will cause an increase in B; (3) an increase in B will cause an increase in A, and a decrease in B will cause a decrease in A; (4) an increase in B will cause a decrease in A, and a decrease in B will cause an increase in A.

For many kinds of research situations, the cross-lagged panel analysis may not enable a researcher to make an unambiguous decision about the competing hypotheses. The technique must be used with extreme caution and its apparent simplicity can be deceptive. For example, other variables may contribute more to a causal relationship than the ones under investigation. Also, the relative stability of a measure, such as the lagged correlations for two panels of one variable, may increase in reliability from the first panel to the second. Thus, this variable may show up as an effect rather than a cause. To rule out a reliability shift explanation, a researcher might have to correct the cross-lagged panel correlations for attenuation when reliability decreases over time. The larger the lag between panels for the time-series (panels one and two), the lower the correlations both for auto-correlations and cross-correlations. In the above discussion of the two panels, it is assumed that the effects of a causal state or event decrease over time. Addition of a third panel might eliminate some rival hypotheses and help to determine if the causal relationship is unidirectional or bidirectional, thus suggesting other causal interpretations.

Threats to Internal Validity. Since the cross-lagged technique does not allow for the manipulation of the independent variable or random assignment of subjects to comparison groups, we cannot say with any certainty that the threats to internal validity of history, maturation, testing, or mortality are controlled. A third variable or event occurring between the first and second measurement of the two variables might influence their relationship or provide a plausible explanation of the relationship under consideration by the researcher. In such cases, a researcher would have difficulty inferring that one variable has a causal effect on a second. The time lag between the measurement of both variables allows for the threat of maturation as a rival explanation. For example, in a study to determine whether intelligence causes achievement (Excerpt 14.6), maturation could account for the relationship between intelligence and achievement test scores. The cross-lagged design does not control for testing as internal threat to the validity of a study. In Excerpt 14.6 the relationship between achievement and intelligence could be explained by the effect of testing. The last possible threat to the internal validity of the cross-lagged design is mortality. If a great number of subjects drop out of the research, mortality could be a plausible explanation of the relationships between the variables rather than the causal influence one variable might have on another.

The threat of selection does not pose a problem for this design. However, if subjects are selected on the basis of extreme scores or dichotomous data, the phenomenon of statistical regression to the mean might be reflected in the correlational coefficients between the lagged and unlagged variables. In the example of Excerpt 14.6 extreme scores on intelligence and achievement were not used; therefore, statistical regression does not threaten the internal validity of the study. Instrumentation should not be a problem in this design if the same procedures and instruments (or alternate forms) are used for both panels of measurement and if the reliability of the instruments or procedures is high. In Excerpt 14.6 the measures of both achievement and intelligence are quite reliable, thus reducing instrumentation as a threat to the internal validity of the study.

Statistical Analysis. A researcher uses a variation of the t test to determine whether two correlations are significantly different. With respect to Figure 14.8 the following correlations are compared: (1) test-retest or lagged autocorrelations ($r_{O_{A_1}O_{A_2}}$ vs. $r_{O_{B_1}O_{B_2}}$), (2) the synchronous or unlagged correlations between variables A and B ($r_{O_{A_1}O_{B_1}}$ vs. $r_{O_{A_2}O_{B_2}}$), and (3) the cross-lagged correlations ($r_{O_{A_1}O_{B_2}}$ vs. $r_{O_{B_2}O_{A_1}}$). The unlagged correlations and the lagged autocorrelations are checks for reliability, and the cross-lagged correlations provide information regarding the direction of the

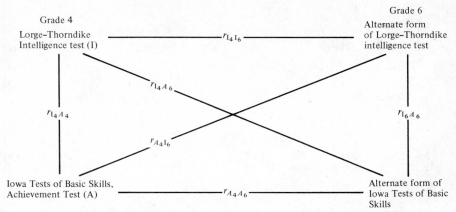

FIGURE 14.9. *Diagram of the cross-lagged panel analysis design presented in Excerpt 14.6.*

possible causal relationship. Thus, if there is a causal sequence between two variables, there will be an inequality between the cross-lagged correlations.

Example of Design. The journal article that we have selected to illustrate the cross-lagged panel analysis design shows an attempt to determine whether intelligence causes achievement. Intelligence and achievement test scores administered during the fourth and sixth grade for 5495 children represented the data for this study. Excerpt 14.6 describes the results and Figure 14.9 is a diagram of the design for this study. Scores on the Lorge-Thorndike Intelligence Test and the Iowa Tests of Basic Skills (achievement test) were the two variables for this study. Note that both tests were administered in the fourth and sixth grades. Figure 14.9 illustrates the six correlation coefficients in this study. The authors used two statistical analyses. They tested the significance of the unlagged IQ and achievement tests. Also, difference between the cross-lagged correlations was statistically tested with a corrected form of the t test between correlations.[4]

Although there are many variations of the five types of quasi-experimental designs, the designs discussed in this chapter provide illustrations of their application to a variety of research situations. In the next chapter we will present several additional forms of the time-series designs that are applied to single subjects ($N = 1$). These extensions of time-series designs are considered because of their increasing popularity and because many investigators consider them legitimate methods of research.

[4] For another recent example of the cross-lagged panel analysis, see J. L. Dyer and L. B. Miller. Note on Crano, Kenny, and Campbell's "Does intelligence cause achievement?" *Journal of Educational Psychology*, 1974, 66, 49–51.

EXCERPT 14.6 PORTION OF THE RESULTS
OF CROSS-LAGGED PANEL ANALYSIS

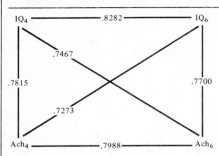

FIGURE 2. *Cross-lagged panel results of the interrelations of intelligence and achievement test composite scores.*

Additional confirmatory information can be obtained by considering the composite score relationship of the synchronous un-lagged IQ and achievement tests. The correlation between these contiguously administered tests is positive and significant at both measurement periods ($r_{I_4A_4} = .7815$, $r_{I_6A_6} = .7700$; $p < .001$, $df = 5493$ for both correlations).[6]

Both of these findings serve to render implausible the rival hypothesis that a negative relationship exists between intelligence and achievement as measured on the tests employed in this investigation. We are thus in a position to investigate the remaining possibilities, namely, that the causal relationship is predominately in the direction of intelligence affecting later achievement, or, of achievement influencing later intelligence.

A number of analytic options is available in this study, but none is completely desirable. One of the most obvious of these consists of a comparison of the crossed and lagged composite score correlations ($r_{I_4A_6}$, $r_{A_4I_6}$). Again, we must emphasize the probable reciprocal causal dependence between these two dimen-

sions. It seems highly probable that both of the possible causal relationships operate to some extent, in a type of feedback system. The test between the cross-lagged coefficients simply enables some estimate concerning the preponderant cause-effect relationship to be made. The pattern of relationships necessary for this comparison is presented in Figure 2. The cross-lagged correlations are both positive and substantial, and suggest a feedback system in which both operations affect one another to a great extent. A comparison of the cross-lagged correlations indicates, however, that the predominant causal sequence is that of intelligence causing later achievement. A test of this inequality revealed that the obtained difference between $r_{I_4A_6}$ and $r_{A_4I_6}$ was statistically significant ($t = 2.941$, $df = 5492$, $p < .01$, two-tailed).[7] For the total group of respondents, then, the preponderant causal sequence is apparently in the direction of intelligence directly predicting later achievement to an extent significantly exceeding that to which achievement causes later intelligence.

[6] The relationship between the intelligence and achievement tests employed in the present investigation appears to be consistent over time. A test of significance between these two correlations disclosed that the null hypothesis that $r_{I_4A_4} = r_{I_6A_6}$ could not be rejected ($z = 1.52$, $p > .05$).

[7] This test was based upon a correction of the usual t test between correlations, suggested by Pearson and Filon (1898), which takes into account the indirect correlation between the arrays under comparison, which are modified by the four other relevant values (see also Peters and Van Voorhis, 1940, p. 185).

SOURCE: W. D. Crano, D. A. Kenny, and D. T. Campbell. Does intelligence cause achievement?: a cross-lagged panel analysis. *Journal of Educational Psychology*, 1972, *63*, 258–275. Copyright 1972 by the American Psychological Association. Reprinted by permission.

REVIEW TERMS

continuous treatment	planned treatment
cross-lagged	quasi-experimental design
ex post facto intervention	self-selected group
generalizability	separate-sample pretest-posttest
group-replication	staggered treatment
group-repetition	subject reactivity
instability of measures	temporary treatment
intact group	time series
integrated moving average model	unlagged measures
lagged measures	
multiple treatment	
nonequivalent control group	
panel	

$$\overline{O\,O}$$
$$O\,O \quad X \quad O\,O \quad Y \quad O\,O$$
$$O\,O \quad X \quad O\,O \quad Y \quad O\,O$$
$$O\,O \quad Y \quad O\,O \quad X \quad O\,O$$
$$XO_5 \, XO_6 \, XO_7$$

REVIEW QUESTIONS

1. What are two variations of the nonequivalent control group design?
2. What threat to internal validity is most probable for unplanned time-series studies?
3. Diagram a time-series design for a single group with continuous and withdrawn single treatment.
4. How many groups are exposed to treatment in the separate-sample pretest-posttest design?
5. How many panels are there in the example study of the correlation analysis design?
6. What possible threat to internal validity is especially associated with time-series designs?
7. What are two types of measurement or observation that a researcher can use with time-series designs?
8. Change in the composition of a group is associated mostly with which source of invalidity?
9. What does $r_{O_{A_1} O_{B_1}}$ represent?
10. What statistical test is used with time-series designs?
11. The degree of "quasi-ness" of a design depends on how much _____ a researcher has.
12. The cross-lagged technique does not manipulate the _____ or _____ a subject to a comparison group.
13. What do the guides for evaluating a research design refer to?
14. What threats to validity are controlled for by using subjects in field or natural settings?
15. Which design discussed in this chapter is least effective for establishing a causal relationship?

CHAPTER 15

DESIGNS FOR APPLIED BEHAVIOR ANALYSIS

The research designs discussed in Chapters 12–14 make comparisons of the dependent variable between two or more groups of subjects. This type of design is sometimes referred to as a *between-group* or *comparative group study* or an *intersubject comparison*. Between-group designs always compare group means or medians by applying inferential statistics to the data of the dependent variable. In this chapter we present designs that compare an individual's behavior under at least two different conditions or time periods. We call this type of design a *within-subject study* or *intrasubject comparison*. As we will see, within-subject designs are similar, and in some cases identical, to the time-series designs presented in Chapter 14. For example, during successive time periods the subject is treated differently. For at least one of the periods, the subject is exposed to the independent variable. In contrast, however, inferential statistical tests are not always applied to the data of applied behavior analysis designs.

The applied behavior analysis designs are based on reinforcement theory. A major principle of this theory is that specific environmental events maintain or change (reinforce) the behavior of an individual. For example, an environmental event might be social approval, grades, playing a game, or tokens. A researcher might be interested in studying the functional relationship between adult social approval (independent variable) and the behavior (dependent variable) of one subject or a small group of subjects. The researcher strives to find out if the independent variable maintains or causes a change in the behavior. Also, researchers using these designs believe that the behavior of two individuals often differs in the way they respond to the same environmental event. For this reason behavior modifiers sometimes study the behavior of only one individual (or small group). The results may indicate that two children do not respond to adult social approval in the same way. One child might be more cooperative after adult social approval and another child might be less cooperative after approval.

All the designs that we present in this chapter may refer to any one of the following kinds of research: contingency management, operant conditioning, behavior modification, functional analysis of behavior, experimental analysis of behavior, or applied behavior analysis. The applied behavior analysis research designs have been used to solve many applied problems in a variety of natural settings. Usually this type of research aims at benefiting the subject of study in his natural setting (home, regular classroom, special classroom, guidance clinic, mental health center, mental hospital, community, or campground).

Applied behavior analysis designs permit a researcher to study the behavior of an individual subject (or small group of subjects) over successive observations rather than to compare the means between groups of individuals. The behavior of the individual is monitored continuously during each time period and between conditions (before and after introducing the independent variable) of the study. With these time-series measurements or observations of behavior, it is possible for a subject to serve as his own control. Thus, the procedures a researcher uses to collect data are very important to the validity of the design. Also, when an inferential statistic is not used to analyze the data, the techniques used to graph the data are important for interpreting the results of a study.

COLLECTING AND GRAPHING DATA

Behaviors

Before a researcher collects the data for a study, he must first define the behavior that he wishes to observe. Simple identification may be a satisfactory definition for some types of behavior under study, such as crawling, climbing, crying, attending school, sitting down, thumbsucking, solving math problems correctly, and bed wetting. However, other types of behavior, such as oppositional, cooperative, or gross motor activities, require a precise definition. For example, the authors of Excerpt 15.1 provide a clear statement

---

EXCERPT 15.1 DEFINITION OF BEHAVIOR TO BE STUDIED

Generally, study behavior was defined as orientation toward the appropriate object or person: assigned course materials, lecturing teacher or reciting classmates, as well as class participation by the student when requested by the teacher.

SOURCE: R. V. Hall, D. Lund, and D. Jackson. Effects of teacher attention on study behavior. *Journal of Applied Behavior Analysis*, 1968, *1*, 1–12. Copyright 1968 by the Society for the Experimental Analysis of Behavior, Inc.

---

of how they define study behavior in their research on first- and third-grade children.

Observation Techniques

Once the behavior under study is identified or defined, then the researcher must decide how to observe the behavior. Three techniques for observing and measuring behavior are commonly reported in journal articles. The first technique is to record the *frequency* of the behavior occurring within a specified time period. For example, a researcher might want to know how many math problems a student can solve correctly during a 50-minute class period. The second technique is to record the *duration* of the behavior, that is, the amount of time a subject engages in a particular behavior during a session or time period. For example, with a stop watch an observer can record the cumulative amount of time a child stays in his seat. The third technique, called *time sampling* observation, is to record the occurrence or nonoccurrence of a behavior once during a predetermined periodic interval. A description of this technique appears in Excerpt 15.2.

EXCERPT 15.2 DESCRIPTION OF TIME SAMPLING OBSERVATION

Each observer had a clipboard, stop watch, and rating sheet. The observer would watch for 10 seconds and use symbols to record the occurrence of behaviors. In each minute, ratings would be made in five consecutive 10-second intervals and the final 10 seconds would be used for recording comments. Each behavior category could be rated only once in a 10-second interval.

SOURCE: C. H. Madsen, Jr., W. C. Becker, and D. R. Thomas. Rules, praise, and ignoring: elements of elementary classroom control. *Journal of Applied Behavior Analysis*, 1968, *1*, 139–150. Copyright 1968 by the Society for the Experimental Analysis of Behavior, Inc.

Depending on the purpose and nature of the study, a time interval may be 10, 20, 30, or 60 seconds, and the technique may be used to observe several behaviors of a subject. Trained observers often use a recording sheet to simplify their data collection (see Excerpt 15.3). The sheet is divided into rows, one for each behavior, and columns for each 10-second interval. Generally, symbols are used to record the observed behavior in the appropriate boxes.

EXCERPT 15.3 OBSERVER RECORDING SHEET

SECONDS

10	20	30	40	50	60																		
N	N	N	N	N	N	N	S	S	S	N	N	S	S	S	S	N	N	N	N	N	N	N	N
		T	T	T							T											T	T
			/	/				/			/												

Row 1 N Non–Study Behavior S Study Behavior

Row 2 T Teacher Verbalization directed toward pupil

Row 3 / Teacher Proximity (teacher within three feet)

FIGURE 1. *Observer recording sheet and symbol key.*

SOURCE: R. V. Hall, D. Lund, and D. Jackson. Effects of teacher attention on study behavior. *Journal of Applied Behavior Analysis*, 1968, *1*, 1–12. Copyright 1968 by the Society for the Experimental Analysis of Behavior, Inc.

Instead of separate rows for each behavior, a recording sheet may have just one box for each time interval in which the symbol of the behavior occurring within that interval is placed. If a researcher is interested in the relationship of two behaviors, he might divide the box for each time interval so that he can record the symbol for a behavior of the subject (dependent variable) in the top half and the symbol for the occurrence of the independent variable in the bottom half. On such a recording sheet, for example, the study behavior (dependent variable) and teacher verbalization (independent variable) would be noted in one box when they occurred within the same time interval.

When several categories of behaviors are used in a study, the researcher will sometimes provide instructions to the observers for recording particular behaviors. The instructions may include time requirements, such as a minimum duration of the behavior during a time interval before it should be recorded. Since some behaviors are more important to the researcher than others, he may provide instructions to give priority to specific behaviors if more than one behavior occurs at the same time or during the same time interval. In an elementary classroom, for example, if a child is rocking his chair (motor behavior) and is talking loudly without permission (disruptive verbalizations) at the same time, the observer might be instructed only to record disruptive verbalizations.

There are many variations of recording techniques. However, defining behaviors, describing the setting, determining experimental procedure(s), devising observational techniques, training observers, and achieving high observer reliability must be completed before the researcher can begin to collect the data. The above illustrations should give you some general notion of what is typically reported in journal articles.

Observer Reliability

Observer reliability is another important part of applied behavior analysis research, but some authors neglect to report observational procedures or observer reliability in their articles. Authors should report *interobserver reliability* (consistency of observation) for the entire study or at least for all conditions of the study. Before the researcher collects the data for his study, he trains observers to use his time sampling techniques with the categories of behavior under study and then assesses reliability. Observation data is collected from two or three observers during the same session in the setting in which the study occurs; then the researcher determines the degree of agreement between observers. The reliability between observers may be reported for each behavior, for all behaviors combined, for several numbers of checks throughout the entire session of the research, or for all the sessions of the research. Examples of reporting interobserver reliability are illustrated in Excerpts 15.4 and 15.5. In Excerpt 15.4 the reliability checks are reported as

---

EXCERPT 15.4 REPORTING OBSERVER RELIABILITY

Average reliabilities (based on 11 checks) for Experiment 1 and Experiment 2 were as follows: attending behavior 92% (Range 89 to 95%), not attending behavior 91% (Range 85 to 99%), disruption 95% (Range 92 to 100%), and token 98% (Range 96 to 100%). Ten reliability checks were also computed on test scoring 99% (Range 98 to 100%) and data recording 99% (Range 97 to 100%).

SOURCE: D. E. Ferritor, D. Bucholdt, R. L. Hamblin, and L. Smith. The noneffects of contingent reinforcement for attending behavior on work accomplished. *Journal of Applied Behavior Analysis*, 1972, 5, 7–17. Copyright 1972 by the Society for the Experimental Analysis of Behavior, Inc.

---

the average percent (with the range) of agreement for each behavior on test scoring and on data recording. In Excerpt 15.5 the consistency of observations is reported as the mean percent (with the range) of agreement for each subject over all 12 sessions and for all behaviors. The author did not report separate reliabilities for each behavior as the authors did in Excerpt 15.4. The authors in the first example did not determine reliability for all sessions (56), but checked on agreement 11 times for the behaviors and 10 times for test scoring and data recording.

EXCERPT 15.5 REPORTING OBSERVER RELIABILITY

For S1, the mean percent of agreement overall 12 duplicated sessions was 86%, with a range of 78 to 96%. For S2, average agreement was 93%, with a range of 84 to 96%.

SOURCE: V. E. Pendergrass. Timeout from positive reinforcement following persistent, high-rate behavior in retardates. *Journal of Applied Behavior Analysis*, 1972, 5, 85–91. Copyright 1972 by the Society for the Experimental Analysis of Behavior, Inc.

The most common method for computing reliability is the percent of agreement, where the number of time intervals of agreements is divided by the total number of agreements plus the total number of disagreements. Researchers sometimes report reliabilities as coefficients of correlation, either as r or rho (see Chapters 2 and 10). It is important for the researcher to report reliability for at least each experimental condition (e.g., baseline, treatment, etc.). Without a relatively high percentage (usually above 85 percent) of interobserver agreement or coefficient of correlation, it is difficult for the researcher to make an accurate statement about the behavior (dependent variable) being measured. For example, if data are compared for two conditions and there is low observer reliability, the researcher will not be certain whether the variation between the conditions is caused by the treatment variable or fluctuations in interobserver agreement. Thus, instrumentation is a threat to applied behavior analysis designs if observer reliability is not very high.

Graphing Data

Although parametric and nonparametric statistics are sometimes used to analyze the data, graphic presentation is the primary analytic tool used by the applied behavior analyst and the most frequently used method to present results. Three graphic forms are commonly used in journal articles. The first form, a line graph, plots the percentage of frequency of the behavior (dependent variable) along the ordinate of the graph as a function of the sessions (days or weeks) along the abscissa. Let us take as a hypothetical example a study involving teacher social approval (experimental variable) and study behavior (dependent variable). Figure 15.1 would be an appropriate line graph used for our hypothetical study. The vertical dash lines indicate when one condition was completed and another started. The first condition, baseline 1, was completed during the third week. The treatment was introduced

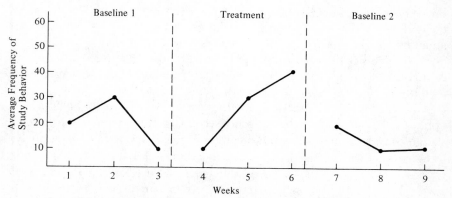

FIGURE 15.1. *Line graph of the average frequency of study behavior of one subject for all conditions of a study.*

during the fourth week. Notice that before the experimental variable was applied, the average frequency of study behavior is 20 for the first week, 30 for the second, and 10 for the third. During the treatment period, study behavior increased. For the third condition (baseline 2), study behavior decreased.

Instead of using line graphs, some authors present the same information with another graphic form, the *histogram*. Figure 15.2 is a histogram based on the same data that was plotted in Figure 15.1.

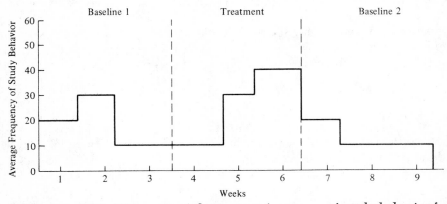

FIGURE 15.2. *Histogram of the average frequency of study behavior for all conditions (data from Fig. 15.1).*

The third form an author can use to present the data is a cumulative curve. This type of graphic representation presents values for the sums of all previous sessions. (Note that since the graph line is plotted on the basis of cumulative figures it continues on an upward angle. Increases within a time period are indicated by an increase of the upward angle (slope) and decreases by a decrease of the upward angle.) For example, in Figure 15.3, an author might plot the average frequency of study behavior as a function of weeks (this is the same data plotted in Figure 15.1). Notice that during the treatment condition, the line has a steeper slope than during baseline 1 or 2.

We have considered some of the procedures and techniques used in applied behavior analysis, but they are secondary to experimental or research design considerations. We will now present several applied behavior analysis designs that may be used by a researcher: (1) AB designs, (2) reversal strategies, (3) multiple baseline strategies, and (4) mixed designs using reversal and multiple baseline strategies. As in our discussions of other experimental designs, we will briefly consider possible threats to the internal validity of each design and a sample article. Again, we remind you that

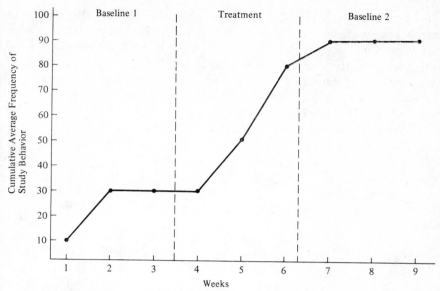

FIGURE 15.3. *Cumulative curve graph of the average frequency of study behavior for all conditions (data from Fig. 15.1).*

internal threats to validity must be assessed in relationship to the particular circumstances associated with a study and that journal articles have been selected for the purposes of illustrating specific designs, not as a reflection of the substantive significance of the research question being investigated. In the last section of this chapter we present a discussion of the variety of statistical procedures that are used with applied behavior analysis designs.

AB DESIGNS

The AB design has two conditions or time periods. The first condition, called the A period, is the *baseline* or pretreatment period. During the baseline period, a subject's behavior is observed for several sessions until it appears that the behavior is relatively stable. The second condition, called the B period, is the *treatment* condition. During this condition, the experimental variable (treatment or intervention) is introduced to see what effect it has on the behavior under investigation. As with the baseline condition, behavior is observed for several sessions until the effects of the treatment can be determined.

There are two variations of the AB design. The first variation is identical to the single group with a continuous single treatment time-series designs illustrated in diagram C of Figure 14.5. In Figure 14.5 X and Y are used to represent the treatment or intervention variables. In this chapter we use the letters typically employed by behavior analysts to represent the treatment variable (e.g., B, C, D, E, etc.). With this variation of the AB design, a researcher can make only one comparison, the series before with the series after treatment. The second variation of the AB applied behavior analysis design has a control group (or control individual). This variation, a multiple group with a continuous single treatment, allows the researcher to make two comparisons, intrasubject and intersubject. Both variations of the AB design can be used with single subject or group research. If the behavior of a group is studied, the researcher usually sums or averages the behavior of the group members, with the group likened to a single subject for purposes of presenting the data graphically. Also, if a comparison group (or individual) is used, a researcher may be able to randomly assign subjects to the control or treatment conditions. As with the time-series designs, group repetition or replication (or individual repetition or replication) can be used with the AB design.

We present two examples of the AB design. In the first study, token reinforcement was made contingent on level of reading for a 14-year-old boy. Excerpt 15.6 shows the boy's reading achievement in terms of grade level during the baseline period (under regular school training) and during the experimental training. After 8½ years of regular school training, the boy was reading on a grade level of 2. At the end of 4½ months of the experimental training in reading, the boy was reading on a grade level of 4.3.

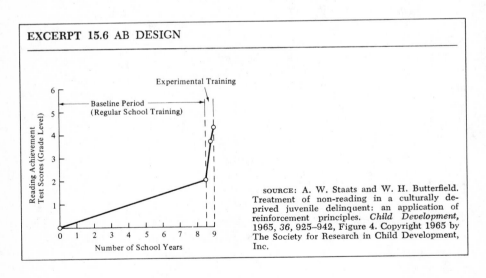

EXCERPT 15.6 AB DESIGN

SOURCE: A. W. Staats and W. H. Butterfield. Treatment of non-reading in a culturally deprived juvenile delinquent: an application of reinforcement principles. *Child Development*, 1965, *36*, 925–942, Figure 4. Copyright 1965 by The Society for Research in Child Development, Inc.

The second study is a good illustration of the use of the AB design with a control group. Teachers were trained in the systematic use of attention and praise (independent variable) in order to assess the effects on reducing disruptive classroom behavior (dependent variable) of four first-grade children. The subjects were problem children selected from three separate classrooms.

Excerpt 15.7 shows the amount of deviant behavior for all four children in the experimental group (Group E). The data presented illustrate the summing and averaging of the behavior for the group. Thus, the group is presented as a single subject, even though three of the subjects were from different classes. Four children were selected at random from the three classes to serve as a control group (Group CI). The data for the control group are also presented in Excerpt 15.7 and provide an additional compari-

---

EXCERPT 15.7 AB DESIGNS WITH A CONTROL GROUP

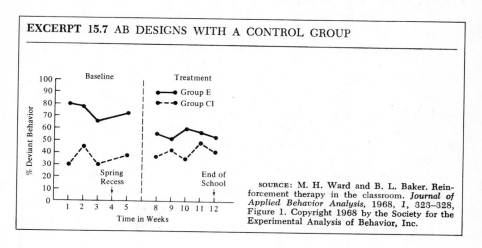

SOURCE: M. H. Ward and B. L. Baker. Reinforcement therapy in the classroom. *Journal of Applied Behavior Analysis*, 1968, *1*, 323–328, Figure 1. Copyright 1968 by the Society for the Experimental Analysis of Behavior, Inc.

---

son. The researchers made an intrasubject comparison of Group E between the baseline and treatment period; the percentage of disruptive behavior decreased. The disruptive behavior increased slightly from baseline to the treatment period for the control group. Although the percent of disruptive behavior was not the same for Group E and Group CI, the treatment appeared to be effective in reducing disruptive behavior. Another unique feature of this study is the use of replication, which is so important for generalization in applied behavior analysis designs. That is, the same procedures were applied to other subjects in different classrooms. This replication produced similar results—a decrease in disruptive behavior.

The first variation of the AB design controls for more threats to internal validity than the one-group pretest-posttest design. As with the single-group time-series designs, the AB design can possibly control for testing and mortality. Obviously mortality cannot pose a threat if one subject is used. Selection cannot logically (by definition) be a problem. Maturation, instrumentation, and regression might be a problem for the reasons mentioned in the discussion of the single-group time-series in Chapter 14. History is the major threat to the internal validity of the single-group or individual applied behavior analysis design.

The threats discussed in the last chapter to the multiple-group time-series design also apply to the AB control group design. Selection may be a possible threat because the groups cannot be considered equivalent unless there is a random assignment of subjects. History, maturation, and regression may also be possible threats. Mortality becomes a threat if there is a loss of subjects from one of the two groups. Tests are infrequently used by applied behavior analysis researchers and therefore testing is not likely to be a threat. If observer reliability is relatively low, instrumentation becomes a source of invalidity of both variations of the AB design. Since repetitive and replicative measurement can be used in both variations of the AB design, a reader should be aware of this distinction in judging the validity of study (see p. 311 for an explanation of why this distinction is so important).

REVERSAL STRATEGIES

Reversal strategies are the most frequently used designs in the professional literature of applied behavior analysis. These strategies are identical to the AB design with the addition of at least one more baseline period, A. The additional A or second baseline period is also referred to as the *reversal* condition or *withdrawal period*. During the reversal period, the researcher attempts to restore or reinstitute the conditions that existed during the first baseline. The researcher reverses the conditions or withdraws the experimental variable or procedure. Thus, the reversal condition tests the accuracy of the original prediction of what the frequency of the behavior would have been at some future time without the treatment variable. The original prediction is verified if the frequency or level of behavior during the second baseline is similar to the level during the first baseline condition. In using the reversal strategies, a researcher is attempting to establish that the treatment procedure was responsible for the change in the behavior under study. Thus, a reversal strategy allows the researcher to demonstrate experimental control if behavior changes only at the point when the treatment is applied. Reversal strategies can be used with an individual subject or a small group of subjects if the group data are treated as a single subject. There are many variations of the reversal strategy. We present the strategies that have most frequently appeared in the professional literature—three variations of the ABA design and five variations of the ABAB design.

ABA Designs

A researcher using the ABA design establishes the baseline condition, introduces the independent variable, and determines whether there is a change in the behavior. If there is a change in behavior (change between the baseline and treatment periods), the researcher withdraws (reverses) the experimental procedure. The independent variable is said to have an effect if the behavior during the reversal period changes to almost the same level as observed during the first baseline period. Thus, the researcher attempts to demonstrate that the behavior under study changes whenever the treatment procedure is applied. The ABA design enables him to compare the treatment period with the before and after periods.

The ABA design can be used by a researcher in three different ways. We have just described the first variation, the extension of the AB design to include the second baseline. To change our hypothetical example of contingent teacher social attention to study behavior to an ABA design, we would add a reversal period during which the teacher withdraws social attention. The graphic presentation of our data might look like Figure 15.1, 15.2, or 15.3. Diagram A in Figure 15.4 illustrates the first variation of ABA design. In the illustrations of the AB designs in Excerpts 15.6 and 15.7, the researchers could have added the withdrawal period by withdrawing the experimental procedures.

Two problems which may arise through the use of reversal are *irreversibility* and *unethical treatment*. Behaviors which do not reverse are said to be irreversible. For example, the higher reading level achieved by the subject in Excerpt 15.6 may later be maintained without token reinforcement, while in Excerpt 15.7 the deviant behavior observed may stay at the same lower level even when the experimental or treatment procedure is withdrawn. In the next section on multiple baseline designs, we present techniques that

A. ABA—TWO BASELINE PERIODS WITH ONE TREATMENT PERIOD.

OOO B $\overline{OOO}$ OOO

B. BCB—TWO TREATMENT PERIODS OF B WITH ONE TREATMENT OF C.

B $\overline{OOO}$ C $\overline{OOO}$ B $\overline{OOO}$

C. BAB—TWO TREATMENT PERIODS WITH ONE BASELINE PERIOD.

B $\overline{OOO}$ OOO B $\overline{OOO}$

D. ABAB—TWO BASELINE PERIODS WITH TWO TREATMENT PERIODS.

OOO B $\overline{OOO}$ OOO B $\overline{OOO}$

E. ABACABAC—MULTIPLE TREATMENTS WITH FOUR BASELINE PERIODS AND TWO TREATMENT PERIODS FOR B AND C.

OOO B $\overline{OOO}$ OOO C $\overline{OOO}$ OOO B $\overline{OOO}$ OOO C $\overline{OOO}$

F. ABABBCCA—MULTIPLE TREATMENTS WITH THREE BASELINE PERIODS, ONE COMBINED B AND C PERIOD, TWO B PERIODS, AND ONE C PERIOD.

OOO B $\overline{OOO}$ OOO B $\overline{OOO}$ BC $\overline{OOO}$ C $\overline{OOO}$ OOO

FIGURE 15.4. *Diagrams of six variations of reversal strategy designs.*

allow a demonstration of control to overcome the problem of irreversible
behaviors. A second problem with reversal designs is that it may be unethical
and undesirable for a researcher to withdraw treatment. That is, if the treatment is beneficial or the withdrawal detrimental to the subject in any way,
it may be unethical for the researcher to withdraw treatment for the sake of
his research findings.

The second variation of the ABA design is the BCB design. The researcher introduces the experimental procedures (B) without a baseline
period, modifies the treatment procedure (C) to see whether he can produce
further changes in the behavior, then withdraws the modified procedure
and restores the conditions (B) that existed during the first period (see
diagram B of Figure 15.4). The first period may still be referred to as a
baseline even if it is a treatment condition because it is the comparison period
for the second treatment period. An illustration of this kind of ABA design
appears in Excerpt 15.8. The authors investigated the effects of points (token

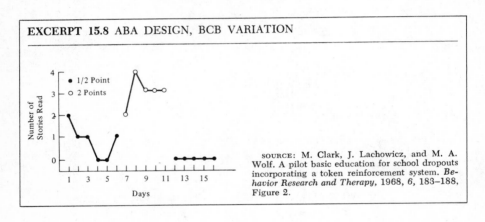

EXCERPT 15.8 ABA DESIGN, BCB VARIATION

SOURCE: M. Clark, J. Lachowicz, and M. A. Wolf. A pilot basic education for school dropouts incorporating a token reinforcement system. *Behavior Research and Therapy*, 1968, 6, 183–188, Figure 2.

system) on reading rate. During the first period, days 1–6, the subject was
given one-half point for every correct answer (treatment procedure, B). During the second period, days 7–11, two points were given for each correct answer (modification of the treatment procedure, C). The one-half point procedure was restored during the last five days of the study (B). Notice that
the increase in the number of stories read was a function of the change to
the two points obtainable for every correct answer. In using the second variation of the ABA design, the authors were attempting to strengthen or increase
the behavior under investigation. In Excerpt 15.8, we have labeled the three
periods BCB. We could have also represented the periods as ABA.

The third variation of the ABA design is the BAB design. The researcher
applies a procedure (B), withdraws the procedure or applies it in ways
which, according to reinforcement principles, should produce an ineffective
outcome (A), and then reinstates the original procedure (B). For example,

if a researcher has a teacher withdraw social attention entirely or change from contingent to noncontingent social attention during the second period (A period of BAB design), this extinction period might provide a control period which could be compared with the first period. Diagram C in Figure 15.4 illustrates the third variation of the ABA design. Excerpt 15.9 describes

---

EXCERPT 15.9 ABA DESIGN, BAB VARIATION

The study, a within-S design, consisted of three conditions: baseline (B), experimental (NT), and return to baseline (B2), in that order. Since a token reinforcement system had been used during the preceding academic year with eight of the Ss, the baseline conditions (B and B2) included dispensing of tokens; the experimental condition (NT) did not. During B and B2, the following conditions were in effect: (1) Social approval and tokens followed cooperative behavior and correct responses to the instructional materials. (2) Social extinction—i.e., no teacher response—followed incorrect responses, and inappropriate, but not disruptive, behavior. (3) A brief time-out period (removal from the classroom) followed disruptive behavior or refusal to comply with instructions. Although no tokens were dispensed during NT, the teachers continued to deliver approval and to administer the time-out procedure in the same fashion as in B and B2.

SOURCE: J. Birnbrauer, M. Wolf, J. Kidder, and C. Tague. Classroom behavior of retarded pupils with token reinforcement. *Journal of Experimental Child Psychology*, 1965, *2*, 219–235.

---

the use of this kind of ABA design. Tokens were given for correct academic responses during the first period (B), removed during the control or baseline period (A), and reinstated during the second treatment period (B). The A condition resulted in a decrease in accuracy on academic tasks. All of the above three variations of the ABA design can be used with a control group or control individual.

ABAB Designs

Two variations of the ABAB design appear frequently in journals: the ABAB with one independent variable and the ABAB with more than one independent variable. As suggested earlier in this chapter, group or individual repetition and group or individual replication can be used with ABAB designs. The ABAB design with one independent variable is an extension of the ABA design (see diagram D of Figure 15.4). After the second baseline period (A), the researcher reinstates the experimental procedure (B). Recall our hypothetical example in which we withdrew teacher social attention (second A) and, now, suppose that we reinstate it to observe the effects on study behavior (second B). As with the ABA design, the researcher is attempting to demonstrate that the dependent variable changes whenever the independent variable is applied. If a researcher can produce these effects by manipulating the independent variable (introducing it for two periods), then evidence for within-subject or intrasubject replication can be provided, that is, the subject's behavior consistently changes from the first baseline to the first treatment period, from treatment to the second baseline, and from the second A to the second B. In other words, the procedures were repeated and the results replicated within one subject (or a small group of subjects). This demonstration of consistent behavioral control is referred to by researchers as *experimental reliability*. The same variations discussed above for the ABA design can be used with the ABAB design (BCBC, BABA).

Two studies from the literature illustrate the use of the ABAB (or $A_1B_1A_2B_2$) design with one treatment procedure. In the first study the effects of individualized reinforcers were contingent on school attendance. "Deals" were made with each of the two subjects for such reinforcers as access to a pool hall, spending time with a girl friend, or weekend privileges. Each subject's reinforcer was provided for each day of class attendance. In Excerpt 15.10 we see that the number of days attending class fluctuated from

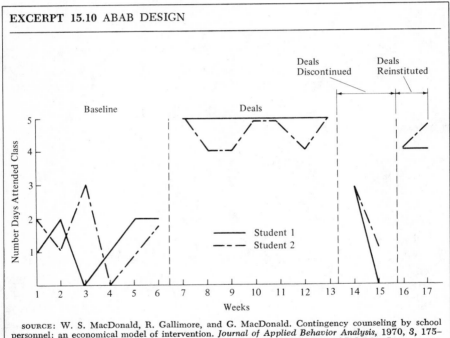

EXCERPT 15.10 ABAB DESIGN

none to three days a week for both students during baseline. During the treatment period (Deals), attendance increased to four and five days for both students. When the deals were discontinued, attendance decreased for both students. Finally, when deals were reinstated, attendance increased.

In our second example of the ABAB design with one treatment procedure, the study evaluated the relationship between two behaviors of a

5-year-old boy, enuresis (bed wetting) and oppositional behavior, as a function of time-out and differential attention by both parents. Note that both
time-out and differential attention by parents constitute a single treatment
procedure. Excerpt 15.11 shows the relationship between the two behaviors
as a function of each condition. Both oppositional and cooperative behaviors
are plotted as well as the number of enuretic episodes between sessions. The
week numbers are listed to correspond with each observation session so that
the time interval between sessions can be easily determined. The data in
Excerpt 15.11 demonstrate oppositional and cooperative behavior changes as
a function of parental differential attention and time-out procedures. Also,
enuretic episodes fluctuated consistently with the presence and absence of
the experimental procedures. The effects of treatment remained stable over
a lengthy observation period during the last condition of the study.

Diagram D in Figure 15.4 shows the ABAB design with one treatment
procedure, and it is also the diagram for our example studies in Excerpts
15.10 and 15.11. Excerpt 15.10 is a good illustration of between-subject
replication because a second subject was used. Excerpt 15.11 is a good illustration of applying treatment to only one behavior (oppositional) and

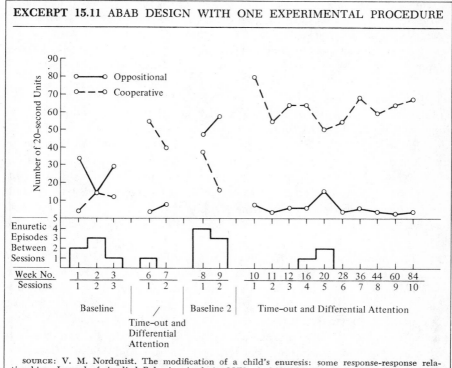

EXCERPT 15.11 ABAB DESIGN WITH ONE EXPERIMENTAL PROCEDURE

SOURCE: V. M. Nordquist. The modification of a child's enuresis: some response-response relationships. *Journal of Applied Behavior Analysis,* 1971, *4,* 241–247, Figure 1. Copyright 1971 by
the Society for the Experimental Analysis of Behavior, Inc.

demonstrating an identical fluctuation with a second behavior (enuresis). Both examples provide demonstration of within-subject replication. Also, in both examples the researchers can compare each period with every other condition or period (e.g., A_1 with B_1, B_2, A_2; B_1 with A_1, A_2, B_2; etc.). Behavior analysts believe that the $A_1B_1A_2B_2$ is an excellent design because evidence for the treatment effect is much stronger when it can be demonstrated twice instead of once. However, this design cannot handle the problem of irreversible behavior.

The second variation of the ABAB design that frequently appears in the literature is the case in which more than one experimental procedure is used. In our hypothetical example of teacher social attention to study behavior, we might add another treatment procedure. For example, we might introduce a point system as one of the experimental conditions. Depending on the purposes of our research questions, the research design could have a couple of variations. If researchers were interested in assessing the effects of each treatment procedure separately, they might use an ABACABAC design with one subject or group, in which C is the symbol for our second treatment procedure. This design is illustrated in diagram E of Figure 15.4. The design has four baseline periods and two treatment periods for B and C. A researcher can form another variation of this design with two subjects or groups. For example, an ABAC design might be used with one subject and an ACAB design used with the second subject. Varying the sequence of treatment for the second subject may increase the strength of design because all treatments have been replicated but not in the same time order or series.

A third variation of the ABAB design is the case in which a researcher wishes to assess the combined and separate effects of two independent variables. Excerpt 15.12 is an example study investigating the combined and separate effects of self-recording and teacher praise for one subject. "No slip" in the figure of Excerpt 15.11 means that slips to record study behavior were not issued to the subject on those days. The diagram for this variation of the ABAB design is shown in F of Figure 15.4. The underlining (BC) indicates that self-recording and teacher praise were combined as one treatment procedure during the fifth time series of the design. Self-recording (B) has two treatment periods and teacher praise (C) has only one period or time series.

Two other variations of the ABAB design are illustrated in Excerpts 15.13 and 15.14. In Excerpt 15.13 after baseline (A) the authors used classroom rules (B), then classroom rules with educational structure (BC), then classroom rules and educational structure combined with teacher's praising appropriate behavior and ignoring inappropriate behavior (BCD), then tokens with all of the above procedures (BCDE), then withdrawal of tokens only (BCD), then reinstatement of tokens (BCDE), and finally a follow-up (a variant of a token program; BCDF). The diagram of this might appear as follows: A B BC BCD BCDE BCD BCDE BCDF. The investigators wanted

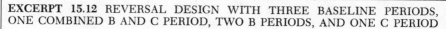

EXCERPT 15.12 REVERSAL DESIGN WITH THREE BASELINE PERIODS, ONE COMBINED B AND C PERIOD, TWO B PERIODS, AND ONE C PERIOD

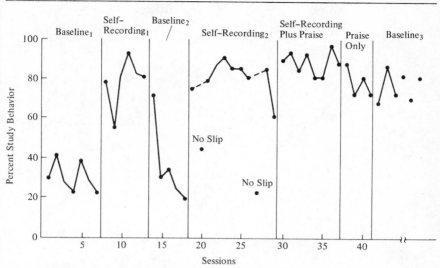

FIGURE 2. *Record of Liza's study behavior and/or teacher attention for study during: baseline₁—before experimental procedures; self-recording₁—Liza recorded study or nonstudy on slips provided by counselor; baseline₂—self-recording slips withdrawn; self-recording₂—self-recording slips reinstated; self-recording plus praise—self-recording slips continued and teacher praise for study increased; praise only—increased teacher praise maintained and self-recording withdrawn; baseline₃—teacher praise decreased to baseline levels.*

SOURCE: M. Broden, R. V. Hall, and B. Mitts. The effect of self-recording on the classroom behavior of two eighth-grade students. *Journal of Applied Behavior Analysis*, 1971, 4, 191–199. Copyright 1971 by the Society for the Experimental Analysis of Behavior, Inc.

to assess the effects of tokens (with B, C, and D) on the disruptive behavior (dependent variable) of 7 children in a second-grade class of 21 children. The researchers can make several comparisons in this study. Perhaps the most interesting contrasts are between the BCD period with the BCDE periods and the BCDF period. Also, contrasts can be made between A (baseline) and B, BC, and BCD periods, and the addition of praise and ignore (BCD) can be compared with the B and BC periods.

The last variation of the ABAB design is shown in Excerpt 15.14. This example study has three treatment variables and uses group-replication observation. The study presents an unusual analysis of three antilitter procedures applied to litter returned by an audience in a theater. The three experimental procedures were: (1) litterbags—a litterbag was given to each person as he entered the theater; (2) litterbags plus instruction—as the previous procedure with the addition of an announcement made during intermission, "Put your trash into the litterbags and put the bag into one of the trash cans

EXCERPT 15.13 REVERSAL DESIGN WITH SIX TREATMENT PROCEDURES

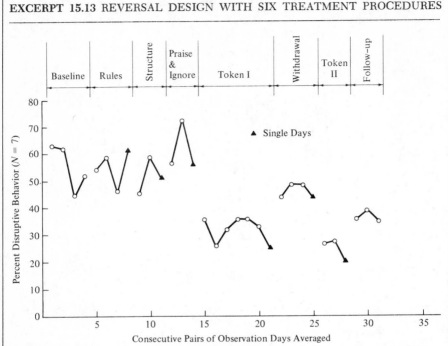

FIGURE 1. *Average percentage of combined disruptive behavior of seven children during the afternoon over the eight conditions: baseline, rules, educational structure, praise and ignore, token I, withdrawal, token II, follow-up.*

SOURCE: K. P. O'Leary, W. C. Becker, M. B. Evans, and R. A. Saudargas. A token reinforcement program in a public school: a replication and systematic analysis. *Journal of Applied Behavior Analysis,* 1969, *2,* 3–13. Copyright 1969 by the Society for the Experimental Analysis of Behavior, Inc.

in the lobby before leaving the theater"; (3) litterbags plus 10¢—Same as the litterbags procedure with the addition that each person was told, "If you bring a bag of litter to the lobby before leaving the theater, you will receive one dime in exchange." The litter collected from the usherettes' container was weighed along with the litter swept from the floor. The dependent variable was the percent of the total litter in the theater deposited in the trash cans. There were two baselines at the beginning of the study and three reversals (one after each of the three experimental conditions). The third procedure (litterbags plus 10¢) was the most effective, litterbags plus instruction next, and litterbags alone the least effective. Also note that during the last three reversal conditions (periods 4, 6, and 8), the percent of litter returned by the audience was almost the same percent as was returned during the first two baselines. The diagram for this variation of a group-replication ABAB design might appear as follows: A_1 A_2 B A_3 $\underline{BC}$ A_4 $\underline{BD}$ A_5. In this

EXCERPT 15.14 ABAB GROUP REPLICATION DESIGN
WITH THREE TREATMENT PROCEDURES

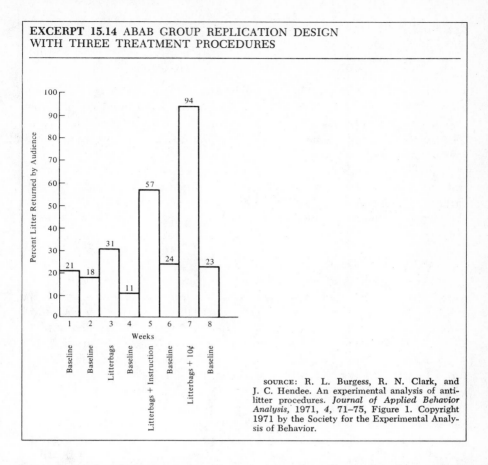

SOURCE: R. L. Burgess, R. N. Clark, and
J. C. Hendee. An experimental analysis of antilitter procedures. *Journal of Applied Behavior
Analysis,* 1971, *4,* 71–75, Figure 1. Copyright
1971 by the Society for the Experimental Analysis of Behavior.

example study, the investigators can compare the effects of each treatment
with any of the four baselines or all four combined. Also, differential effects
of each of the three treatments can be compared with each other.

The internal threats to the validity of reversal designs are the same as
the threats discussed for the AB designs. A single-group or subject reversal
design will probably not control for as many threats to validity as a multiple-
group or subject reversal design. Also, the single-subject multiple treatments
presented in this section are not as strong for controlling threats as the single
or multiple group (individual) with a single-treatment design. For the reasons mentioned in the last chapter, readers should be alert to what type of
measurement is used in evaluating an ABAB study (e.g., repetitive or replicative). One feature that the ABAB design has that the AB designs do not have
is within-subject replication. Within-subject replication (ABAB) may be a
stronger design for demonstrating the effects of treatment (two times), but
it does not guarantee control for all of the internal sources of invalidity. The
extent to which a study controls for the threats to validity depends a great

deal on the particular circumstances of the investigation, the specific design, and the measures that are used.

As indicated earlier, the use of reversal strategies may present problems. The behaviors produced may be irreversible so that withdrawal will have little or no effect. That is, after the experimental variable has been applied for a number of days, the behavior may be well established and may not change during the reversal or withdrawal period. For example, in Excerpt 15.6 the reading achievement test scores would not be likely to reverse to a lower level. Also, natural reinforcers in the subject's environment may maintain his reading skills even if the treatment is withdrawn. In such cases it is difficult for the researcher to establish a functional relationship. If there are problems of irreversibility or ethical considerations, or if it is undesirable to withdraw treatment, an appropriate alternative may be a multiple baseline strategy.

MULTIPLE BASELINE DESIGNS

The multiple baseline designs require concurrent observation (two or more baselines) across situations, behaviors, or individuals. For example, the treatment variable is applied to only one situation, behavior, or individual after baseline. If a change in behavior is demonstrated after the first application of the treatment variable, then the treatment variable is applied to a second behavior (or in another situation or for another individual), later to a third, and possibly to a fourth. Consider a hypothetical example. Suppose a researcher observes the study behavior of a subject during social studies class, language arts class, and math class each day for two weeks. The researcher could then apply the independent variable (teacher social approval) to study behavior first in the social studies class, later in the language arts class, and finally in the math class. The researcher can establish a relatively strong causal inference about the relationship between the independent and dependent variables if the behavior in each class changes from the baseline at the point when the independent variable was introduced.

As with the reversal designs, a multiple baseline design can be used with one individual or with a small group. Again, if a researcher is studying the behavior of a group of subjects, the behaviors of all the members of the group are usually summed or averaged and the group is considered as a single subject. Repetitive or replicative measurement can also be used with multiple baseline designs.

We present three types of multiple baseline designs: (1) across two or more situations, (2) across two or more behaviors or responses, and (3) across two or more subjects. Diagram A in Figure 15.5 shows the multiple baseline design that can be used with three different situations, individuals, or behaviors. The multiple baseline design is identical to the multiple group with staggered and continuous single treatment discussed in the last chapter (Figure 14.7, diagram D).

A. MULTIPLE BASELINE DESIGN APPLIED ACROSS THREE SITUATIONS, INDIVIDUALS,
OR BEHAVIORS.

Three different situations
or individual or behaviors
{
I   O O O B $\overline{\text{O O O}}$ $\overline{\text{O O O}}$ $\overline{\text{O O O}}$

II O O O   O O O B $\overline{\text{O O O}}$ $\overline{\text{O O O}}$

III O O O   O O O   O O O B $\overline{\text{O O O}}$

B. MIXED DESIGN—ABABC REVERSAL APPLIED ACROSS THREE SITUATIONS.

Situation I   O O O B $\overline{\text{O O O}}$ $\overline{\text{O O O}}$ $\overline{\text{O O O}}$ O O O B $\overline{\text{O O O}}$ C $\overline{\text{O O O}}$

Situation II O O O   O O O B $\overline{\text{O O O}}$ $\overline{\text{O O O}}$ O O O B $\overline{\text{O O O}}$ C $\overline{\text{O O O}}$

Situation III O O O   O O O   O O O B $\overline{\text{O O O}}$ O O O B $\overline{\text{O O O}}$ C $\overline{\text{O O O}}$

C. MIXED DESIGN—ABAB APPLIED TO BEHAVIOR I AND AB APPLIED TO BEHAVIOR II.

Behavior I   O O O B $\overline{\text{O O O}}$ O O O B $\overline{\text{O O O}}$

Behavior II O O O   O O O O O O B $\overline{\text{O O O}}$

FIGURE 15.5. *Diagrams of multiple baseline design and two mixed designs.*

Multiple Baseline Across Situations

In an across-situations multiple baseline design, the same behavior(s) or response(s) for the same individual(s) is observed across two or more situations (settings or environmental conditions). Baseline measures are obtained for the behavior in all situations. Then, the independent variable is applied first in one situation, later in a second, then in a third, and possibly in a fourth. The situation can vary in many different ways, for example, different time periods, types of activities, classes or teachers, locations, or composition of groups.

A portion of the record of the pupils returning late to their fifth-grade classroom after morning, noon, and afternoon recess is presented in Excerpt 15.15. The independent variable involved listing the names of students who were not late to class from recess on a chart entitled "Today's Patriots." The children were studying colonial patriotism and they discussed how they could be patriots living in the 1960s. The procedure was applied to the noon recess about session 14, to the morning recess about session 23 (note the dotted line), and to the afternoon recess about session 27. Excerpt 15.15 shows that the number of pupils late to class decreased for each situation at the point when the patriots chart was introduced.

Multiple Baseline Across Individuals

The second multiple baseline design involves measuring concurrently the same behavior, in the same situation, across two or more individuals. For

EXCERPT 15.15 MULTIPLE BASELINE ACROSS SITUATIONS

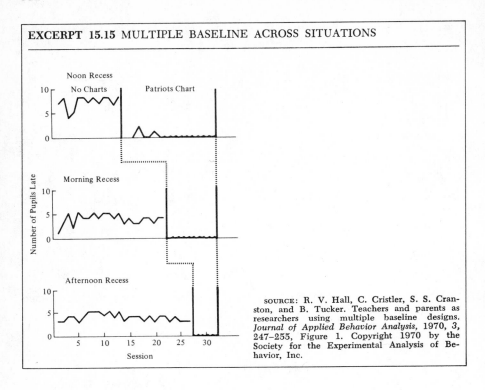

SOURCE: R. V. Hall, C. Cristler, S. S. Cranston, and B. Tucker. Teachers and parents as researchers using multiple baseline designs. *Journal of Applied Behavior Analysis*, 1970, *3*, 247-255, Figure 1. Copyright 1970 by the Society for the Experimental Analysis of Behavior, Inc.

example, Excerpt 15.16 shows the record of daily quiz grades for three high school French class students. The independent variable consisted of telling the subjects that whenever they earned a score of D or F on a daily quiz, they would be required to come after school for a tutoring session. Notice that the dependent variable in this design was letter grades. Changes in grades occurred for each student at the point when the independent variable was introduced (first to Dave, later to Roy, and finally to Debbie).

Multiple Baseline Across Behaviors

An across-behaviors design involves measuring concurrently two or more behaviors of the same individual or group in the same situation. For example, the authors in Excerpt 15.17 measured the percent of four undesirable meal-time behaviors of a hospital cottage of retardates. After a baseline condition was obtained for all behaviors, the independent variable (a time-out procedure) was applied first for stealing food, then to eating food with fingers, later to improper use of utensils, and finally to pigging (eating food by placing the mouth directly on it without use of fingers or utensils). The data plotted in the line graphs of Excerpt 15.7 are for 16 subjects. The

EXCERPT 15.16 MULTIPLE BASELINE ACROSS INDIVIDUALS

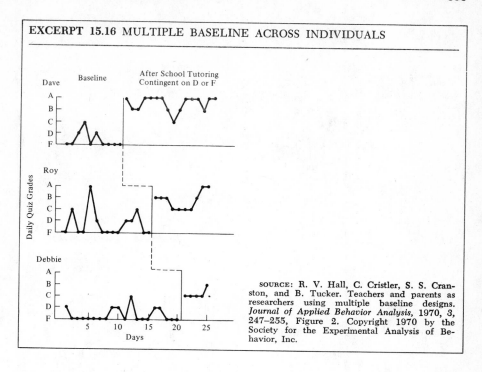

SOURCE: R. V. Hall, C. Cristler, S. S. Cranston, and B. Tucker. Teachers and parents as researchers using multiple baseline designs. *Journal of Applied Behavior Analysis*, 1970, *3*, 247–255, Figure 2. Copyright 1970 by the Society for the Experimental Analysis of Behavior, Inc.

graphs show a decline for each behavior at the point when the time-out from the meal was applied. Also, a graph of "total disgusting behaviors" is presented in the excerpt.

There may be a problem of generalization for a researcher using an across-behaviors multiple baseline design.[1] For example, the application of the treatment variable to one behavior may generalize to another behavior. Generalization can be good if the researcher desires this effect. However, generalization may not be the desired effect if the researcher wants to demonstrate experimental control of the independent variable.

The multiple baseline designs provide a within (behavior, individual, or situation) comparison in addition to an across (behavior, individual, or situation) comparison. Also, a before-and-after treatment comparison for each behavior, individual, or situation under study is possible. Thus, the multiple baseline designs that we have considered may be stronger designs than the reversal strategies discussed in the previous section because there are more comparisons. Selection can be a possible threat to internal validity for the multiple baseline across individuals, but it is doubtful that it could

---

[1] Used in this context, generalization does not mean the same thing as external validity.

EXCERPT 15.17 MULTIPLE BASELINE ACROSS BEHAVIORS

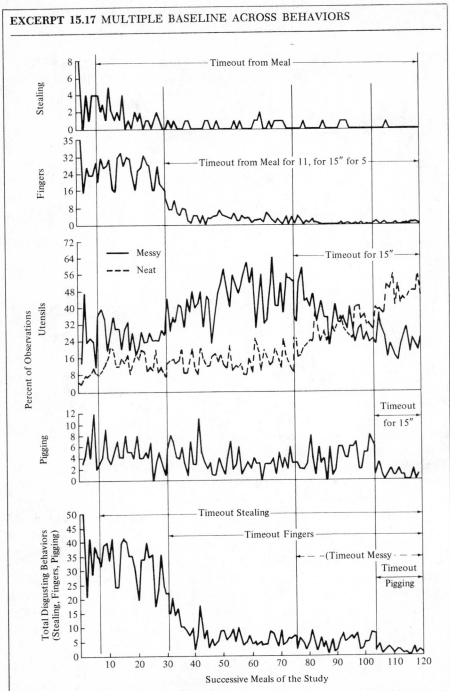

SOURCE: E. S. Barton, D. Guess, E. Garcia, and D. M. Baer. Improvement of retardates meal-time behaviors by timeout procedures using multiple baseline techniques. *Journal of Applied Behavior Analysis*, 1970, 3, 77–84. Copyright 1970 by the Society for the Experimental Analysis of Behavior, Inc.

be a threat for across situations or behaviors. History, maturation, and regression may also be sources of invalidity. However, these threats may be less likely with multiple baseline strategies than with the reversal strategies because there are more across and within comparisons. Unless a subject or subjects drop out, mortality is not a threat to the multiple baseline designs. Instrumentation and testing may be controlled by the multiple baseline strategies.

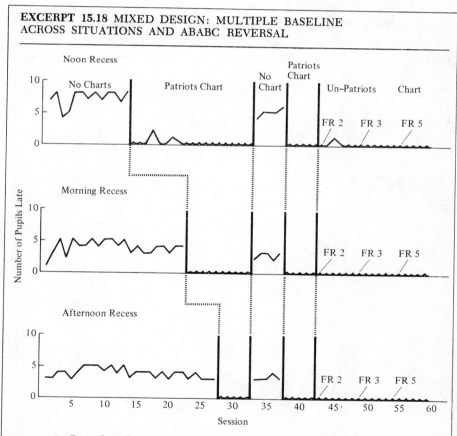

EXCERPT 15.18 MIXED DESIGN: MULTIPLE BASELINE ACROSS SITUATIONS AND ABABC REVERSAL

FIGURE 1. *Record of the number of pupils late in returning to their fifth-grade classroom after noon, morning, and afternoon recess: no charts—baseline, before experimental procedures; patriots chart—posting of pupil names on "Today's Patriots" chart contingent on entering class on time after recess; no chart—posting of names discontinued; patriots chart—return to patriots chart conditions; un-patriots chart—posting of names on "Un-Patriots" chart contingent on being late after recess (FR 2) every two days, (FR 3) every three days, and (FR 5) every five days.*

SOURCE: R. V. Hall, C. Cristler, S. S. Cranston, and B. Tucker. Teachers and parents as researchers using multiple baseline designs. *Journal of Applied Behavior Analysis,* 1970, 3, 247–255. Copyright 1970 by the Society for the Experimental Analysis of Behavior, Inc.

MIXED DESIGNS: REVERSAL AND MULTIPLE
BASELINE STRATEGIES

Researchers sometimes use a mixed design that combines reversal and multiple baseline strategies in one study.[2] As with the other designs, repetitive or replicative measurement can be used with a mixed design. There are many variations of mixed designs. We will present three studies to illustrate the designs most frequently reported in the literature.

The first study is an example of multiple baseline strategy and reversals applied to each situation. A portion of the study was presented in Excerpt 15.15, and we saw on that chart that the design was a multiple baseline across three situations (three different time periods). Now, looking at the complete chart in Excerpt 15.18, we see that the study used a reversal design in addition to the multiple baseline. A reversal period (no chart) was introduced from about session 32 to 37. The independent variable was restored between sessions 38 and 42 for all three situations. A second independent or treatment

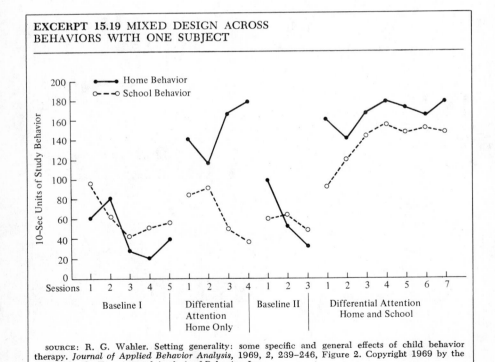

EXCERPT 15.19 MIXED DESIGN ACROSS
BEHAVIORS WITH ONE SUBJECT

source: R. G. Wahler. Setting generality: some specific and general effects of child behavior therapy. *Journal of Applied Behavior Analysis,* 1969, 2, 239–246, Figure 2. Copyright 1969 by the Society for the Experimental Analysis of Behavior, Inc.

[2] Mixed designs should not be confused with the mixed designs of repeated measures ANOVAs discussed in Chapter 6 or with the mixed models presented in Chapter 13.

variable (Un-patriots chart) was introduced during the last period of the study (between session 42 and 60). Thus a multiple baseline design was used (independent variable staggered across the three situations), and a reversal design (ABABC) was also used for each situation. Diagram B in Figure 15.5 illustrates the design for this study.

In the above example the authors collected data from the entire class. In our second example, the author used a multiple baseline and ABAB design with one subject, an 8-year-old boy. The independent variable was applied first in the home setting and later in the school situation. After the baseline for both situations, parental differential social attention (approval) to their son was made contingent upon study behavior in the home setting only. As we see from Excerpt 15.19, the experimental procedure increased study behavior at home but not at school. During the reversal (second baseline), study behavior at home decreased to about the same level as during the first baseline. Differential attention was reinstated (second B condition) for study behavior at home and also applied to study behavior at school during the last condition of the study. When the procedure was applied to study behavior at school by the teacher, study behavior increased to almost the same level as that behavior at home. This study is illustrated in diagram C of Figure 15.5.

The third example of a mixed design applied the treatment variable to

---

EXCERPT 15.20 DESCRIPTION OF EXPERIMENTAL PHASES FOR A MIXED DESIGN

The experimental design included both reversal and multiple baseline phases. The data were recorded separately during the reading and math periods providing the two baselines. The study was divided into four corresponding phases. A session in one class period corresponded to a session in the other class period in that they were recorded consecutively and on the same day.

I. MATH-Baseline, READING-Baseline.
For 10 sessions, the normal (baseline) rates of out-of-seat and talking-out behaviors of the class were recorded during the math and reading periods. The teacher carried out her classroom activities in her usual manner.

II. MATH-Game$_1$, READING-Baseline.
During the second phase the game was introduced during math but not during reading.

III. MATH-Reversal, READING-Game. In the third phase, the game was introduced during reading and withdrawn during math.

IV. MATH-Game$_2$, READING-Game.
Lastly, the game was reintroduced in math period and remained in effect during reading period. Both periods were treated as one extended period, thus using the same initial criteria of the last number of marks or five or fewer marks.

SOURCE: H. H. Barrish, M. Saunders, and M. M. Wolf. Good behavior game: effects of individual contingencies for group consequences on disruptive behavior in a classroom. *Journal of Applied Behavior Analysis*, 1969, *2*, 119–124. Copyright 1969 by the Society for the Experimental Analysis of Behavior, Inc.

two behaviors (talking-out in class and out-of-seat) across two situations (math period and reading period). The treatment was a good behavior game played by a fourth-grade classroom of 24 students and first applied to the math period. After the baseline rate of the inappropriate behaviors was obtained, the class was divided into two teams. Each out-of-seat and talking-out response by an individual child resulted in a mark being placed on the chalkboard, which meant a possible loss of privileges by all members of the student's team. In this way a contingency was arranged for the inappropriate behavior of each child, while the consequence (possible loss of privileges) of

EXCERPT 15.21 MIXED DESIGN: MULTIPLE BASELINE ACROSS SITUATIONS AND ABAB REVERSAL

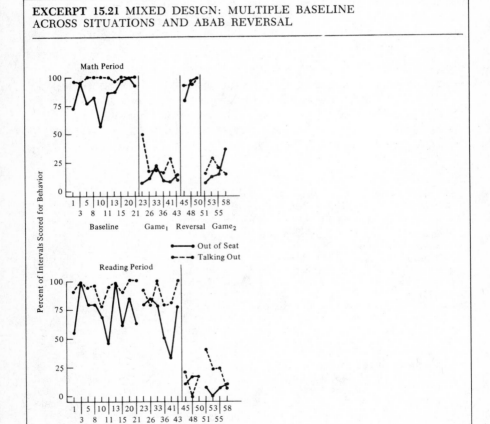

SOURCE: H. H. Barrish, M. Saunders, and M. M. Wolf. Good behavior game: effects of individual contingencies for group consequences on disruptive behavior in a classroom. *Journal of Applied Behavior Analysis,* 1969, *2,* 119–124, Figure 1. Copyright 1969 by the Society for the Experimental Analysis of Behavior, Inc.

the child's behavior was shared by all members of his team as a group. The privileges were activities which were available in the classroom. The experimental phases (periods) of this study are described in Excerpt 15.20 and the results are presented in Excerpt 15.21. Notice that during the math period the ABAB design was used and that later the game was introduced in the reading period. Both disruptive behaviors (talking-out and out-of-seat) decreased as a function of the treatment procedure. The behaviors in the reading period did not decrease until the treatment was applied at session 44.

This study also provides a good illustration (in the math period) of when researchers should change from one treatment condition to the next. After a relatively long baseline period or time series, the treatment procedure was applied for several sessions. After its effects had been demonstrated, the treatment was withdrawn. Then, when the effects of the reversal (withdrawal of treatment or second baseline) had been demonstrated (both behaviors increased to about the same level as the first baseline), treatment (Game$_2$) was reinstated. Diagram C in Figure 15.5 shows the design of this example study, but note that for this study the labels Behavior I and Behavior II would read Situation I (math period) and Situation II (reading period).

The number of threats to internal validity controlled by the mixed applied behavior analysis designs are possibly similar to the number controlled by the multiple-group time-series designs. The mixed designs have the potential for controlling several sources of invalidity. Selection may be a problem if two groups are used which are not randomly assigned to staggered treatment conditions and for research using replication measurement with groups (see Chapter 14). Of course, selection could not be a threat in the three examples described above because the same group or subject was used across different situations or behaviors. Neither history nor maturation is likely to be a source of invalidity when a single treatment procedure is staggered across two or more baselines. Regression may become a plausible hypothesis if extreme scores or measures of behavior are used. Mortality is certainly not a threat for the above three examples, but it could pose a problem if students were to drop out in examples one and three (Excerpts 15.18 and 15.21). Instrumentation and testing can be controlled by the mixed applied behavior analysis designs, but instrumentation may become a problem when there is relatively low observer reliability.

Mixed designs can provide within-subject replication (reversal strategy), which makes the study stronger in demonstrating the effects of treatment (twice instead of once). Thus, for researchers who want relatively good control for the sources of invalidity and also want within-subject replication, the mixed designs may be a good choice. However, the sources of invalidity must be evaluated with respect to each specific study. As we discussed with the time-series designs, the applied behavior analysis designs may have greater external validity (generalization) than true experimental designs because they are usually conducted in natural settings.

STATISTICAL ANALYSIS FOR APPLIED
BEHAVIOR ANALYSIS DESIGNS

Inferential statistical tests are not always used in applied behavior analysis studies. Graphic presentation of the data is the most frequently used technique for interpreting the results of a study. However, some behavior analysts use a variety of inferential statistical procedures, such as the Mann-Whitney U test, the independent samples t test, the one-way repeated measures ANOVA, and follow-up tests. In most cases, these statistical procedures are applied to studies that have group data, but some researchers feel that a one-way ANOVA can be used with one subject exposed to four conditions or time series $(A_1B_1A_2B_2)$. In this case the ANOVA is applied to determine if there is a significant F between the means of the conditions. If the F is significant, a follow-up test is used.

Although some behavior analysts believe that such tests as the ANOVA the independent samples t test, or the Mann-Whitney U test can aid in the interpretation of the treatment effects, others feel that these statistical procedures are inappropriate to use with time-series or applied behavior analysis designs. They consider the procedures inappropriate for two reasons. First, it may not be correct to compare the pre- and posttreatment means in order to determine the effects of treatment. For example, the t test and ANOVA do not consider (in the calculations) differences between successive observations in a time series. Thus, when a statistically significant difference is obtained with these tests, the significant difference may be a consequence of the statistical procedure rather than the effect of the treatment. A second reason for considering the statistical tests inappropriate is that the assumption of independent data cannot be met when using time-series or applied behavior analysis designs. Instead, time-series data should be analyzed with the integrated moving average model (see Chapter 14) because of the dependency in the observations of time-series data.

The time-series statistical procedures (e.g., the integrated moving average) have not been used very frequently with applied behavior analysis designs, but their popularity as alternatives to the t test and ANOVA may increase. We cannot resolve the issue of which statistical procedures to use with applied behavior analysis designs. Graphic presentation of the data for a study may be one of the most satisfactory techniques for helping to interpret the results. If behavior analysts seek additional aid for data interpretation from statistical procedures, they should be well acquainted with the available statistical techniques and the assumptions related to those techniques.

REVIEW TERMS

across behaviors
across individuals
across situations
baseline
between-group comparison
cumulative frequency curve
duration of behavior
experimenter reliability
frequency count
group replication
histogram
intersubject comparison
intrasubject comparison

mixed design
multiple baseline
observer reliability
reversal strategy
time sample
within-subject comparison
within-subject replication
withdrawal
AB
ABA
ABAB
B, C, D

REVIEW QUESTIONS

1. When a researcher uses a multiple baseline and a reversal strategy, the study may be referred to as being a _____ design.
2. The letter A represents which time period in an applied behavior analysis design?
3. What three observational techniques can a researcher use to collect data?
4. Diagram in letters a behavior analysis design that has within-subject replication.
5. What source of invalidity is observer reliability related to?
6. After observing three children daily for a period of two weeks, an intervention technique is applied first to one child, later to the second, and finally to the third child. What type of design was used in this hypothetical study?
7. For the hypothetical study in question 6, how many intrasubject comparisons can be made? How many intersubject comparisons?
8. When a researcher says that a reversal strategy was used, what was reversed?
9. What type of graph does not show decreases in the curve?
10. A researcher staggers the following sequence of conditions in two situations for six hospitalized patients: baseline, feedback, baseline, feedback and tokens, baseline, and tokens. Describe and diagram the design for this study.
11. How many independent variables are there for the hypothetical study in question 10?

12. What are two considerations that a behavior analyst must explore before using a reversal design in an investigation?
13. Two examples of AB designs were presented in this chapter. Which design has fewer threats to internal validity?
14. What is a reason for using a single subject in the experimental analysis of behavior?
15. One of the example studies discussed in this chapter used litter behavior as the dependent variable. What type of measurement was used in this study?

FOUR

SOME
POST
HOC
COMMENTS

In the previous three parts of this book, we have presented the typical format for journal articles or research reports, a variety of statistical techniques used with experimental research, and several types of experimental designs that a researcher can use. In this part of the book we present some reasons investigators may have for conducting experimental research and explore some questions that a reader can ask in evaluating a research report or journal article.

Some of these questions are related to the research report per se. Other equally important questions about the nature and direction of behavioral science research in general are also considered. These considerations are discussed in the context of two types of validity: ecological setting and replicative validity. Finally, on the basis of these questions we discuss possible future trends in behavioral science methodology. Our look into the future represents our bold idealism or plea for more systematic and cooperative efforts in behavioral science research.

CHAPTER 16

CURRENT AND FUTURE TRENDS IN BEHAVIORAL SCIENCE RESEARCH

WHY DO RESEARCH?

The statistical procedures and research designs presented in this book are the techniques for conducting experimental investigations of behavioral phenomena. However, they are merely tools and do not represent or explain the motives or reasons for engaging in experimental research. A behavioral scientist might have many reasons for conducting experimental research, and we present a few possibilities that might be most representative. These reasons are not mutually exclusive. Some researchers may do experimental studies for any one of the following reasons or for any combination of these and other reasons.

First, researchers may perform experimental studies to satisfy their scientific curiosity about behavior. For instance, a researcher might ask, "I wonder what would happen if juvenile delinquents were taught to use transcendental meditation?" or "What would happen to the quality of production if automobile workers were allowed to work in small groups and to decide their work schedule and working arrangements?" or "What would happen to the patterns of EEG waves if people were trained to experience positive thoughts?" While such questions could be guided by hypotheses and theories, most researchers engage in this type of experimental investigation without any formulated hypothesis based on a theoretical position.

Second, behavioral scientists may perform experimental studies in order to evaluate a hypothesis that is based on a theory. Researchers engaged in this type of experimental investigation often tend to emphasize the importance of systematized hypothesis testing and theory building, and their research usually elaborates on previous experimental work. For example, a researcher might test hypotheses that are based upon a particular theory of child development. Research based on hypothesis testing can provide insightful assumptions and behavioral evidence to support a theoretical position, but it can also result in a series of experimental studies that only generate a vast amount of trivia. Hypothesis testing to support a theory, like any other form of research, can be performed at many levels of contribution. Behavioral scientists who rigidly follow this approach to experimental research may underestimate the possible importance of research performed to satisfy the researcher's curiosity. Some important discoveries about behavioral phenomena have been made accidentally or serendipitously by curious researchers.

Third, researchers may perform experiments to explore the effects of a new method or technique on some dependent variable. Thus, a researcher might be interested in assessing the technique of self-monitoring on smoking or eating, or a researcher might examine the effects of a personalized system of instruction on academic achievement and later job success. Most of the hypothetical examples presented in Part Three of this book were examples of research to explore the effects of a method or technique on certain kinds of behavior (e.g., particular diet on weight reduction, human relations training on sales, etc.). Research performed for this reason can, at the same time, be based on the desire to satisfy curiosity, to test a hypothesis, or to elaborate on previous work.

Fourth, behavioral scientists may perform experimental studies to establish the existence of a behavioral phenomenon and to explore conditions under which this phenomenon occurs. This approach is sometimes referred to as exploratory research. For example, a researcher might set up an experiment to establish that a covert cognitive process called thought stopping occurs and then investigate under what conditions thought stopping could be used to facilitate change in thought patterns. Another investigator might wish to determine what behaviors are most typical at different levels of population density. Or what effect varying levels of population density have on delinquency, crime, suicide, or mental illness. Again, this reason may be combined with any one or all of the other reasons for doing experimental research.

Fifth, a researcher may perform experimental research in response to the demands of the sociological process itself in the behavioral sciences. Views of the role of the behavioral scientist, the structure of the organizations and institutions, the influence of the special interest groups, the policies of the professional journals, all contribute to motivating the behaviorial scientist to engage in research and, almost more important, to publish results. Academic institutions often focus on the number of articles published by a behavioral scientist rather than on the relevance and quality of his thought and investigations. The academic training for most behavioral scientists fosters individual experiments without providing satisfactory models for cooperative and systematic research.

Another contributing factor to this process is the institutionalization of the various disciplines within the behavioral sciences (e.g., anthropology, education, psychology, and sociology). These disciplines have their own professional organizations and their own professional journals. Even within one discipline there are divisions representing a variety of professional interests and often creating their own professional journals to publish their research. Quantity of research sometimes seems to reflect a greater concern for maintaining the identity and bolstering the prestige of a discipline or group than for expanding knowledge. The editorial policy of some professional journals also contributes to this sociological process. For example, some journals must devote most of the number of alloted pages for each issue to experimental research. The consequence of this sociological process is the proliferation of unsystematic, repetitive, and irrelevant experimentation. There seems to be little cooperative or systematic effort to investigate behavioral phenomena either across disciplines of the behavioral sciences, within various interest groups of a particular discipline, or among individual scientists. This situation can have a direct influence on the validity of a study.

OTHER KINDS OF VALIDITY

In the third part of this book we discussed internal and external validity to assist readers of journal articles or research reports in evaluating experimental studies. A reader may evaluate a study by asking a series of four questions about the experiment. First, is the research question or the hypothesis stated clearly? A description of the question under investigation should leave little doubt in the reader's mind about the purpose and intent of the study.

Second, does the research design for this study help to provide a reasonably clear answer to the problem under investigation? As we have seen, different designs are appropriate for different types of problems, and the researcher must choose the design which answers the question he poses. For example, a researcher may be more concerned about the effects of an independent variable on some behavior than about generalization of the results to a similar population. In such a case a quasi-experimental design may not provide a satisfactory answer to the research question because the results may be confounded with internal threats to valid inference. In other words, whether or not the treatment variable made a difference may be difficult to determine with a quasi-experimental design.

Third, is the statistical procedure appropriate for the research design used in the study? For example, multiple t tests as follow-up tests for an ANOVA used with one of the true experimental designs considered in Chapter 13 may not be appropriate.

Fourth, does the analysis of the data or the results provide a reasonably clear answer to the research question? At first glance, this may seem to be a rather simple question to ask of a journal article or research report because the answer may appear to be quite obvious. However, as Tukey suggests, bending the research question to fit the analysis should be avoided.[1] Therefore, if the research question cannot be answered by an analysis of variance, it is improper to insist on using the ANOVA.

All of the above questions refer to experimentation reported in most professional journals that contain experimental studies. These studies can also be evaluated in terms of the kinds of internal and external threats to validity that we presented in this book and that are implicit in the above questions. These threats to validity are helpful in evaluating one-shot experimentation. There are other, equally important kinds of validity of research. If researchers in the behavioral sciences seriously consider the following kinds of validity, some changes may occur in the methodology of behavioral science research.

Ecological Setting Validity

Ecological setting validity, a special case of ecological validity (see Chapter 12), focuses on the need to study behavioral phenomena in naturalistic settings. Some phenomena may be investigated satisfactorily in the laboratory, but generalizations from laboratory results may not be valid because of distinctive features of the laboratory setting. Furthermore, it may be impossible to study other phenomena outside their natural setting. As Brunswick suggests, it may be more important to sample situations or settings than subjects.[2] It is important for us to find out what the consequences are of such environmental conditions as poverty, population density, transient populations, communal living, or public education. It appears that society is asking such questions with a sense of urgency and the behavioral sciences have not provided satisfactory answers.[3] There seems to be a public plea for accountability in providing solutions to social problems. Tradi-

[1] J. W. Tukey. Analyzing data: sanctification or detective work? *American Psychologist*, 1969, *24*, 83–91.

[2] E. Brunswick. *Perception and the Representative Design of Psychological Experiments.* Berkeley: University of California Press, 1956.

[3] R. G. Barker. Wanted: an eco-behavioral science. In E. P. Willems and H. L. Raush (Eds.), *Naturalistic Viewpoints in Psychological Research.* New York: Holt, Rinehart and Winston, 1969, pp. 31–43.

tionally, behavioral scientists have assumed that these questions can be answered with existing and popular methodological procedures (paradigms). Possible future trends for behavioral science research might include new methods and tactics that will be conducted in a variety of environmental settings. The new methods will be derived from a variety of disciplines in the behavioral sciences such as sociology, anthropology, psychology, and education. These methods will also include a variety of techniques that focus on systematic observation as well as qualitative data from self-report and introspection of subjects in the naturalistic settings.

Shulman suggests that the basis for more ecological research should be task validity.[4] This kind of validity specifies that the operations required to perform a particular task in the laboratory must be similar to those required in the applied settings to which the results are to be generalized. For example, some investigations of verbal learning, concept formation, and problem solving conducted in laboratory settings may require entirely different kinds of mental processes and operations than are required in classrooms or other applied settings. There is often discontinuity between laboratory research and field research when the same behavioral phenomenon is supposedly examined by both. This is not to disparage laboratory research. However, there should perhaps be more cooperation between basic laboratory and applied research. Generalizations or inferences about the results obtained from laboratory research may not be valid because of the "sterile" setting in which the data were collected. We must be aware of features which distinguish the laboratory setting for a particular study from the field setting to which the operations of the independent variable are to be generalized.

McClelland alludes to this type of validity when he discusses the invalidity of intelligence tests and ability tests to predict job success.[5] The grades given by high school teachers, elementary teachers, and college professors fall into this type of classification of validity. Grades or the intelligence tests may not be valid because they do not measure the operations required for performance on the job or for job success. The mental operations, the kinds of skills, and the types of intellectual competencies that a learner may be required to perform for a grade or to demonstrate for an achievement or intelligence test may be incongruent with the type of behaviors, skills, and mental operations that are required in terms of job performance and success.

Ecological research may increase in the future also because of the growing acceptance of the suggestion that the population that has a problem may possess the best resources for describing and dealing with that problem. For example, ex-convicts, ex-drug addicts, or ex-mental patients

---

[4] L. S. Shulman. Reconstruction of educational research. *Review of Educational Research*, 1970, *40*, 371–396.

[5] D. C. McClelland. Testing for competence rather than for "intelligence." *American Psychologist*, 1973, *28*, 1–14.

may be the most appropriate personnel for successfully helping people with these problems in their environments. To explore this possibility, behavioral science research must examine these effects in naturalistic settings.

Ecological research may increase in popularity if behavioral scientists focus on preventive approaches for solving problems, rather than remediation. For example, behavioral scientists have not developed or focused on research for future environments (preventive) but have instead devised techniques for the repairing of existing settings (remediation). Preventive research would focus on designing an environment to decrease the number of problems typically created or generated by social factors within the environments. This kind of research endeavor would require that behavioral scientists anticipate what problems are generated by particular environmental settings. In terms of ecological research the question facing behavioral scientists is whether they will conduct research that uses a preventive or remedial approach.

Replicative Validity

Replication is one of the basic principles of competent research. Yet, if we were to survey the professional journals or textbooks about research in the behavioral sciences, we would find that behavioral scientists have paid little attention to this principle. Sidman is one of the few researchers who has treated the subject of replication at any length.[6] He distinguishes four methods or techniques for replication that involve varying the use of subjects. The first technique, *intergroup replication,* is the application of the same research efforts to two or more different groups. The second technique, *intersubject replication,* is the application of the same experimental design and independent variable to two or more subjects or individuals. The third technique, *intragroup replication,* is a repeated application of different processes or experimental conditions to the same group. The last technique, *intrasubject replication,* is the repeated measurement of the same subject for at least two or more experimental conditions.

Sidman also distinguishes two types of replication that can be incorporated into a research endeavor, direct and systematic.[7] *Direct replication* is the exact repetition of procedures in a replicated study. In this type of replication the researcher is attempting to demonstrate the generality of his procedures. Direct replication may also be applied to any one of the four techniques of replication described above. The other type of replication that can be incorporated into a particular research design is *systematic replication,* which is the replication of a study with slight modification of the independent variable. Systematic replication gives the researcher the same advantage as direct replication in demonstrating generality. Also, each successive replicative study or investigation yields additional bits of information

[6] M. Sidman. *Tactics of Scientific Research.* New York: Basic Books, 1960.
[7] Ibid.

that would not have been obtained if only direct replication were incorporated into the design. In other words, systematic replication might also be referred to as quasi-replication in which other techniques or research strategies are employed to investigate the same behavioral phenomenon.

Other possible future trends with respect to replication might be the use of staggered replication built into the same investigative effort. This is similar to the time-series staggered treatment discussed in Chapter 14. For instance, treatment could be staggered across different facets or segments of the environment (e.g., socioeconomic level, cities, ghettos, rural communities, rehabilitation centers) in which treatment is applied to one setting and later applied to another. In other words, staggered replication is an attempt to ascertain whether or not social or cultural changes have an influence on the dependent variable over time.

There is also a statistical rationale for either direct or systematic replication. Almost all research endeavors reported in behavioral science journals today rely heavily on statistical hypothesis testing. Tukey remarked that the use of statistical significance testing was never meant to serve as a substitute for replication.

> *The modern test of significance, before which so many editors of psychological journals are reported to bow down, owes more to R. A. Fisher than to any other man. Yet Sir Ronald's standard of firm knowledge was not one very extremely significant result, but rather the ability to repeatedly get results significant at 5%. Repetition is the basis for judging variability and significance and confidence. Repetition of results, each significant, is the basis, according to Fisher, of scientific truth.*[8]

It is common practice to assume that if results are significant at a given level, usually .05 or even at .01 or better, as they must be if they are to be accepted by many behavioral science journals, there is no need for replication of the experimental study. However, this may or may not be true. In the future studies may involve more replication across researchers, laboratories, populations, socioeconomic levels, and ecological settings. It is also possible to include replication in a study in terms of minor variations of not only the independent variable, but also the research techniques employed to investigate a specific behavioral phenomenon. If behavioral science research is to have any validity and to make any impact on some of our social problems, then replicated studies will have to be conducted across ecological or environmental settings and become an integral part of the research design.

Thus, replicative validity refers to the degree that an experimental study is repeated either directly or systematically. Replication can be per-

---

[8] Tukey. Analyzing data, p. 85.

formed across environmental settings, across populations, across research designs or methodologies, or across laboratories. The possibility of replicating a study across designs or methodologies has some advantages over the traditional one-shot experimental studies. This is not to disparage the traditional and somewhat rigid experimental and control group design, but to encourage more systematic attempts at data verification and behavioral phenomena verification in terms of a variety of methodologies, either experimental or descriptive. The systematic approach could focus on a variety of related behavioral phenomena. To conduct this kind of research—a systematic attempt to use a variety of methodologies or designs and to replicate these paradigms over varying tasks and conditions—would require a tremendous amount of cooperative effort. The use of a variety of paradigms to focus on a single behavioral phenomenon has more replicative validity than the typical experimental and control group study that is so popular.

The physicist Werner Heisenberg proposed half a century ago the principle of uncertainty that points out that the very act of observing disturbs the system.[9] The Heisenberg principle also suggests that rational science may be limited in its ability to comprehend nature and at best it can only arrive at certain statistical probabilities about behavioral phenomena. Therefore, the incorporation of a variety of methodologies (paradigms) on a common investigative effort of a particular behavioral phenomenon may provide more replicative evidence than isolated experimental studies. If behavioral scientists seriously consider the above kinds of validity and reconsider some of the reasons for conducting experimental research, some changes in the methodology of behavioral science research may be imminent.

FUTURE TRENDS IN METHODOLOGY

In an address to the American Psychological Association in 1956 Robert Oppenheimer gave a warning that might be heeded by all disciplines in the behavioral sciences. He said, "The worst thing that psychology might do would be to model itself after a physics which is not there any more and which has been outdated."[10] Most experimental researchers in behavioral science have clung to the habit of conducting research in the classical experimental and control group mode. This research mode was modeled on the so-called pure or physical sciences. However, the alleged objectivity of these methodological procedures is questionable. Observation in the behavioral sciences is an active, not a passive, event. The events being observed are also active and dynamic. Failure to recognize other procedures could retard development of a more inclusive behavioral science methodology. Behavioral scientists should liberate themselves from the old conceptualization of research and existing methodological biases that are current in the behavioral sciences today.

[9] W. Heisenberg. *Physics and Philosophy*. New York: Harper & Row, 1958.
[10] R. Oppenheimer. Analogy in science. *American Psychologist*, 1956, *11*, 134.

What is needed in the behavioral sciences is for experimental research to relinquish its dependence on the safe one-shot studies and to emphasize more systematic replicative effort in the appropriate ecological settings. We do not mean to say that the controlled experimentation that we considered in Chapters 12 and 13 should be discarded. However, an overall strategy of research is necessary to direct more integrated and productive studies and to reduce fragmented and irrelevant experimentation. This approach would mean a longitudinal series of investigative efforts by applying a variety of designs to a single behavioral phenomenon (dependent variable).

There also must be a closing of the gap between the methodologists and the substantive researchers. Some methodologists disdain knowing anything about substantive research areas, that is, they do not wish to engage in understanding the possible implications of a particular theory or approach for the subjects. They instead limit their thought to the development and choice of methods, that is, the tools of research, without consideration of the substantive areas of investigation. On the other hand, substantive researchers often use old methodological designs and statistical procedures without keeping up to date with the more current research designs and statistical techniques. Research on a larger scope requires a synthesis of the contributions made by both the methodologist and the substantive researcher.

If behavioral scientists do not use a variety of investigative or methodological techniques, they may retard research efforts and may become sterile in their investigative procedures. Thus, to use classical experimental paradigms may be restricting efforts to investigate dynamic and flexible behavioral phenomenon. The point is, as Smith indicates, that behavioral science ". . . must turn its attention to working out the philosophy of science which is most appropriate for its subject matter."[11] There must be developed, according to Smith,

> research techniques indigenous and appropriate to the enterprise upon which it is engaged. Aping the techniques of other established sciences whether in research for "respectability" or because indigenous techniques have not been developed can retard the mature development of psychology as a science in its own right and can, of course, lead to the compilation of data, either of less value or of data less revealing for the questions at hand than otherwise might be the case.

In the following we present some bold speculations and suggestions about future trends in research design or methodology and in statistical procedures.

[11] N. C. Smith, Jr. Replication studies: a neglected aspect of psychological research. *American Psychologist,* 1970, 25, 970–975.

Research Design

Future trends in research for the behavioral sciences might involve more cooperative and systematic approaches for investigating a single behavioral phenomenon. Ideally, these investigative efforts would involve cooperation not only across disciplines but also cooperation between "camps" and "schools" within a behavioral science discipline. We can construct a model of a new design, a grand strategy for conducting research, which incorporates the methodology and techniques for observing and collecting data from different disciplines and from groups within the behavioral sciences. In other words, a variety of designs and observational procedures would be employed to investigate a single behavioral phenomenon.

To construct such a model of an expanded research design, we should consider six factors that are related to any experimental research endeavor. First, any research effort can be classified with respect to the type of design used. A research report can use a true, quasi-, or pseudoexperimental design, or a variety of descriptive techniques which do not manipulate an independent variable but use some kind of observational scheme to collect data. Second, a research project can have one of two forms of data, either quantitative data or qualitative data. Qualitative data can represent nominal kinds of data or descriptions of events or processes. Third, research can be performed in a laboratory setting or in a field setting. Fourth, a research project can use either a group of subjects or one subject. Fifth, the dependent variable(s) of a research project can be classified with respect to who measures or observes. In most experimental studies, either the researcher or trained observers are the ones who measure the dependent variable or record these data. On the other hand, the subjects (or subject) can record the dependent measures (self-monitoring of behavior, self-reporting, etc.). Sixth, a research project can be a one-shot study or an integral part of a series of replicated studies. Research that is replicated or repeated can be either direct or systematic. Thus there are six factors (one factor with four levels and each of the other five with two levels) that can apply to any study. For example, a researcher may study the treatment of self-monitoring for one subject on the number of cigarettes smoked each day for a two-week period. This study can be classified as descriptive, quantitative data, field research, $N = 1$, subject record, and one-shot.

The conceptualization of this model is illustrated in Figure 16.1. The large cube at the bottom of the figure shows three factors: type of design, form of data, and where the research is performed. Each smaller block of this cube can be further divided in terms of the three other factors: type of subject, data recorder, and one-shot or replicated. The model suggests that investigation of any one behavioral phenomenon should include a variety of research strategies with many combinations of the levels for each factor. Also, the specific technique that is generally associated with any one be-

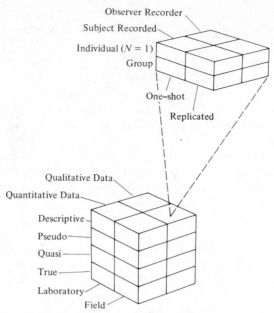

FIGURE 16.1. *Six factors for classifying research design.*

havioral science discipline could be incorporated in the grand strategy of this model.

The model provides a systematic structure classifying behavioral science research and hopefully can encourage examination of a single behavioral phenomenon from several methodological designs and possibly across a variety of behavioral science disciplines. The rationale for this model is simply that no one design or methodology is best in the investigative process and no one approach is most powerful. Kessel stated that "the analogy drawn by Eddington[12] between the scientist and the ichthyologist conveys the point best: Having used a fishing net with a certain size mesh, the ichthyologist proclaims that there are no fish in the universe smaller than the mesh size."[13] The size of the fish net represents acceptable experimental methodology and proper statistical procedures. The catch is, to a large extent, the function of the size of the mesh. For perhaps a more accurate perspective of behavioral phenomena, scientists should use many "nets of varying mesh size" or methodological approaches. Assumptions or inferences made about a behavioral phenomenon that have been derived from one methodological procedure without being replicated may limit the perspective about that phenomenon.

[12] A. S. Eddington. *The Nature of the Physical World.* New York: Wiley, 1928.
[13] F. S. Kessel. The philosophy of science as proclaimed and science as practiced: identity or "dualism"? *American Psychologist,* 1969, *24,* 1004.

Statistical Procedure

The flexibility that we have described for research design would also have to apply to statistical analysis. Researchers would have to employ a variety of techniques, and a probable future trend could be a decrease in the distinction between nonparametric and parametric statistical analysis. In other words, a variety of statistical procedures would be used on any particular investigative effort. Again, this would require a team effort and more systematic analyses of the data. A systematic approach to analysis similar to the systematic approach to research design would require more complicated types of statistical tests and techniques. For example, multivariate procedures will perhaps increase in their use in terms of both qualitative and quantitative data. It is essential that researchers always employ the particular statistical procedure that answers the particular question under investigation.

Every researcher has the obligation to admit that each part or set of his data could have come about quite randomly given the oddest state of nature.[14] Therefore, conjoint measurement and analysis in terms of the kinds of data or the kinds of statistical analysis employed would mean a more liberalized approach to data analysis and the use of statistical procedures rather than rigid and rigorous adherence to the typical, for example, univariate analysis of variance. Also, because of the complexity of designs and the necessity for replicated studies, repeated measures designs, either in terms of simultaneous replication or staggered replication, might also increase in the future. Multivariate statistical procedures that can accommodate repeated measures would probably increase in popularity because there would be a variety of independent variables and dependent variables used in a single study.

Statisticians have been concerned with the type of errors that can be made with respect to a statistical analysis. In Chapter 3 we mentioned Type I and Type II errors. There are other types of errors that can be made by a researcher. For example, there are three different versions of a Type III error. One version of a Type III error that might be committed in testing statistical hypotheses is selecting a statistical test falsely to suit the significance of the particular sample data available. Another version was alluded to by Kaiser and is defined as an incorrect decision of direction following a rejected two-tailed test of hypothesis.[15] This error would occur if one concluded that the difference favors one condition or treatment even though the true state of affairs would reveal a difference in parameters in the opposite direction. A final definition of a Type III error, noted by Kimball, is the error committed by giving the right answer for the wrong problem.[16] It is made whenever a

[14] Tukey. Analyzing data, p. 85.

[15] H. F. Kaiser. Directional statistical hypotheses. *Psychological Review*, 1960, 67, 160–167.

[16] A. W. Kimball. Errors of the third kind in statistical consulting. *Journal of the American Statistical Association*, 1957, 52, 133–142.

statistician supplies the correct answer to a problem other than the one proposed by the client.

A Type IV error has been described by Marascuilo and Levin. They define Type IV error as the incorrect application of follow-up (post hoc) procedures. Type IV errors, they say, "may be likened to a physician's correct diagnosis of an ailment followed by the prescription of the wrong medicine."[17]

TYPE V ERROR

All of the above types of errors refer to statistical errors, but recall our discussion of statistics and research design. If good design is the prerequisite of good statistical analysis, a researcher cannot have satisfactory statistical analysis with a poor experimental design. In view of this, we would like to define Type V error as the error committed by a researcher by rigidly adhering to one investigative design without a systematic approach. As a result of this error, data or results are sanctified without being verified through replication by other kinds of research designs or investigative efforts. In other words, Type V error is the error committed by a researcher relying on a single research design without replication. Results of one-shot experimentation, no matter what the level of significance, should not be accepted as the absolute. The attempt to avoid Type V errors will require a tremendous effort of cooperation between the methodologists and substantive researchers, among a variety of design and statistical specialists, and across and between disciplines in the behavioral sciences.

To achieve this cooperation will also require a variety of models for examining a behavioral phenomena. As Coulson and Rogers said,

> the model of a precise beautifully built and unassailable science which most of us hold consciously or unconsciously becomes then a limited and distinctively human construction incapable of precise perfection. Openness to experiences can be seen as being fully as important a characteristic of the scientist as the understanding of a research design. And the whole enterprise of science can be seen as but one portion of a larger field of knowledge in which truth is pursued in many equally meaningful ways, science being one of these ways.[18]

Again, the implication is that no one approach for viewing behavior phenomena is most powerful or most valid. A variety of methods are perhaps

---

[17] L. A. Marascuilo and J. R. Levin. Appropriate post hoc comparisons for interaction and nested hypotheses in analysis of variance designs: the elimination of Type IV errors. *American Educational Research Journal*, 1970, 7, 397–421.

[18] W. R. Coulson and C. R. Rogers (Eds.). *Man and the Science of Man.* Columbus, Ohio. Merrill, 1968, p. 8.

more appropriate than any one single approach. Flexible methodology is required by behavioral scientists because the phenomena they seek to investigate are dynamic.

SUGGESTED READINGS

We have ended each previous chapter with a few short questions on the content of that chapter. However, the goal of this chapter differs from that of the others. While the others provide the reader with the basic information needed to understand statistics and research design in the literature, we hope this chapter will give the reader an idea of current and future issues and will stimulate further reading in the area. In short, we hope we have raised some questions rather than answered them. To further this goal we are providing a list of books in a variety of areas for the reader who would like a more advanced view about statistics, research design, and philosophy of behavioral science than this book could provide.

Campbell, Donald T., and Stanley, Julian C. *Experimental and Quasi-Experimental Designs for Research.* Chicago: Rand McNally, 1966.

Hays, William L. *Statistics for the Social Sciences.* New York: Holt, Rinehart and Winston, 1973.

Kaplan, Abraham. *The Conduct of Inquiry: Methodology for Behavioral Sciences.* San Francisco: Chandler, 1964.

Kerlinger, Fred N. *Foundations of Behavioral Research.* New York: Holt, Rinehart and Winston, 1973.

Kuhn, Thomas S. *The Structure of Scientific Revolutions.* Chicago: The University of Chicago Press, 1970.

Morrison, D. F. *Multivariate Statistical Methods.* New York: McGraw-Hill, 1967.

Siegel, Sidney. *Nonparametric Statistics for the Behavioral Sciences.* New York: McGraw-Hill, 1956.

Stanley, Julian C. (Ed.). *Improving Experimental Design and Statistical Analysis.* Chicago: Rand McNally, 1967.

Tatsuoka, Maurice M. *Multivariate Analysis.* New York: Wiley, 1971.

Winer, B. J. *Statistical Principles in Experimental Design.* New York: McGraw-Hill, 1971.

INDEX

(The numbers in parentheses refer to excerpts in which the term is used by the author of a journal article. Note: In some cases the excerpt references may not be all-inclusive.)